Gas Sensors for Monitoring Environmental Changes

Gas Sensors for Monitoring Environmental Changes

Editors

Kai Xu

Zhong Li

MDPI

Basel • Beijing • Wuhan • Barcelona • Belgrade • Novi Sad • Cluj • Manchester

Editors
Kai Xu
RMIT University
Melbourne
Australia

Zhong Li
Southwest Jiaotong University
Chengdu
China

Editorial Office
MDPI AG
Grosspeteranlage 5
4052 Basel, Switzerland

This is a reprint of articles from the Special Issue published online in the open access journal *Chemosensors* (ISSN 2227-9040) (available at: https://www.mdpi.com/journal/chemosensors/special_issues/5099967663).

For citation purposes, cite each article independently as indicated on the article page online and as indicated below:

Lastname, A.A.; Lastname, B.B. Article Title. *Journal Name* **Year**, *Volume Number*, Page Range.

ISBN 978-3-7258-2459-5 (Hbk)
ISBN 978-3-7258-2460-1 (PDF)
doi.org/10.3390/books978-3-7258-2460-1

Contents

chemosensors

Article

Preparation and Hydrogen-Sensitive Property of WO_3/Graphene/Pd Ternary Composite

Lin Wang [1,2], Fei An [2], Xinmei Liu [1,*], Dongzhi Zhang [3] and Zhe Yang [2,*]

1 College of Chemistry and Chemical Engineering, China University of Petroleum (East China), Qingdao 266580, China; wangl.qday@sinopec.com
2 State Key Laboratory of Safety and Control for Chemicals, SINOPEC Research Institute of Safety Engineering Co., Ltd., Qingdao 266071, China; anf.qday@sinopec.com
3 College of Control Science and Engineering, China University of Petroleum (East China), Qingdao 266580, China; dzzhang@upc.edu.cn
* Correspondence: lxmei@upc.edu.cn (X.L.); yangzhe.qday@sinopec.com (Z.Y.); Tel.: +86-532-86980908 (X.L.); +86-532-83786640 (Z.Y.)

Abstract: Hydrogen (H_2) is a renewable energy source that has the potential to reduce greenhouse gas emissions. However, H_2 is also highly flammable and explosive, requiring sensitive and safe sensors for its detection. This work presents the synthesis and characterization of WO_3/graphene binary and WO_3/graphene/Pd (WG-Pd) ternary nanocomposites with varying graphene and Pd contents using the microwave-assisted hydrothermal method. The excellent catalytic efficacy of Pd nanoparticles facilitated the disintegration of hydrogen molecules into hydrogen atoms with heightened activity, consequently improving the gas-sensing properties of the material. Furthermore, the incorporation of graphene, possessing high conductivity, serves to augment the mobility of charge carriers within the ternary materials, thereby expediting the response/recovery rates of gas sensors. Both graphene and Pd nanoparticles, with work functions distinct from WO_3, engender the formation of a heterojunction at the interface of these diverse materials. This enhances the efficacy of electron–hole pair separation and further amplifies the gas-sensing performance of the ternary materials. Consequently, the WG-Pd based sensors exhibited the best gas-sensing performance when compared to anther materials, such as a wide range of hydrogen concentrations (0.05–4 vol.%), a short response time and a good selectivity below 100 °C, even at room temperature. This result indicates that WG-Pd ternary materials are a promising room-temperature hydrogen-sensing materials for H_2 detection.

Keywords: WO_3; graphene; Pd; microwave; hydrothermal; H_2; sensors; room temperature

Citation: Wang, L.; An, F.; Liu, X.; Zhang, D.; Yang, Z. Preparation and Hydrogen-Sensitive Property of WO_3/Graphene/Pd Ternary Composite. *Chemosensors* **2023**, *11*, 410. https://doi.org/10.3390/chemosensors11070410

Academic Editor: Michele Penza

Received: 12 June 2023
Revised: 10 July 2023
Accepted: 14 July 2023
Published: 21 July 2023

1. Introduction

Hydrogen, an eco-friendly and renewable energy source, has gained considerable interest in its wide range of applications, including automobile, fuel cell and space rocket applications [1,2]. However, hydrogen exhibits a colorless, odorless and explosive nature with a low ignition energy (0.02 mJ) and a wide flammable range of 4–74 vol.% in air, leading to many safety issues [3,4]. Therefore, rapid and accurate detection of hydrogen leakage is necessary for large-scale utilization of hydrogen. In recent years, chemiresistive hydrogen sensors based on metal oxide semiconductors (MOS), such as ZnO, SnO_2, MoO_3, WO_3, etc., have been widely applied to hydrogen detection due to their low cost, high sensitivity and short recovery time [5–8]. Among these MOS, WO_3, with its unique morphological structure, gasochromic properties, and high diffusion coefficient of oxygen vacancies, has proven to be one of the most attractive sensing materials [9]. However, the pure WO_3 sensors need a high working temperature (200–400 °C) to detect hydrogen, which may lead to power consumption and source of ignition [10,11].

Based on recent research findings, incorporating graphene or its derivatives (such as graphene oxide or reduced graphene oxide) has emerged as a promising strategy to lower the operating temperature of MOS-based sensors. This is attributed to the remarkable properties of graphene, including excellent conductivity, high carrier mobility at room temperature, low electrical noise and a large surface area [12–14]. Moreover, graphene and its derivatives have various active sites (such as oxygen functional groups, defects, vacancies and π-π covalent bonds) that can selectively adsorb target gases, thereby enhancing the sensitivity and selectivity of the sensors. Furthermore, a heterogeneous structure (p–n or p-p junction) can be formed at the interface between graphene and metal oxides. This structure can facilitate the adsorption or resistance modulation of target gas molecules, which can improve the sensing performance of the material [15]. Consequently, chemiresistive sensors based on WO_3 sheets, hemispheres, aerogels, etc., combined with graphene have demonstrated lower working temperatures (<150 °C) towards target gases. For instance, Chu et al. reduced the optimal operating temperature of WO_3 from 180 °C to 100 °C by adding 0.1 wt% graphene [16]. Gui et al. fabricated hemispherical WO_3/graphene nanocomposites with hollow structures to achieve triethylamine sensing at room temperature [17]. Zhao et al. synthesized mesoporous WO_3@graphene aerogel nanocomposites, which exhibited a good response to acetone at 150 °C [18].

The hydrogen-sensing properties (e.g., sensitivity, response time and selectivity) of MOS can also be enhanced by decorating them with noble metals (such as Au, Ag, Pt or Pd). This enhancement results from the chemical and electronic sensitization of noble metals, where the former is achieved by the dissociation of hydrogen molecules and the latter by the change of electron depletion layers generated by heterogeneous structures [19]. Among these noble metals, Pd is considered to have the greatest effect on improving the hydrogen-sensing performance. It can spill over hydrogen molecules into atoms, which combine with Pd to form palladium hydride, resulting in a significant change in resistance. This reaction is reversible at room temperature. For example, Zhu et al. synthesized PdNPs@WO_3NPs, which showed a fast response (1.2 s) to hydrogen at 50 °C [20]. Le et al. fabricated a fast and efficient hydrogen gas sensor using $PdAu_{alloy}$@ZnO core–shell nanoparticles [21].

In this study, we examined the morphology and distribution of WO_3/graphene binary materials and WO_3/graphene/Pd ternary materials (WG-Pd) fabricated by microwave-assisted hydrothermal method. The structure of both binary and ternary materials was influenced by the content of graphene and Pd. Moreover, we investigated the relationship between the content of graphene in WO_3/graphene binary materials and their hydrogen-sensing properties to determine the optimal graphene content for hydrogen sensing. Additionally, we evaluated the working temperature, repeatability and selectivity of WG-Pd ternary materials. Finally, we discussed the sensing mechanism of WG-Pd.

2. Materials and Methods

2.1. Materials

GO suspension (GO-1, Hangzhou Gaoxi Technology Co., Ltd, Hangzhou, China); absolute ethanol (99.7%, Sinopharm, Shanghai, China); HCl (99.5%, Sinopharm, Shanghai, China); Na_2WO_3 (Sigma-Aldrich, St. Louis, MO, USA); $PdCl_2$ (Sigma-Aldrich, St. Louis, MO, USA). All reagents were of analytical grade without further purification, and the deionized water was used in all experiments.

2.2. Fabrication of WO_3/Graphene/Pd Ternary Materials

A certain amount of GO was dispersed in 190 mL deionized water followed by ultrasonication. Next, 1.06 g of Na_2WO_3 was added into GO dispersion, and 50 mL of HCl (2 M) was added drop-by-drop into the mixture under vigorous stirring for 30 min. The suspension was poured into autoclaves and subjected to microwave heating at 180 °C for 1 h using a Multiwave PRO oven (Anton Paar, Graz, Austrian). The precipitate was then collected and washed by centrifugation in deionized water and absolute ethanol, followed by freeze-drying under vacuum at room temperature. The WO_3/graphene nanomaterials were synthesized with different GO/WO_3 mass ratios of 0.5, 1, 2 and 3:50, and denoted as W0.5G, WG, W2G and W3G, respectively. Pure WO_3 was also prepared in the same way without GO for comparison. A quantity of 50 mg of WG and a specific amount of $PdCl_2$ were dispersed in 5 mL of absolute ethanol using ultrasonication for a duration of 30 min. The resulting mixture was then dried in an oven at a temperature of 80 °C for a period of 12 h, followed by thermal annealing in a nitrogen atmosphere at a temperature of 300 °C for 2 h. The samples obtained from this process, with $PdCl_2$ masses of 1, 5 and 10 mg, respectively, were designated as WG-1Pd, WG-5Pd and WG-10Pd.

2.3. Characterization

Scanning electron microscopy (SEM, JEOL JSM-7610F, Tokyo, Japan) and transmission electron microscopy (TEM, FEI Talos, Columbia, SC, USA) were used to observe the morphologies of ternary nanomaterials. The phases of samples were analyzed by X-ray diffraction (XRD, Bruker D8 Advance, Billerica, MA, USA). The chemical compositions of ternary nanomaterials were measured using Thermo Fisher (Waltham, MA, USA) ESCALAB 250 XI X-ray photoelectron spectroscopy (XPS).

2.4. Fabrication and Test of Gas Sensors

The WG and WG-Pd powders were blended with terpineol in a 1:2 mass ratio, while finely grinding them in a mortar. The resultant paste was uniformly coated on an Al_2O_3 tube with a pair of Pt wires. To heat the gas sensor, a Ni-Cr heating wire was inserted into the tube. The sensors were pre-treated at 100 °C for 5 days to enhance their stability before the tests. The gas-sensing measurements were carried out using a CGS-8 Gas Sensing Measurement System (Beijing Elite Tech Company Limited, Beijing, China) with a 500 mL test chamber. The sensors' resistance was stabilized at the desired temperature before a known volume of gas was injected into the chamber. The tests were performed under ambient conditions of 25 ± 5 °C and 40 ± 5% relative humidity.

The gas response was calculated as $(R_{air} - R_{gas})/R_{air}$, where R_{air} and R_{gas} are the sensor resistance in air and target gas, separately. The response time and recovery time are the time intervals required for the sensor response to reach 90% of its total change upon exposure to the target gas or air, respectively.

3. Results

3.1. Morphology and Structure

In order to characterize the surface morphology and distribution of WO_3, WG and WG-Pd, SEM and EDS were utilized. As shown in Figure 1a, WO_3 obtained by the microwave-assisted hydrothermal method showed a nanoribbon-like structure with a length of about 200 nm and a thickness of about 30 nm. The two-dimensional structure had a high surface-to-volume ratio and abundant surface functional groups, which offer a large number of adsorption sites for the target gas and oxygen molecules. This leads to effective physisorption and chemisorption of gas molecules at low temperature. Hence, two-dimensional materials have some advantages in providing fast and complete diffusion of H_2 gas throughout the structure and are widely used as gas-sensing materials [22–24]. Following the incorporation of GO, WG was interconnected by delicate folded RGO layers amid multiple WO_3 nanoribbons. The folded configuration of RGO primarily resulted from lamellar distortion induced by partial reduction of GO during the microwave-assisted hydrothermal process (Figure 1b–e) [25,26]. In the case of W0.5G with a lower graphene content, only a limited number of WO_3 nanoribbons were linked by graphene (Figure S1a). Conversely, with a higher graphene content in W2G, a significantly larger proportion of WO_3 nanoribbons were connected by graphene, albeit some were enveloped by it (Figure S1b). As the graphene content increased, porous structures gradually formed within the composite due to graphene stacking, enveloping the WO_3 nanoribbons and leaving only a few exposed (Figure S1c). Subsequent to Pd loading onto the surface of WG, a substantial amount of Pd was uniformly distributed, as evidenced by the EDS image (Figure 1f–h).

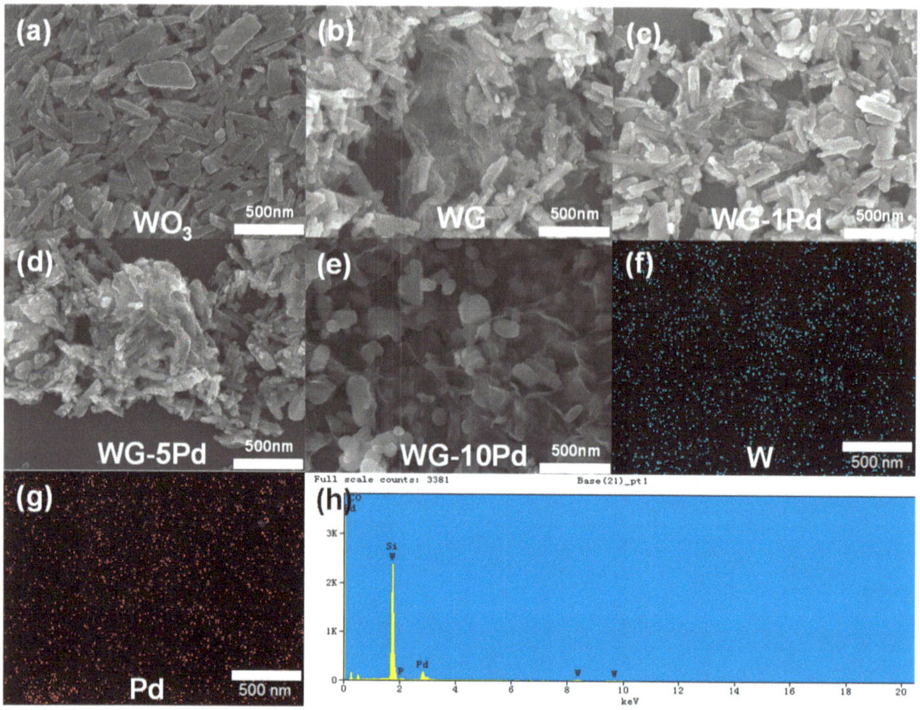

Figure 1. (**a**–**e**) SEM images of WO_3, WG and WG-Pd, (**f**–**h**) corresponding EDS elemental mapping and EDS spectra of WG-10Pd.

The hierarchical structures of WG and WG-Pd were characterized by TEM. As shown in Figure 2a–c, in addition to large nanoribbons, there were a few nanosheets of WO_3 around 10–20 nm of WG, which should be grown on the surface of graphene. There were a few nanosheets of WO_3 of around 10–20 nm on the surface of WG, indicating that they were grown on the graphene surface. After loading Pd onto WG, spherical Pd nanoparticles appeared on the surface of graphene (Figure 2d–f). With a Pd content of 1%, Pd nanoparticles were sporadically distributed on graphene nanosheets with a size of 3 nm. When increasing the Pd content, the number and size of Pd nanoparticles also increased. When the Pd content increased to 10%, Pd nanoparticles were more densely distributed and had a larger size of 10 nm. Moreover, even for the WG-Pd with the highest content of Pd, no significant aggregation of Pd nanoparticles was observed. This phenomenon stemmed from the mutual interaction between the abundant oxygen-containing functional groups on the surface of graphene oxide and Pd ions, which led to their adsorption and uniform growth on the surface of graphene nanoflakes and prevented the aggregation of Pd nanoparticles. This phenomenon can be attributed to the interaction between the abundant oxygen-containing functional groups on the surface of graphene oxide and Pd ions. This interaction leads to the adsorption and uniform growth of Pd nanoparticles on the surface of graphene nanoflakes, preventing their aggregation.

Figure 3 shows XRD patterns of WO_3 and WG-Pd. WO_3 exhibited distinct diffraction peaks of orthorhombic phase of $WO_3{\cdot}0.33H_2O$ (JCPDS 72-199) [27,28], indicating the presence of crystal H_2O before thermal annealing. For the WO_3/graphene composite nanomaterials, the characteristic peaks at 22.7°, 28.1°, 36.5°, 49.9° and 55.3° were slightly different from those of pure-phase WO_3. This was attributed to the interaction between the functional groups (hydroxyl, carbonyl and carboxyl groups) on the graphene surface and Na_2WO_3, which influenced the WO_3 growth. Moreover, slight deviation from stoichiometry (WO_{3-x}) could also affect the peak position. They also exhibited a broad characteristic peak around 20°, which was caused by the imperfect restacking due to the curvature of graphene sheets [29] (Figure S2). After loading with Pd and thermal annealing, WG-1Pd, WG-5Pd and WG-10Pd had different XRD patterns with WG. As shown in Figure 3b–d, there were diffraction features matching hexagonal phase of WO_3 due to the dehydration after annealed [30]. Furthermore, characteristic peaks of Pd (JCPDS 46-1043) and PdO (JCPDS 43-1024) appeared in WG-Pd, applying the presence of Pd and PdO attributed to the particle oxidization of Pd nanoparticles in air [31].

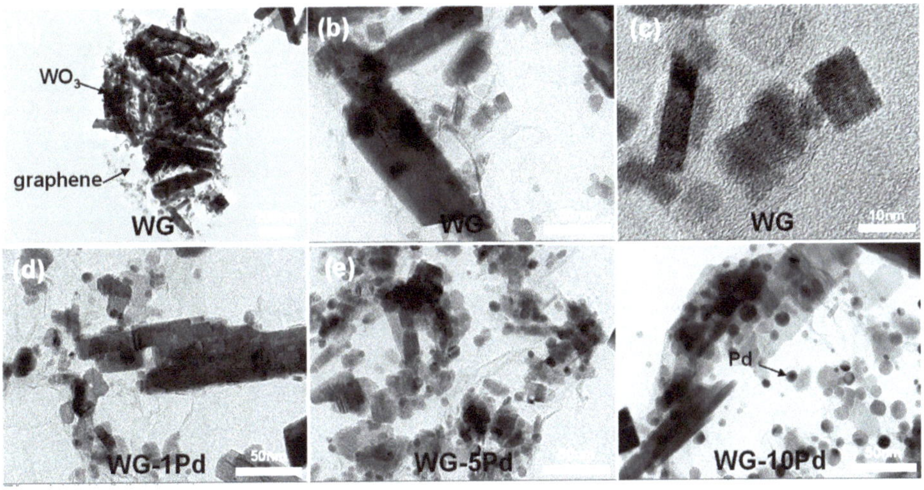

Figure 2. TEM images of (**a–c**) WG and (**d–f**) WG-Pd.

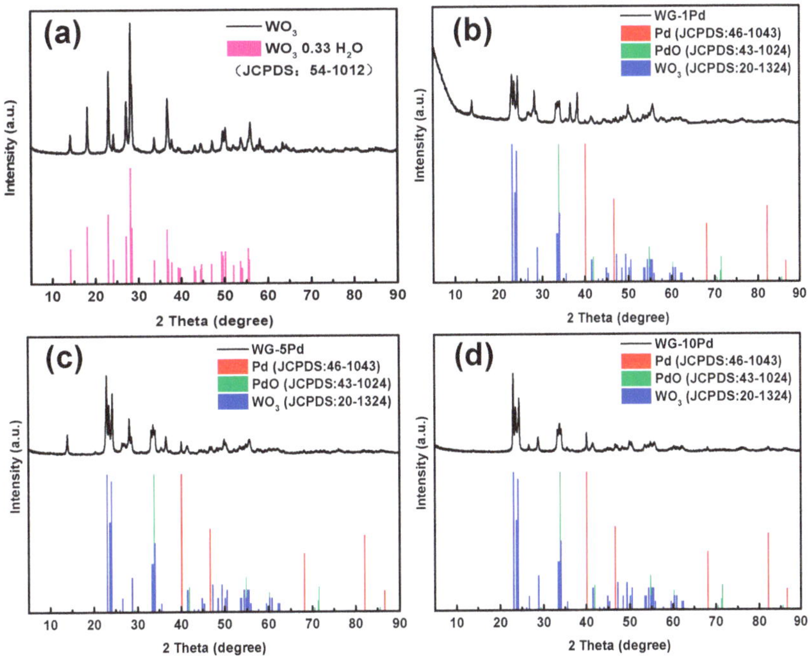

Figure 3. XRD patterns of (**a**) WO_3 and (**b–d**) WG-Pd.

In order to accurately analyze the content and valence state of elements in WG-10Pd, XPS analysis was utilized (Figure 4). As shown in Figure 4a, the material was mainly composed of W, C, O and Pd. Figure 4b demonstrates the deconvoluted Pd (3d) XPS spectra of WG-10Pd. It can be found that Pd element was mainly metallic Pd^0 located at 335.3 and 340.55 eV. Additionally, there were other peaks of oxidized Pd^{2+} located at 336.6 and 341.85 eV and Pd^{4+} located at 338.2 and 343.45 eV [32]. This also corresponds to the XRD spectra. A small amount of oxidized Pd came from the partial oxidation of surface atoms of Pd nanoparticles by air due to their high activity.

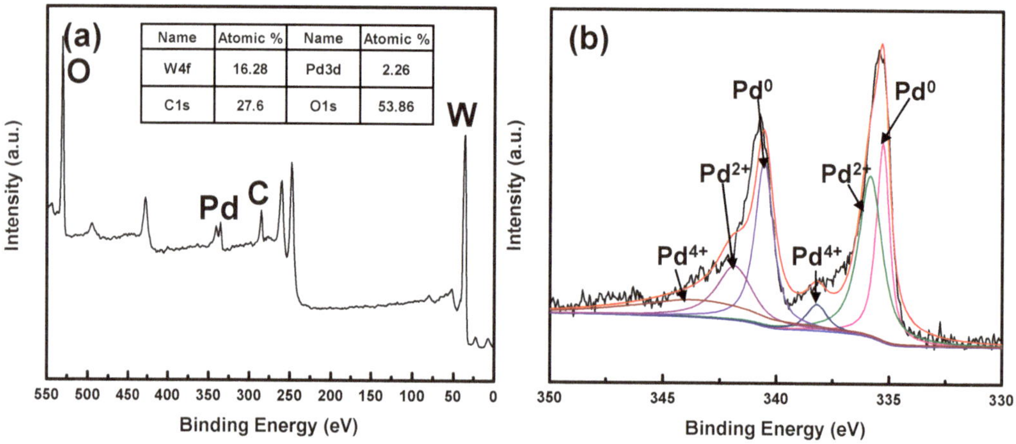

Name	Atomic %	Name	Atomic %
W4f	16.28	Pd3d	2.26
C1s	27.6	O1s	53.86

Figure 4. (**a**) XPS survey of WG-10Pd; (**b**) high-resolution XPS spectra of Pd 3d of WG-10Pd.

3.2. Hydrogen Gas Sensing

The effects of graphene content and working temperature on the hydrogen response performance of WO_3/graphene nanomaterials are shown in Figure 5. As can be seen from Figure 5a, at 50 °C, WO_3, W0.5G and W3G had no response to hydrogen. Among them, WO_3 and W0.5G had large resistances (7130 MΩ and 2500 MΩ, respectively) that exceeded the instrument resistance range (500 MΩ) and could not be measured. This was because WO_3, as a metal oxide semiconductor, had a large resistance at low temperatures, and even with a small amount of graphene added, it still could not meet the measurement requirements. WG had the best response performance, but due to the low working temperature, it only responded to hydrogen concentrations above 0.4%. Although W2G and W3G had higher graphene contents than WG, their response performance was reduced. This was attributed to the excess graphene that could cover or even wrap around WO_3 and provide an additional path for current flow, resulting in a significant increase of the conductivity of the sample, which made resistance variation in reducing environment less noticeable. As the working temperature increased to 100 °C and 150 °C, the response performance of WG, W2G and W3G were all improved. The response of WG, W2G and W3G to 1% H_2 at 100 °C was 30.8, 28.9 and 2.5%, respectively. After heating at 150 °C, the corresponding responses increased to 90.5, 56.3 and 7.9%, respectively. This result indicated that WG had the highest response performance. At the same time, as the working temperature increased from 50 °C to 150 °C, WG's response time to 1% hydrogen was shortened from 223 s to 67 s (Figure 5d,e). Meanwhile, WG showed p-n switching phenomenon at different temperatures, which was mainly caused by the combination of different semiconductor materials and the change of charge transfer path at different working temperatures. Graphene (p-type semiconductor) and WO_3 (n-type semiconductor) formed a composite, in which the carriers of both graphene and WO_3 participated in the conduction path, showing a mixed p-n characteristic. At low temperature, WO_3 transferred electrons to GO, showing p-type behavior. At high temperature, WO_3 received electrons from GO, showing n-type behavior. However, WG still had no response to hydrogen at room temperature and needed to be composited with Pd to further improve its response performance. Therefore, we used WG as the base material and composite different contents of Pd nanoparticles to study the effects of Pd content and working temperature on the hydrogen response performance.

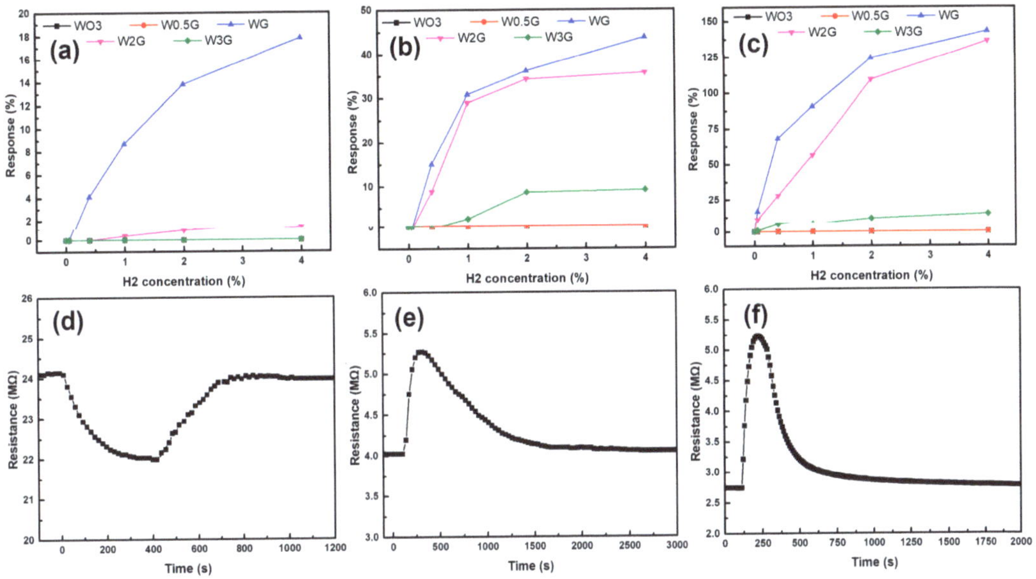

Figure 5. The response of WO_3/graphene nanomaterials at 50 (**a**), 100 (**b**) and 150 °C (**c**); The resistance of WG for 1% H_2 at 50 (**d**), 100 (**e**) and 150 °C (**f**).

The effect of Pd content and working temperature was studied by testing these gas sensors at different temperature ranging from 25 to 100 °C as shown in Figure 6. For WG-1Pd and WG-5Pd, there was no evident response to 500 ppm of H_2 at 25 °C, applying low content of Pd was not sufficient for H_2 sensing at room temperature. However, WG-10Pd with larger content of Pd showed response to 500 ppm of H_2 at 25 °C with value of 2.3%. WG-1Pd, WG-5Pd and WG-10Pd were more sensitive to H_2 at higher working temperature. The response to 500 ppm H_2 of WG-1Pd, WG-5Pd and WG-10Pd increased to 86.1%, 92.6% and 94.6% at a working temperature of 100 °C, respectively. This phenomenon stemmed from the lower gas activation energy of adsorption/desorption at higher temperature. Meanwhile, WG-10Pd still showed a higher response to H_2 than the others. Compared with the WO_3/graphene binary materials without Pd, the hydrogen-sensing performance of the WO_3/graphene/Pd ternary composite improved significantly. WG exhibited no noticeable response to 500 ppm hydrogen, while the response value increased to 86.1% with the addition of 1% Pd nanoparticles, indicating that Pd had a remarkable effect on enhancing the hydrogen response.

The dynamic response of WG-10Pd was studied at various H_2 concentrations (ranging from 0.05–4%) and at temperatures of 25, 50, 75 and 100 °C. As shown in Figure 7a, WG-10Pd showed excellent response to hydrogen in the range of 0.05–4% at room temperature. The response value increased gradually with the increase of the concentration of H_2. At 4% of H_2, the response value reached 96.1%. On the contrary, the response time was gradually shortened. The response time was reduced to within 40 s at 4% of H_2. After heating to 50, 75 and 100 °C, the response value was further improved and the response time was shorter. At 100 °C, the response value of 0.05% H_2 could reach 94.6%, and the response time was shortened to 6 s. However, the response curves reached a plateau at relatively low concentrations, proving its insensitivity to high concentration of H_2. Therefore, WG-10Pd was more suitable for the measurement of low-concentration hydrogen below 0.05% after heating. The saturation phenomenon was ascribed to the increased catalytic effect of Pd at high temperature, which generated abundant chemisorbed water and occupied the active sites for oxygen adsorption.

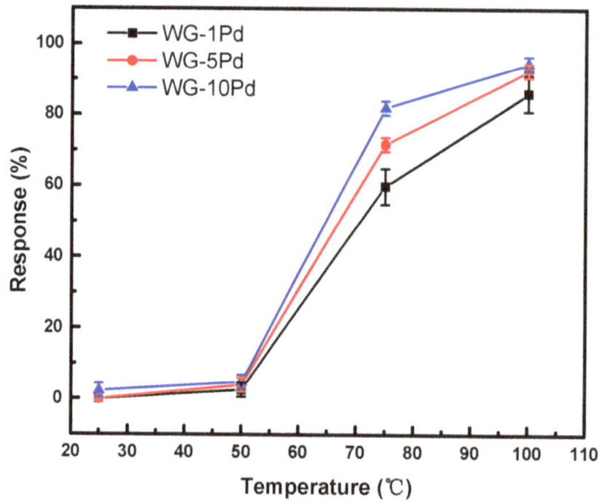

Figure 6. The response of WG-Pd for 500 ppm H_2 at different temperatures ranging from 25 to 100 °C.

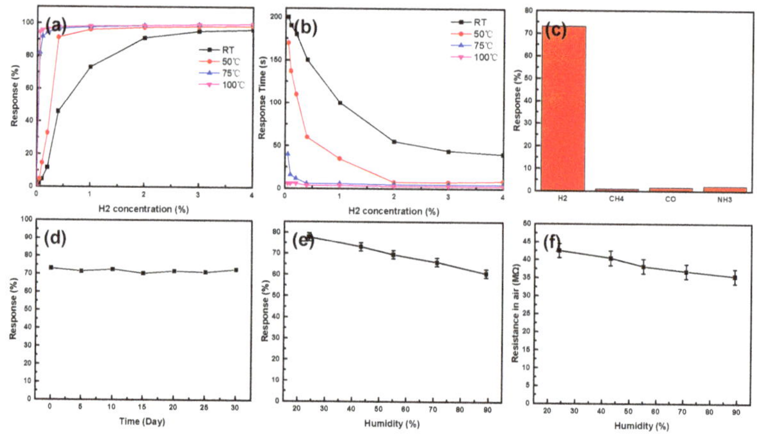

Figure 7. (**a**) The response value and (**b**) response time of WG-10Pd to 0.05–4% H_2 at different temperature ranging from 25 to 100 °C; (**c**) the response of WG-10Pd to 1% H_2, CH_4, CO and NH_3; (**d**) the response of WG-10Pd to 1% H_2 for 30 days; (**e**) the response and (**f**) the resistance in air of WG-10Pd under different RH.

WG-10Pd hydrogen-sensing material had excellent selectivity and stability. As demonstrated in Figure 7c, the response value of WG-10Pd to 1% H_2 at room temperature was significantly higher than that of CH_4, CO and NH_3 with the same concentration, showing high selectivity of H_2, due to the catalytic properties of Pd to hydrogen dissociation. After 30 days of uninterrupted testing, the change of response value was still maintained within 5%, implying the high stability of WG-10Pd.

The effect of relative humidity (RH) on the sensing performance of WG-10Pd at room temperature was also investigated. Figure 7e,f shows that the response of WG-10Pd decreased with increasing RH. This phenomenon was caused by the water vapor occupying the adsorption sites and hindering the oxygen adsorption. The response values of WG-10Pd to 1% H_2 were 77.8%, 73.1%, 69.4%, 65.9% and 60.6% under 24%, 43%, 55%, 71% and 89% RH, respectively. The response variation (calculated as 1 − Response (89RH%)/Response (24RH%)) was 22%, indicating that RH had a significant influence on the hydrogen sensing. Meanwhile, the resistance in air decreased from 42.5 to 35.4 MΩ as RH changed from 24% to 89%. Therefore, the resistance variation was about 16.7%.

Similarly, the response of WG-1Pd and WG-5Pd at 50–100 °C were studied (Figure 8). The variation trend of response value and response time was basically the same as that of WG-10Pd, but the response value was slightly reduced and the response time was extended. The responses to 0.05% H_2 at 100 °C of WG-1Pd and WG-5Pd were 86.1% and 92.6%, respectively, while the response times were 25 s and 9 s, respectively.

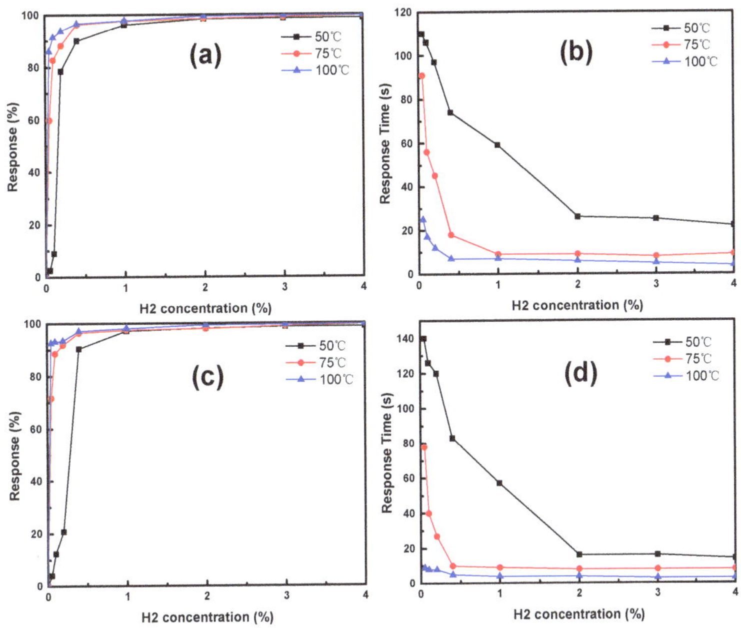

Figure 8. (**a**) The response value and (**b**) response time of WG-1Pd to 0.05–4% H_2 at different temperature ranging from 25 to 100 °C; (**c**) the response value and (**d**) response time of WG-5Pd to 0.05–4% H_2 at different temperatures ranging from 25 to 100 °C.

4. Discussion

The WG-Pd ternary materials had a good response to H_2 from room temperature to 100 °C due to their chemical composition and morphological structure (Figure 9). According to the surface depletion layer model, when the material contacts oxygen in the air, it forms oxygen negative ions (O_{2^-}, O^-, O^{2-}, etc.) on the surface by adsorbing electrons [33]. This increases electrical resistance. At different temperatures, oxygen anions have different valence states. Below 100 °C, it is mainly O_2^- ions. Between 100–300 °C, O^- dominates. Above 300 °C, oxygen anions are mostly O^{2-} [34,35]. Thus, the adsorbed oxygen ions on the surface of WG-Pd material are mainly O_2^-, as the operating temperature is below 100 °C. When it contacts hydrogen, O_2^- reacts with hydrogen to form water vapor and releases electrons to the material, causing a drop in resistance [36]. The resistance decrease is proportional to the hydrogen concentration, allowing measurement of hydrogen concentration through resistance change. The reaction process is as follows:

$$2H_2 + O_2^- \rightarrow 2H_2O + e^- \quad (1)$$

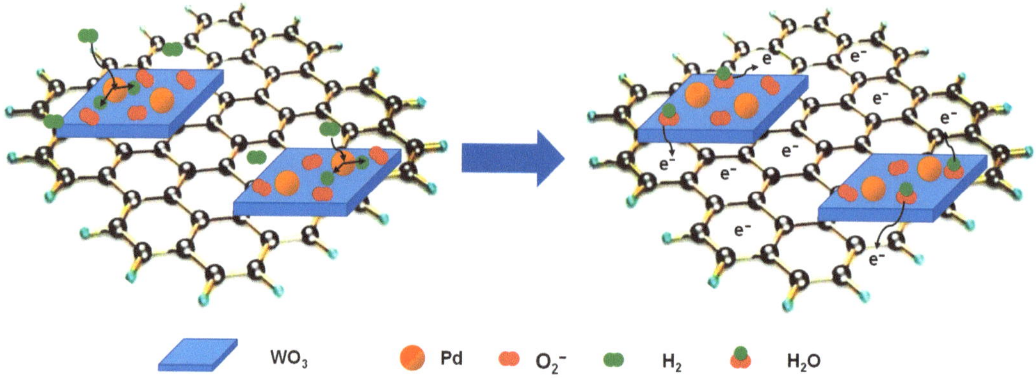

Figure 9. The schematic illustration of WG-Pd during exposure to H_2.

In the WG-Pd ternary composites, the three components showed different functions. As the material with the highest content, WO_3 played the main role in gas sensing. The abundant oxygen defects on its surface provided active sites for oxygen adsorption [37]. Pd nanoparticles could dissociate H_2 molecules into highly active hydrogen atoms, thus accelerating the reaction of O^{2-} with H_2. H_2 could also be directly dissolved into metal Pd to form Pd-H complex, thus significantly reducing the Schottky barrier on WO_3-Pd interface, enabling more electrons to enter the conduction band of the material and improving the electrical conductivity of the material [6]. However, electron transfer between different WO_3/Pd nanosheets was still hindered by the Schottky barrier between interfaces, while graphene sheets could significantly reduce that, providing multiple paths for electron transfer, thus further reducing the resistance [38]. Graphene itself had a high specific surface area and oxygen-containing functional groups, which improved response performance [39]. Moreover, the hydrogen-sensing performance could also be enhanced by the heterogeneous structure formed at the interface between graphene and WO_3. WO_3 is an n-type semiconductor material with more electrons than holes as the main carriers. Graphene sheets are p-type semiconductor materials with more holes than electrons as the main carriers, due to the intrinsic properties of graphene. Thus, there are many heterojunctions (p-n junctions) at the interface between p-type graphene and n-type WO_3. Because of the different work functions of graphene and WO_3, the holes transfer from graphene to WO_3 and the electrons transfer from WO_3 to graphene at the p-n junction. After the carrier exchange, the holes and electrons with opposite charges recombine. This leads to the decrease of concentration of effective carriers and formation of electron depletion layer and hole depletion layer at the interface of WO_3 and graphene, respectively. The depletion layer can increase the resistance of WG-Pd ternary composites in air and facilitate the adsorption or resistance modulation of H_2 gas molecules, which can improve the hydrogen-sensing properties at room temperature. Through the synergistic effect among the three components of WG-Pd, a sensitive response to hydrogen is achieved at temperatures lower than 100 °C.

5. Conclusions

In this paper, we fabricated WO_3/graphene binary materials and WG-Pd ternary materials with different contents of graphene and Pd using the microwave-assisted hydrothermal method. The sensing results showed that WG with 1% graphene exhibited the best hydrogen performance at 100 and 150 °C compared to other WO_3/graphene binary materials with different contents of graphene. However, WG still exhibited no noticeable response to hydrogen below 100 °C. For WG-Pd ternary materials, WG-10Pd showed the highest response to H_2 at room temperature. Moreover, it showed a wide range of hydrogen concentrations (0.05–4 vol.%), a short response time (40 s) and a good selectivity. Its sensing properties were further enhanced after heating to 50, 75 and 100 °C. The excellent sensing properties resulted from the synergistic effect among the three components of WG-Pd. In summary, WG-Pd ternary materials are promising sensing materials for detection of H_2 leakage at temperatures lower than 100 °C, even at room temperature.

Supplementary Materials: The following supporting information can be downloaded at: https://www.mdpi.com/article/10.3390/chemosensors11070410/s1, Figure S1: SEM images of (a) W0.5G, (b) W2G and (c) W3G; Figure S2: XRD patterns of (a) W0.5G, (b) WG, (c) W2G and (d) W3G.

Author Contributions: Conceptualization, L.W.; methodology, F.A.; software, F.A.; validation, F.A.; formal analysis, L.W.; investigation, L.W.; resources, Z.Y.; data curation, F.A.; writing—original draft preparation, L.W.; writing—review and editing, L.W.; visualization, F.A.; investigation, formal analysis, supervision, X.L. and D.Z.; project administration, L.W.; investigation, methodology, supervision, funding acquisition, Z.Y. All authors have read and agreed to the published version of the manuscript.

Funding: This research was funded by The National Natural Science Foundation of China (52003297) and the National Key R&D Program of China (2022YFB3205504).

Institutional Review Board Statement: Not applicable.

Informed Consent Statement: Not applicable.

Data Availability Statement: All data supporting the conclusions of this article are included within this article.

Conflicts of Interest: The authors declare no conflict of interest.

References

1. Jabbr, A.I.; Gaja, H.; Koylu, U.O. Multi-objective optimization of operating parameters for a H_2/diesel dual-fuel compression-ignition engine. *Int. J. Hydrogen Energy* **2020**, *45*, 19965–19975. [CrossRef]
2. Robledo, C.B.; Leeuwen, L.; Wijk, A. Hydrogen fuel cell scooter with plug-out features for combined transport and residential power generation. *Int. J. Hydrogen Energy* **2019**, *44*, 29648–29657. [CrossRef]
3. Wang, F.; Hu, K.; Liu, H.; Zhao, Q.; Zhang, Y. Low temperature and fast response hydrogen gas sensor with Pd coated SnO_2 nanofiber rods. *Int. J. Hydrogen Energy* **2020**, *45*, 7234–7242. [CrossRef]
4. Zhang, Z.; Yin, C.; Yang, L.; Jia, W.; Zhou, J.; Xu, H.; Cao, D. H_2 response characteristics for sol–gel-derived WO_3-SnO_2 dual-layer thin films. *Ceram. Int.* **2017**, *43*, 6693–6699. [CrossRef]
5. Cheng, I.K.; Lin, C.Y.; Pan, F.M. Gas sensing behavior of ZnO toward H_2 at temperatures below 300 °C and its dependence on humidity and Pt-decoration. *Appl. Surf. Sci.* **2021**, *541*, 148551. [CrossRef]
6. Cai, Z.; Park, S. Synthesis of Pd nanoparticle-decorated SnO_2 nanowires and determination of the optimum quantity of Pd nanoparticles for highly sensitive and selective hydrogen gas sensor. *Sens. Actuators B Chem.* **2020**, *322*, 128651. [CrossRef]
7. Xu, K.; Liao, N.; Xue, W.; Zhou, H. First principles investigation on MoO_3 as room temperature and high temperature hydrogen gas sensor. *Int. J. Hydrogen Energy* **2020**, *45*, 9252–9259. [CrossRef]
8. Rahmani, M.B.; Yaacob, M.H.; Sabri, Y.M. Hydrogen sensors based on 2D WO_3 nanosheets prepared by anodization. *Sens. Actuators B Chem.* **2017**, *251*, 57–64. [CrossRef]
9. Chen, M.; Zou, L.; Zhang, Z.; Shen, J.; Li, D.; Zong, Q.; Gao, G.; Wu, G.; Shen, J.; Zhang, Z. Tandem gasochromic-Pd-WO_3/graphene/Si device for room-temperature high-performance optoelectronic hydrogen sensors. *Carbon* **2018**, *130*, 281–287. [CrossRef]
10. Boudiba, A.; Zhang, C.; Umek, P.; Bittencourt, S.R.; Olivier, M.G.; Debliquy, M. Sensitive and rapid hydrogen sensors based on Pd-WO_3 thick films with different morphologies. *Int. J. Hydrogen Energy* **2013**, *38*, 2565–2577. [CrossRef]
11. Mozalev, A.; Calavia, R.; Vazquez, R.M.; Grdcia, I.; Cane, C.; Correig, X.; Vilanova, X.; Gispert-Guirado, F.; Hubalek, J.; Llobet, E. Mems-microhotplate-based hydrogen gas sensor utilizing the nanostructured porous-anodic-alumina-supported WO_3 active layer. *Int. J. Hydrogen Energy* **2013**, *38*, 8011–8021. [CrossRef]
12. Tung, T.T.; Nine, M.J.; Krebsz, M.; Pasinszki, T.; Coghlan, C.J.; Tran, D.; Losic, D. Recent advances in sensing applications of graphene assemblies and their composites. *Adv. Funct. Mater.* **2017**, *27*, 1702891. [CrossRef]
13. Sajjad, S.; Leghari, S.; Iqbal, A. Study of graphene oxide structural features for catalytic, antibacterial, gas sensing and metals decontamination environmental applications. *ACS Appl. Mater. Interfaces* **2017**, *9*, 43393–43414. [CrossRef]
14. Chatterjee, S.G.; Chatterjee, S.; Ray, A.K.; Chakraborty, A.K. Graphene–metal oxide nanohybrids for toxic gas sensor: A review. *Sens. Actuators B Chem.* **2015**, *221*, 1170–1181. [CrossRef]
15. Lonkar, S.; Singh, K.K.; Swaminathan, S.; Abdala, A. Recent advances in graphene based gas sensors. *Sens. Actuators B Chem.* **2015**, *218*, 160–183.
16. Chu, X.; Tao, H.; Feng, G.; Dong, Y.; Sun, W.; Bai, L. Gas sensing properties of graphene–WO_3 composites prepared by hydrothermal method. *Mater. Sci. Eng. B* **2015**, *193*, 97–104. [CrossRef]

17. Gui, Y.; Zhao, J.; Wang, W.; Tian, J.; Zhao, M. Synthesis of hemispherical WO_3/graphene nanocomposite by a microwave-assisted hydrothermal method and the gas-sensing properties to triethylamine. *Mater. Lett.* **2015**, *155*, 4–7. [CrossRef]
18. Tao, Z.; Yuan, R.; Jia, G.; Zhao, Y.; Fan, Y.; Yang, J.; Zhang, X.; Jiang, W.; Wang, L.; Luo, W. Facile synthesis of mesoporous WO_3 @graphene aerogel nanocomposites for low-temperature acetone sensing. *Chin. Chem. Lett.* **2019**, *30*, 2032–2038.
19. Hashtroudi, H.; Atkin, P.; Mackinnon, I.; Shafiei, M. Low-operating temperature resistive nanostructured hydrogen sensors. *Int. J. Hydrogen Energy* **2019**, *44*, 26646–26664. [CrossRef]
20. Zhu, Z.; Xing, X.; Feng, D.; Li, Z.; Tian, Y.; Yang, D. Highly sensitive and fast-response hydrogen sensing of WO_3 nanoparticles via palladium reined spillover effect. *Nanoscale* **2021**, *13*, 12669–12675. [CrossRef]
21. Le, H.J.; Dao, D.V.; Yu, Y.T. Superfast and efficient hydrogen gas sensor using Pd Au alloy@ZnO core–shell nanoparticles. *J. Mater. Chem. A* **2020**, *8*, 12968–12974. [CrossRef]
22. Yang, S.; Wang, Z.; Zou, Y.; Luo, X.; Pan, X.; Zhang, X.; Hu, Y.; Chen, K.; Huang, Z.; Wang, S.; et al. Remarkably accelerated room-temperature hydrogen sensing of MoO_3 nanoribbon graphene composites by suppressing the nanojunction effects. *Sens. Actuators B Chem.* **2017**, *248*, 160–168. [CrossRef]
23. Yang, S.; Zhao, W.; Hu, Y.; Luo, X.; Gu, H. Highly responsive room-temperature hydrogen sensing of α-MoO_3 nanoribbon membranes. *ACS Appl. Mater. Interfaces* **2015**, *7*, 9247–9253. [CrossRef]
24. Kwak, T.D.; Wang, M.; Koski, K.J.; Zhang, L.; Lei, Y. Molybdenum trioxide (α-MoO_3) nanoribbons for ultrasensitive ammonia (NH_3) gas detection: Integrated experimental and density functional theory simulation studies. *ACS Appl. Mater. Interfaces* **2019**, *11*, 10697–10706. [CrossRef]
25. Peng, F.; Wang, S.; Yu, W.; Huang, T.; Dai, N. Ultrasensitive ppb-level H_2 gas sensor at room temperature based on WO_3/rGO hybrids. *J. Mater. Sci. Mater. Electron.* **2020**, *31*, 5008–5016. [CrossRef]
26. Hu, K.; Xie, X.; Szkopek, T.; Cerruti, M. Understanding hydrothermally reduced graphene oxide hydrogels: From reaction products to hydrogel properties. *Chem. Mater.* **2016**, *13*, 1756–1768. [CrossRef]
27. Perfecto, T.M.; Zito, C.A.; Volanti, D.P. Room-temperature volatile organic compounds sensing based on $WO_3{\cdot}0.33H_2O$, hexagonal-WO_3, and their reduced graphene oxide composites. *RSC Adv.* **2016**, *6*, 105171–105179. [CrossRef]
28. Perfecto, T.M.; Zito, C.; Mazon, T.; Volanti, D.P. Flexible room-temperature volatile organic compound sensors based on reduced graphene oxide–$WO_3{\cdot}0.33H_2O$ nano-needles. *J. Mater. Chem. C* **2018**, *6*, 2822–2829. [CrossRef]
29. An, F.; Li, X.; Min, P.; Liu, P.; Jiang, Z.; Yu, Z. Vertically aligned high-quality graphene foams for anisotropically conductive polymer composites with ultrahigh through-plane thermal conductivities. *ACS Appl. Mater. Interfaces* **2018**, *10*, 17383–17392. [CrossRef]
30. Shi, J.; Cheng, Z.; Gao, L.; Yuan, Z.; Zhao, H. Facile synthesis of reduced graphene oxide/hexagonal WO_3 nanosheets composites with enhanced H_2 sensing properties. *Sens. Actuators B Chem.* **2016**, *230*, 736–745. [CrossRef]
31. Lee, J.; Kim, S.Y.; Yoo, H.S.; Lee, W. Pd-WO_3 chemiresistive sensor with reinforced self-assembly for hydrogen detection at room temperature. *Sens. Actuators B Chem.* **2022**, *368*, 132236. [CrossRef]
32. Kumar, N.; Haviar, S.; Zeman, P. Three-layer $PdO/CuWO_4/CuO$ system for hydrogen gas sensing with reduced humidity interference. *Nanomaterials* **2021**, *11*, 3456–3471. [CrossRef] [PubMed]
33. Walker, J.M.; Akbar, S.A.; Morris, P.A. Synergistic effects in gas sensing semiconducting oxide nano-heterostructures: A review. *Sens. Actuators B Chem.* **2019**, *286*, 624–640. [CrossRef]
34. Li, Z.; Huang, Y.; Zhang, S.; Chen, W.; Kuang, Z.; Ao, D.; Liu, W.; Fu, Y. A fast response & recovery H_2 gas sensor based on α-Fe_2O_3 nanoparticles with ppb level detection limit. *J. Hazard. Mater.* **2015**, *300*, 167–174.
35. Chang, S.J.; Hsueh, T.J.; Chen, I.C.; Huang, B.R. Highly sensitive ZnO nanowire CO sensors with the adsorption of Au nanoparticles. *Nanotechnology* **2008**, *19*, 175502. [CrossRef] [PubMed]
36. Zhou, R.; Lin, X.; Xue, D.; Zong, F.; Zhang, J.; Duan, X.; Li, Q.; Wang, T. Enhanced H_2 gas sensing properties by Pd-loaded urchin-like $W_{18}O_{49}$ hierarchical nanostructures. *Sens. Actuators B Chem.* **2018**, *260*, 900–907. [CrossRef]
37. Al-Hashem, M.; Akbar, S.; Morris, P. Role of oxygen vacancies in nanostructured metal-oxide gas sensors: A review. *Sens. Actuators B Chem.* **2019**, *301*, 126845. [CrossRef]
38. Kathiravan, D.; Huang, B.; Saravanan, A. Self-assembled hierarchical interfaces of ZnO nanotubes/graphene heterostructures for efficient room temperature hydrogen sensors. *ACS Appl. Mater. Interfaces* **2017**, *9*, 12064–12072. [CrossRef]
39. Zhang, L.S.; Wang, W.D.; Liang, X.Q.; Chu, W.S.; Song, W.G.; Wang, W.; Wu, Z.Y. Characterization of partially reduced graphene oxide as room temperature sensor for H_2. *Nanoscale* **2011**, *3*, 2458–2460. [CrossRef]

chemosensors

MDPI

Article

The Human Nose as a Chemical Sensor in the Perception of Coffee Aroma: Individual Variability

Roberto Crnjar, Paolo Solari and Giorgia Sollai *

Department of Biomedical Sciences, University of Cagliari, 09042 Monserrato, Italy
* Correspondence: gsollai@unica.it; Tel.: +39-070-6754160

Abstract: The flavor of foods and beverages is generally composed of a mixture of volatile compounds, however not all the molecules that form an aroma are sensorially relevant. The odor-active compounds present in a mixture are different for each subject, both in quantitative and qualitative terms. This means that the ability of the human nose to act as a chemical sensor varies among individuals. In this study, we used the headspace of roasted coffee beans as a complex olfactory stimulus and, by means of the coupled Gas Chromatography-Olfactometry (GC-O) technique, the single components of coffee flavor were separated. Each subject, previously classified for his/her olfactory status (normosmic, hyposmic or anosmic) by means of the Sniffin' Sticks battery (composed of Threshold, Discrimination and Identification subtests), had to identify and evaluate each smelled molecule. The results show that the individual ability to detect individual compounds during the GC-O experiments and the odor intensity reported during the sniffing of pen #10 (the pen of the identification test) containing coffee aroma were related to TDI olfactory status (based on the score obtained from the sum composed of Threshold, Discrimination and Identification scores). We also found that the number of total molecules and of molecules smelling of coffee is linearly related to the TDI olfactory score. Finally, the odor intensity reported when sniffing pen #10 containing coffee aroma is positively correlated with the number of molecules detected and the average intensity reported. In conclusion, our findings show that the human perception of both individual compounds and complex odors is strongly conditioned by the olfactory function of subjects.

Keywords: inter-individual physiological variations; olfactory function; VARU intensity; Sniffin' Sticks test; olfaction

Citation: Crnjar, R.; Solari, P.; Sollai, G. The Human Nose as a Chemical Sensor in the Perception of Coffee Aroma: Individual Variability. *Chemosensors* **2023**, *11*, 248. https://doi.org/10.3390/chemosensors11040248

Academic Editor: Andrea Ponzoni

Received: 8 March 2023
Revised: 31 March 2023
Accepted: 15 April 2023
Published: 17 April 2023

1. Introduction

All living organisms are able to sense the chemicals present in the environment where they live to obtain useful information on the availability of energy-rich food sources, potential mating partners and on the presence of predators [1–8]. All odorous molecules, whether natural or synthetic and whether perceived as pleasant or unpleasant, are highly volatile and can activate the olfactory receptors present in the human nose or in the olfactory organs of all animals [8–14]. In particular, in humans the perception of odors has been observed to affect the quality of life, exerting a relevant influence on eating habits of individuals and consequently on their body weight, on the ability to perceive molecules that signal the presence of dangers (e.g., toxic and/or harmful gases, smoke and spoiled food) and in interpersonal relationships [15–20].

Most odors found in nature, such as those of food and drink, are rarely formed by individual compounds; more commonly they are complex mixtures, composed of multiple molecules, only some of which are sensorially relevant [21–23]. An important challenge is therefore represented by the understanding of which molecules within a mixture are perceived, and thus constitute the odor-active compounds, and which remain irrelevant from a sensory point of view. This problem can be solved using the gas chromatography-olfactometry (GC-O) technique which simultaneously uses the chromatographic column to separate the individual compounds within a mixture and the human nose as a chemical sensor capable of perceiving and evaluating the volatiles coming out of the olfactometric port [24–28]. In addition to understanding olfaction at a physiological level, many attempts have been recently made to detect odorants and volatile organic compounds by means of artificial olfactory sensor technology, which was widely introduced in many fields, such as environmental monitoring, detection of food conditions and clinical diagnostics [29–31]. Semiconductor metal oxides such as SnO_2 and TiO_2, for instance, are largely employed as gas sensors for a number of favorable functional properties like good stability and sensitivity combined with low cost [32,33].

An important aspect that should not be underestimated is that the ability to perceive individual compounds is directly correlated with the olfactory function of individuals who, as known, present a great variability [34–36]. In fact, as a result of physiological, genetic, behavioral and environmental factors, humans can be classified as normosmic (normal olfactory function), hyposmic (reduced olfactory function) or functionally anosmic (olfactory blindness, which can be general or specific) [16,37–50]. It is known that the human olfactory system has a high potential in terms of discrimination and sensitivity even if, at present, the number of stimuli that humans are able to perceive has not yet been quantified [51]. Furthermore, the information available on the potential of the human nose as a chemical sensor in relation to the olfactory function of individuals is still lacking, in terms of odor-active compounds, in the intensity at which they are perceived and in our ability to identify and/or associate them with the aroma of the mixture to which they belong [26,36,52].

Based on these considerations, this study had three different but interconnected objectives, with the aim of better understanding the relationship between the olfactory function of individuals, in terms of normosmia or hyposmia, and the ability of the human nose to act as a chemical sensor of individual molecules belonging to a complex mixture, both in quantitative (number of molecules perceived) and qualitative (intensity of perception) terms. The first objective regards the effect of the olfactory function of individuals on their ability to smell the individual compounds of the complex aroma of coffee as they are eluted from the chromatographic column and the correlation between the number of odor-active compounds and the TDI olfactory score obtained by each subject. With the second objective we evaluated whether the intensity reported by subjects for the odor of coffee contained in the pen #10 of the identification test (one of the subtests of the Sniffin' Sticks battery) was correlated with the number of odor-active compounds and with the average intensity with which they were perceived. Finally, we investigated the effect of olfactory status (normosmic vs. hyposmic) on perceived intensity for pen #10 and its correlation with the TDI olfactory score obtained by each participant.

2. Materials and Methods

2.1. Subjects

Thirty-eight Caucasian healthy non-smoking volunteers (24 females, 14 males, 28.8 ± 1.61 y), recruited in Cagliari (Sardinia, Italy), took part in this study. To estimate the ability to detect the odor-active compounds, during the GC-O analysis we used the detection frequency method, which does not require the presence of qualified evaluators and the results it produces represent the inter-individual variability [24,28,53–55].

All participants fasted for at least 90 min prior to testing and were free of perfume. Before starting the olfactory tests, the experimental protocol approved by the local Ethical Committee was read to them and they were asked to sign an informed consent (Prot. PG/2018/22 of 2 January 2018).

2.2. Olfactory Sensitivity Screening

The olfactory function of each individual was evaluated by means of the TDI olfactory score given by the sum of the scores obtained with the tests of Threshold (T-test), Discrimination (D-test) and Identification (I-test), which represent the subtests of the Sniffin' Sticks test (Burghart Instruments, Wedel, Germany), based on odor-containing felt-tip pens [56]. According to the reference values reported by Hummel et al. [57], based on the total TDI olfactory score obtained, and on sex and age, each participant was classified as normosmic, hyposmic or functionally anosmic. For each olfactory test the score is between 0 and 16. For the T-test, the score is given by the average of the last 4 reversals out of 7, while for the D-test and I-test the score is given by the number of correct discriminations and identifications (for details visit: https://www.uniklinikum-dresden.de/de/das-klinikum/kliniken-polikliniken-institute/hno/forschung/interdisziplinaeres-zentrum-fuer-riechen-und-schmecken/neuigkeiten/downloads).

For the coffee-odor of pen #10 presented during the I-test, the subjects must also give a personal evaluation of the perceived intensity, marking a sign on the "Visual Analogue Rating Units" (VARUs) scale, ranging from 0 to 20 VARUs [58].

2.3. Dynamic Headspace Sampling

The dynamic headspace method, as described by Rizzolo et al. [59] and Nuzzi et al. [27], was used to collect the volatile compounds. In terms of volatiles, the dynamic headspace method is considered the most appropriate for obtaining an extract whose composition is closely linked to the quality of the scent as assessed by the consumer [60]. In addition, it has the ability to acquire extracts for GC-MS and sensory assessment via GC-O analysis by a human assessor [27].

In detail, approximately 100 g of roasted coffee beans were placed in a 0.5 L airtight glass vessel, with a flow-through system fitted to a Porapak Q (150/75 mg, 50/80; Supelco; St. Louis, MO, USA) in a glass adsorption tube (5 mm Ø) inserted into the collection port at the top of the vessel. By flushing the system with purified air for three hours at a rate of 30 L/h (500 mL/min), volatiles were recovered at room temperature. Using 1.5 mL of 1-hexane, trapped volatiles were released from the Porapak Q tube, resulting in a solution containing the isolated volatile chemicals. Samples were then stored at −20 °C until used. By performing three GC runs 24 h after sample preparation and before they were used for GC-O experiments, the effectiveness of the extraction and the reproducibility of the chromatogram were confirmed. Besides, to verify that the sample was not altered, before each section of the GC-O experiment, a GC-run without any sniffing session was made. The fact that the volatile chemical profile observed in this study is remarkably comparable to the headspace volatile profiles published in the literature data provides evidence of its validity [61–69].

2.4. Mass Spectrometry/Gas Chromatography–Olfactometry (MS/GC-O) Analysis

An Agilent 6890N gas chromatograph (GC; Agilent technologies; Santa Clara, CA, USA) was simultaneously connected with an Agilent model 5973 series mass spectrometer (MS) and with an olfactometry detection system (Gerstel ODP3; Mülheim an der Ruhr, Germany) to perform the analyses. A constant flow of 1.2 mL/min of He was used as carrier gas. The flow was split 1:1 between the olfactometry and the MS detector at the outlet of the chromatographic column and the injection volume was 1 µL [36].

The chromatographic column was a 30 m HP-INNOWax, 0.25 mm internal diameter × 0.50 µm film thickness (Agilent 19091N-233; Agilent technologies, Santa Clara, CA, USA). The temperature of the injector and the MS interface temperature were set at 250 °C and 260 °C respectively. The oven temperature was maintained at 40 °C (0.2 min), 40 °C/min to 90 °C (0.50 min), 2 °C/min to 150 °C, 30 °C/min to 230 °C (12 min). The injector mode was splitless; the temperatures for the ion source and the quadrupole mass filter were 230 °C and 150 °C, respectively. Chromatograms were recorded by monitoring the total ion current in a 40–550 mass range. The transfer line to the GC-ODP3 sniffing port was held at 220 °C [36].

To identify the volatiles, we used the mass spectrum found in the MS Standard Library NIST2014 (US National Institute of Standards and Technology; Gaithersburg,

MD, USA). In accordance with Gonzales-Kristeller et al. [70], the Good Scents Company Information System (www.thegoodscentscompany.com; 20 January 2023) was used to obtain information regarding odorant natural occurrence, "odor type" (i.e., roasted, floral, woody, etc.) and "odor descriptors" (i.e., coffee, fruit, cheese, wood, etc.).

The "Sniffin' Sticks" test, as previously described, was used to characterize the olfactory function of each panelist, prior to testing. Participants were asked to assess the volatile strength and duration during elution for the GC-O analyses [27,71] by using a PC-connected audio recorder and digital signaling system (GERSTEL ODP recorder 3 for Windows 7). The signaling system is characterized by the presence of 4 keys that represent a 4-point intensity scale: 1 = weak odor, 2 = distinct odor, 3 = intense odor, 4 = very intense odor. The participant pressed one of the signaling system keys whenever he/she detected an odor to express his/her subjective assessments of the aroma intensity (based on which button was pressed), the stimulus duration, the degree of pleasantness or unpleasantness and the description of the odor-active compound. The chromatograms were overlaid with the obtained olfactograms after the PC automatically recorded the retention time and sniffing time of each odor-active compound. The samples were presented completely blind to avoid psychological conditioning.

2.5. Statistical Analysis

The Pearson correlation test was used to evaluate the relationship between: (a) the total number of odor-active compounds (hereafter, total-compounds) or the number of odor-active compounds smelling of coffee (hereafter, coffee-compounds) smelled by each subject and his/her TDI olfactory score; (b) the intensity perceived by each subject for the pen of the identification test containing the coffee aroma (hereafter, the coffee-odor pen) and his/her TDI olfactory score, (c) the number of total- and coffee-molecules smelled and the intensity perceived for the coffee-odor pen by each subject, (d) the reported average intensity for total- and coffee-molecules and the perceived intensity for the coffee-odor pen by each subject. The correlation coefficient "r" was considered to measure the strength of the linear relationship or straight-line between two variables: $r < 0.3$ means lower correlation, $0.3 < r < 0.7$ means medium correlation, $r > 0.7$ means higher correlation [72,73]. Statistical analyses were performed using GraphPad Prism 6 (GraphPad Software, San Diego, CA, USA). A statistically significant correlation was defined with a p-value < 0.05.

One-way ANOVA was used to analyze the effect of the TDI olfactory status of the subject on: (a) his/her ability to smell individual compounds during the GC-O analysis and (b) the intensity referred for the coffee-odor pen. Post-hoc comparisons, following one-way ANOVA, were conducted with the Fisher's test of least significant difference (LSD). Statistical analyses were performed using STATISTICA for WINDOWS (version 7.0; StatSoft Inc., Tulsa, OK, USA). p values < 0.05 were considered significant.

3. Results

3.1. Volatile Compounds of Coffee Aroma

The 50 volatile compounds obtained from the extract of the roasted coffee beans by means of the dynamic headspace method are listed in Table 1. Using the information from "The Good Scent Company System" on the organoleptic properties, we classified each volatile for its odor type and odor descriptors. On the basis of the odor type information, we found nine nutty compounds, five fruity, two buttery, two bready, two fishy and one for each of the following odor type: coffee, chocolate, sweet, potato, creamy, caramellic, cheesy, roasted, winey, citrus, green, vegetable, waxy, musty, balsamic, aromatic, phenolic, acidic and terpenic. Based on odor descriptors, 21 compounds were identified as smelling of coffee, eight of which are of the nutty odor type, three are of the fruity odor type, two are of the bready odor type, one is of the phenolic, fishy, coffee, balsamic buttery, caramellic, roasted or vegetable odor type. In the Good Scent Company System, we did not find any information regarding odor type for 11 compounds (no. 1, 2, 7, 18, 33, 35, 36, 38, 44 and 47 in Table 1) and for 7 regarding odor descriptor (no. 1, 6, 33, 35, 36, 44 and 47 in Table 1).

Table 1. Volatile compounds found in the headspace of roasted coffee beans.

N.	Compound	RT [a]	Odor Type [b]	Odor Descriptors [b]
1	Octane, 3,5-dimethyl-	8.10	-	-
2	Oxalic acid, isobutyl nonyl ester	8.33	-	Bland, mild, caramellic
3	Toluene	8.54	Phenolic	Solventy, woody, roasted, coffee
4	β-Pinene	10.20	Terpenic	Sweet, fresh, pine, woody, hay, green
5	Ethylbenzene	10.69	-	Petroleum-like odor
6	p-Xylene	10.90	Aromatic	-
7	Oxalic acid, isobutyl pentyl ester	11.85	-	Bland, mild, caramellic
8	Pyridine *	12.55	Fishy	Sour, sickening, putrid, coffee
9	D-Limonene *	12.81	Citrus	Citrus, orange, fresh, sweet
10	Furan,2-pentyl- *	13.30	Fruity	Fruity, green, earthy, beany, vegetable, metallic
11	Pyrazine, methyl- *	15.65	Nutty	Coffee, cocoa, roasted, chocolate, peanut, green, nutty brown, musty, earthy
12	Acetoin	16.50	Buttery	Sweet, creamy, green, butter, dairy, milk, fatty, buttery, creamy, sour, fatty, vanilla
13	2-Propanone, 1-hydroxy-	17.28	Sweet	Pungent, sweet, caramellic, ethereal
14	Pyrazine, 2,5-dimethyl- *	18.11	Chocolate	Cocoa, roasted nutty, beefy roasted, beefy, woody, grassy, medicinal
15	Pyrazine, ethyl- *	18.67	Nutty	Peanut, butter, musty, woody, roasted, cocoa, coffee
16	Pyrazine, 2,3-dimethyl- *	19.30	Nutty	Musty, nut skin, cocoa, powdery, caramellic, roasted, potato, coffee, peanut, butter,
17	DL-2,3-Butanediol *	19.81	Creamy	Fruity, creamy, buttery
18	Vinyl butyrate	20.02	-	Organic solvent
19	Hex-4-yn-3-one, 2,2-dimethyl-	20.67	Winey	Chemical, winey, fruity, fatty, terpenic, cauliflower
20	Pyrazine, 2-ethyl-6-methyl- *	21.08	Potato	Roasted, potato
21	Pyrazine, 2-ethyl-3-methyl- *	22.09	Nutty	Nutty, peanut, musty, corn, raw, coffee, bready
22	Pyrazine, 2-(n-propyl)- *	22.89	Green	Green, vegetable, nutty
23	Pyrazine, 2,6-diethyl- *	23.59	Nutty	Nutty, potato, cocoa, roasted, coffee
24	Pyrazine, 3-ethyl-2,5-dimethyl- *	24.08	Nutty	Potato, cocoa, roasted, nutty, coffee
25	2-Propanone, 1-(acetyloxy)-	24.70	Fruity	Fruity, buttery dairy, nutty
26	Pyrazine, 2-ethyl-3,5-dimethyl- *	25.01	Nutty	Peanut, caramellic, coffee, cocoa
27	Furfural *	25.27	Bready	Sweet, brown, woody, caramellic, bread baked, coffee, almond
28	Pyrazine, tetramethyl-	25.71	Nutty	Nutty, musty, chocolate, coffee, cocoa, brown, lard, burnt, dry, vanilla
29	Pyrazine, 3,5-diethyl-2-methyl- *	26.66	Nutty	Nutty, meaty, vegetable
30	Pyrazine, 2-ethenyl-5-methyl-	27.10	Coffee	Coffee
31	Furan, 2-acetyl- *	27.67	Balsamic	Sweet, balsamic, almond, cocoa, coffee, caramellic, nutty, brown, toasted, milky, lactonic

Table 1. *Cont.*

N.	Compound	RT [a]	Odor Type [b]	Odor Descriptors [b]
32	2,3-Pentanedione *	28.21	Buttery	Pungent, sweet, buttery, creamy, nutty, caramellic, cheesy, coffee
33	2-Butanone, 1-(acetyloxy)-	28.52	-	-
34	2-Furanmethanol, acetate *	28.81	Fruity	Coffee, sweet, fruity, banana, horseradish, roasted, cocoa
35	Pyrazine, 2-methyl-6-(2-propenyl)-	29.37	-	-
36	2-Cyclopenten-1-one, 2,3-dimethyl-	30.12	-	-
37	Acetic acid, diethyl- *	30.56	Acidic	Acidic, fruity, whiskey, dry berry, dairy, tropical
38	Pentanoic acid, 4-oxo-, methyl ester	31.05	-	Caramellic, flavouring agent
39	2-Furancarboxaldehyde, 5-methyl- *	31.54	Caramellic	Sweet, caramellic, bready, brown, coffee, spicy, maple
40	2-Furanmethanol, propanoate *	32.16	Fruity	Sweet, fruity, green, banana, oily, coffee, spicy
41	Furan, 2,2′-methylenebis- *	32.81	Fruity	Rich, roasted, coffee
42	2-Furanmethanol *	34.17	Bready	Sulfurous, estery, chemical, musty, sweet, brown, caramellic, bready, coffee, alcoholic
43	Butanoic acid, 3-methyl- *	34.35	Cheesy	Cheesy, dairy, acidic, sour, pungent, ripe, fatty, fruity, stinky feet, sweaty, tropical
44	Furan, 2-(2-furanylmethyl)-5-methyl- *	34.68	-	-
45	Pyrazine, 2-acetyl-6-methyl	35.33	Roasted	Roasted, coffee, cocoa, popcorn
46	4(H)-Pyridine, N-acetyl- *	35.76	Fishy	Sour, fishy, putrid, ammoniacal
47	Octaethylene glycol monododecyl ether	36.21	-	-
48	2-Hexadecanol	36.38	Waxy	Waxy, clean, greasy, floral, oily
49	N-Furfurylpyrrole *	37.95	Vegetable	Plastic, green, waxy, fruity, coffee, vegetable
50	2-Acetylpyrrole *	40.68	Musty	Musty, nut, skin, cherry, maraschino, cherry, bready, coumarinic, licorice, walnut

[a] RT = retention time in I-Wax column. [b] Odor type and odor descriptors from the Good Scent Company Information System. (www.thegoodscentscompany.com). Asterisks indicate the molecules that have also been found in other coffee extracts [56–64].

Table 2 shows that 47 compounds found in the extract were odor-active for at least two of the participants; in fact, the "ethylbenzene" and "furan, 2-(2-furanylmethyl)-5-methyl" (indicated with no. 5 and 44 in Table 1) was found to be active for just one individual; while, the "2-Butanone, 1-(acetyloxy)-" (indicated with no. 33 in Table 1) it was not perceived by any of the participants. Furthermore, the panelists described 21 of the odor-active compounds as smelling of coffee, even though the odor descriptors reported in Table 1 define only 17 of them as actually having a coffee odor.

Table 2. GC-O analysis: odor-active compounds and odor descriptions by subjects.

N.	Odor-Active Compound	Odor Description	df
1	Octane, 3,5-dimethyl-	Woody, unknown	2
2	Oxalic acid, isobutyl nonyl ester	Burnt, unknown	2
3	Toluene	Coffee, smoked, solvent, roasted	8
4	β-Pinene	Sweet, floral, vanilla	8
5	Ethylbenzene	Petrol	1
6	p-Xylene	Vanilla, medicinal, floral	5
7	Oxalic acid, isobutyl pentyl ester	Floral, fruity, vanilla	6
8	Pyridine *	Coffee, smoked, roasted, cheese	3
9	D-Limonene *	Sweet, sour, citrus	6
10	Furan,2-pentyl- *	Smoked	2
11	Pyrazine, methyl- *	Coffee, nutty, roasted, smoke	3
12	Acetoin	Coffee, sweet, roasted, parfum	10
13	2-Propanone, 1-hydroxy-	Sweet, pungent, fish, solvent, wet	10
14	Pyrazine, 2,5-dimethyl- *	Coffee, citrus, medicinal, sweet, cocoa	7
15	Pyrazine, ethyl- *	Coffee, nutty, egg	3
16	Pyrazine, 2,3-dimethyl- *	Coffee, burnt, caramellic, fruity	5
17	DL-2,3-Butanediol *	Sweet, caramellic, rose, wet	5
18	Vinyl butyrate	Floral, parfum, bitter, solvent, pungent	7
19	Hex-4-yn-3-one, 2,2-dimethyl-	Sweet, solvent	4
20	Pyrazine, 2-ethyl-6-methyl- *	Coffee, sweet, smoked, medicinal, solvent, parfum, roasted	19
21	Pyrazine, 2-ethyl-3-methyl- *	Coffee, cocoa, solvent, bitter, nutty, roasted	25
22	Pyrazine, 2-(n-propyl)- *	Green, musty, woody, earthy, wet, herbs, floral	22
23	Pyrazine, 2,6-diethyl- *	Coffee, roasted, earthy, musty, burnt, mushrooms	25
24	Pyrazine, 3-ethyl-2,5-dimethyl- *	Coffee, nutty, roasted, floral, bitter	20
25	2-Propanone, 1-(acetyloxy)-	Pungent, parfum	6
26	Pyrazine, 2-ethyl-3,5-dimethyl- *	Coffee, musty, roasted, wet	21
27	Furfural *	Coffee, sweet, solvent, floral, pungent	13
28	Pyrazine, tetramethyl-	Coffee, roasted, burnt, vanilla	13
29	Pyrazine, 3,5-diethyl-2-methyl- *	Floral, musty, wet, solvent, fresh	15
30	Pyrazine, 2-ethenyl-5-methyl-	Coffee, nutty, bitter, plastic	14
31	Furan, 2-acetyl- *	Parfum	2
32	2,3-Pentanedione *	Floral, herbs, earthy, sweat, musk, cheese, pungent	24
34	2-Furanmethanol, acetate *	Roasted, fruit, earthy, herb, woody, coffee	21
35	Pyrazine, 2-methyl-6-(2-propenyl)-	Pungent, sour, bitter	6
36	2-Cyclopenten-1-one, 2,3-dimethyl-	Sweet, floral, lavender	4
37	Acetic acid, diethyl- *	Roasted, solvent, rotten, musty, herbs, wet earth	22
38	Pentanoic acid, 4-oxo-, methyl ester	Sweet	4
39	2-Furancarboxaldehyde, 5-methyl- *	Coffee, sweet, parfum	4
40	2-Furanmethanol, propanoate *	Coffee, pungent, floral, musty, herb, sweet, burnt	14
41	Furan, 2,2′-methylenebis- *	Coffee, nutty, popcorn, roasted, fish, sour	21
42	2-Furanmethanol *	Coffee, smoke, popcorn, nutty, roasted	24
43	Butanoic acid, 3-methyl- *	Cheese, smoke, stinky feet, acidic, fruity, putrid	16
44	Furan, 2-(2-furanylmethyl)-5-methyl- *	Unknown	1
45	Pyrazine, 2-acetyl-6-methyl	Putrid, musty, cheese	6
46	4(H)-Pyridine, N-acetyl- *	Shoes, wet, sweat	7
47	Octaethylene glycol monododecyl ether	Sweat, acidic	4
48	2-Hexadecanol	Cheese, musty, putrid, plastic	29
49	N-Furfurylpyrrole *	Solvent, cheese, musty	15
50	2-Acetylpyrrole *	Coffee, roasted, almond, sweet, burnt, parfum, fresh	29

Odor-active compounds: list of compounds eluted from the chromatography column and smelled by at least one participant. Odor description: personal description given by each subject for the odor smelled. df = detection frequency (number of participants who smelled the compound). Asterisks indicate the molecules that have also been found in other coffee extracts [56–64].

3.2. Olfactory Function and Odor-Active Compounds

One-way ANOVA showed that the number of molecules perceived for both total and smelling of coffee is significantly higher for normosmic individuals than for hyposmic ones (total-molecules: F (1,36) = 16.19, $p = 0.0003$; coffee-molecules: F (1,36) = 24.25, $p < 0.0001$) (Figure 1).

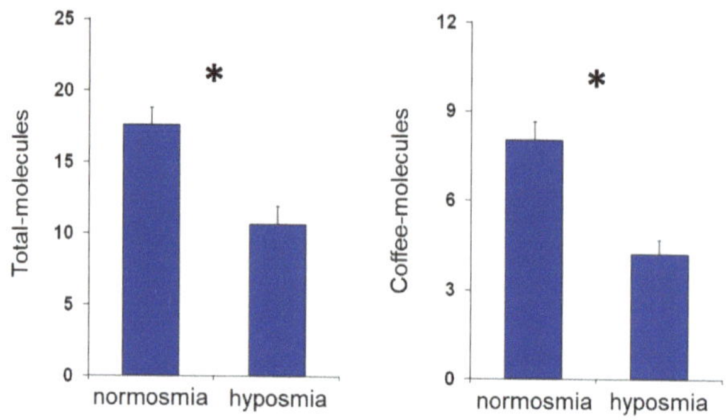

Figure 1. Mean (±SEM) values on the number of total- and coffee-molecules smelled during the GC-O experiments by each subject, according to their TDI olfactory status. Asterisk indicates significant differences between individuals with normosmia or hyposmia ($p < 0.0005$; Fisher's LSD test subsequent to one-way ANOVA).

The Pearson correlation test was used to investigate for a correlation between the number of odor-active compounds and the TDI olfactory score reached by each subject. In detail, the results shown in Figure 2 indicated that TDI olfactory score was positively correlated with both the number of total-molecules (Pearson r = 0.56, $p = 0.0002$) and of coffee molecules smelled by each subject (Pearson r = 0.55, $p = 0.0004$).

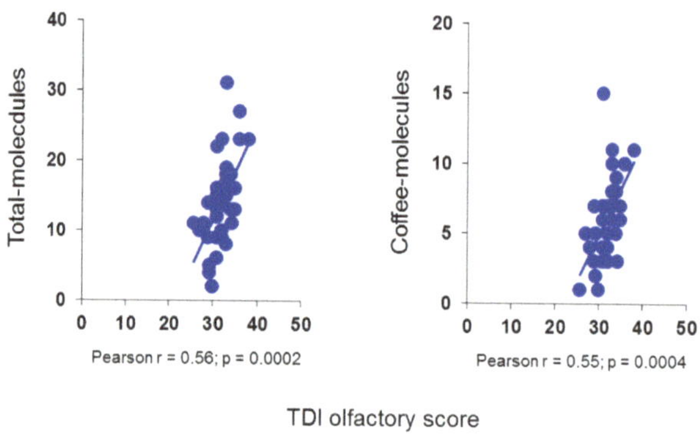

Figure 2. Correlation analysis between the number of total- and coffee-molecules smelled by each subject and his/her TDI olfactory score.

Besides, a positive correlation was found between the intensity value that each subject attributed to the coffee-odor pen and the number of odor-active compounds (total-molecules: Pearson r = 0.67, $p < 0.0001$; coffee-molecules: Pearson r = 0.65, $p < 0.0001$) (Figure 3) and the average intensity referred for total- and coffee-molecules (total-molecules: Pearson r = 0.56, $p = 0.0003$; coffee-molecules: Pearson r = 0.62, $p < 0.0001$) (Figure 4).

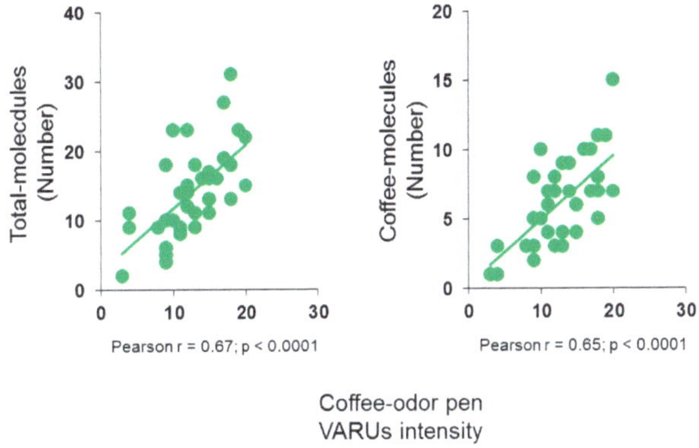

Figure 3. Correlation analysis between the number of total- and coffee-molecules smelled, and the intensity perceived for the coffee-odor pen by each subject.

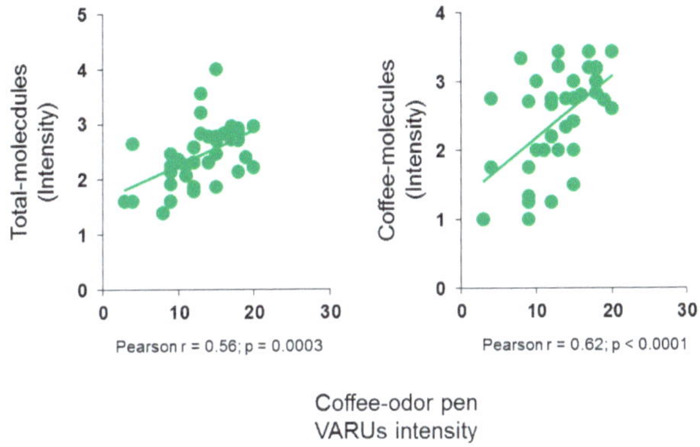

Figure 4. Correlation analysis between the average intensity referred for total- and coffee-molecules and the intensity perceived for the coffee-odor pen by each subject.

The mean values ± SEM of the intensity perceived for the coffee-odor pen by panelists classified by their TDI olfactory status are shown in Figure 5. One-way ANOVA revealed that the intensity perceived by normosmic individuals was significantly higher than that of hyposmic individuals ($F(1,36) = 11.63$, $p = 0.0016$).

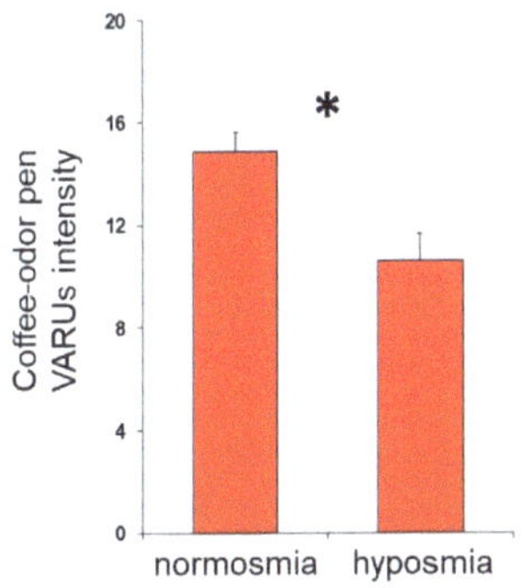

Figure 5. Mean (±SEM) values of the intensity perceived for the coffee-odor pen by subjects, according to their TDI olfactory status. Asterisk indicates significant differences between individuals with normosmia or hyposmia ($p < 0.005$; Fisher's LSD test subsequent to one-way ANOVA).

The Pearson correlation test also revealed that the intensity of the coffee-odor pen reported by each subject was positively correlated with his/her TDI olfactory score (Pearson r = 0.49, p = 0.0016) (Figure 6).

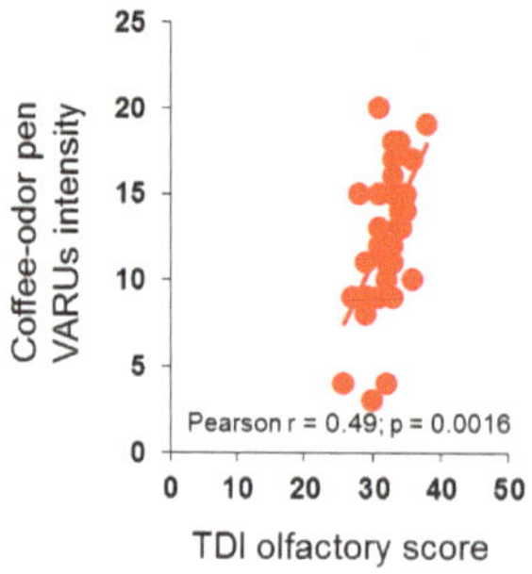

Figure 6. Correlation analysis between the intensity perceived for the coffee-odor pen by each subject and his/her TDI olfactory score.

4. Discussion

The main objective of this study was to evaluate the role of the human nose as a chemical sensor, particularly, in relation to its ability to perceive the single odorous molecules that make up a complex odor. It is known that the human nose has a vast sensitivity and a high discriminative power, but the number of odorants it can perceive is still unknown [51], especially when considering that among humans there is a great inter-individual variability due to multiple factors: physiological, genetic, environmental, cultural and behavioral [18,20,37,38,40,42–50]. Most of the odors that surround us, and in particular those of food and drink, are composed by a mixture of volatile compounds that can be separated by means of a chromatographic column and used as single olfactory stimuli [24,25,53,71,74–77].

In this study, by means of the coupled Gas Chromatography-Olfactometry (GC-O) technique, the single components of the coffee aroma were separated, identified and verbally evaluated by each subject, using their own nose as a chemical sensor. The number of odor-active compounds (i.e., the number of sensory active molecules for the subject during the GC-O experiment) was evaluated by means of the frequency detection method [36,54,71], which has the advantage of not requiring qualified participants and of highlighting inter-individual variability [28,54]. The results show that all compounds eluted from the chromatographic column were perceived by the participants and that 17 of the 21 compounds commonly defined as smelling of coffee, were described as having coffee odor. This is remarkable considering that the participants were unaware of the mixture injected into the chromatographic column, so they did not have a mental representation of the odor, known to exert a great influence on the formation of the perceived odor quality [78].

Since little is known about the ability of the human nose to perceive individual compounds as they elute from the chromatography column, we evaluated the effect of subjects' olfactory function on their ability to smell the individual molecules that make up the complex odor of coffee. Our results show that the number of odor-active compounds smelled by each individual depends on his/her olfactory status. In fact, for individuals classified as normosmic the number of odor-active compounds was significantly higher than that of those classified as functionally hyposmic. The TDI olfactory status represents the general olfactory status of the individual attributed on the basis of the score obtained from the sum of the olfactory threshold, discrimination and identification scores. This means that a condition of hyposmia can be determined by reduced ability in all three olfactory performances, or in two of them or only in one. We believe that the number of odor-active compounds for hyposmic individuals is lower than for normosmic ones, due to a reduced ability to perceive and discriminate odors. The olfactory threshold represents the minimum concentration that an odor must have in order to be perceived: hyposmic individuals could present an increased olfactory threshold (i.e., to perceive odors, they must have a higher concentration than that required by normosmic individuals) and consequently they may not perceive those odors that are eluted at subthreshold concentrations for them. Discrimination represents the ability to recognize different odors: hyposmic subjects show difficulty in recognizing different odors that are similar, and this could reduce the number of single compounds that they smell during GC-O experiments. This is compatible with the fact that in our sample the general state of hyposmia is mainly determined by a low score obtained with the threshold and discrimination tests. In addition, correlation analyses showed that the number of odor-active compounds is directly correlated with the TDI olfactory score achieved by each individual. Furthermore, results are similar whether we consider the number of total perceived molecules or of molecules commonly defined as coffee odorants. Within each complex odor there are both molecules that even individually smell like the complex odor and molecules whose odor is completely different. This aspect is noteworthy if we consider that the more sensorially active molecules are also those that contribute more to the odor of the mixture [14,26,53]. Therefore, the individual perception of a complex odor is strongly conditioned by its intensity, number and type of individual compounds perceived, making the odor unique and characteristic for each person. This means that the idea that each individual has of a complex odor, formed by a set of many molecules, may be different from that of other individuals for whom the odor-active compounds are different both quantitatively and qualitatively. This also explains, at least in part, why the intensity with which an odor is perceived differs between individuals.

Based on these considerations, the second aim of our work was to evaluate whether a correlation exists between the perceived intensity of the coffee-odor pen and both the number of odor-active compounds and the intensity with which they are perceived. The results we obtained show that the reported intensity for the coffee-odor pen is positively correlated with the number of odor-active compounds, both total and coffee-smelling. Moreover, a positive correlation was also found for the reported intensity of individual compounds perceived during the GC-O experiments: the greater the intensity with which each individual perceives each molecule, the greater the intensity with which the mixture is perceived. These results are in agreement with the fact that in our sample the condition of hyposmia of individuals is mainly determined by a reduced ability of odor discrimination and an increased odor threshold. Therefore, on the one hand the sensorially active molecules are less numerous and on the other, the odor-active compounds are perceived with less intensity, making the perception of the complex coffee blend less intense.

Finally, given the correlation between the number of odor-active compounds and the TDI olfactory score and between the number of odor-active compounds and the intensity of the coffee-odor pen, the last objective of this study was to look for a correlation between the perceived intensity of the coffee-odor pen and the olfactory function of each individual. The results show that the values of the coffee-odor pen intensity and those of TDI are linearly correlated and that normosmic individuals report perceiving the odor of coffee more intensely than hyposmic ones. These findings are in accordance with a previous study in which a positive correlation was found between the subjects' olfactory function and their ability to detect individual compounds eluted from a chromatographic column and between the perceived intensity of the complex odor of banana and the number of odor-active compounds smelling of banana sniffed by each subject [36].

5. Conclusions

In conclusion, the results of this study show that the ability of the human nose as a chemical sensor is strongly conditioned by the individual olfactory function and that the intensity with which a complex odor is perceived depends on the number of odor-active compounds and on the intensity at which they are perceived. Furthermore, the knowledge of which compounds of a complex odor are odor-active could be of great interest not only for the food and perfume industry, but also for developing electronic noses capable of identifying specific volatile molecules even in complex mixtures and/or reproducing the functional organization of the olfactory system [14,53].

Author Contributions: Conceptualization, R.C. and G.S.; methodology, G.S.; formal analysis, G.S.; investigation, P.S. and G.S.; resources, R.C. and G.S.; data curation, G.S.; writing—original draft preparation, G.S.; writing—review and editing, R.C., P.S. and G.S.; supervision, G.S.; funding acquisition, R.C. and G.S. All authors have read and agreed to the published version of the manuscript.

Funding: This work was supported by a grant from the University of Cagliari (Fondo Integrativo per la Ricerca, FIR 2019–2020).

Institutional Review Board Statement: The study was conducted in accordance with the Declaration of Helsinki and approved by the Ethics Committee of the University Hospital of Cagliari (Prot. PG/2018/22 del 2 January 2018).

Informed Consent Statement: Informed consent was obtained from all subjects involved in the study. Written informed consent has been obtained from the patient(s) to publish this paper" if applicable.

Data Availability Statement: The data presented in this study are available on request from the corresponding author. The data are not publicly available due to restrictions (e.g., privacy or ethical).

Acknowledgments: The authors thank the volunteers, without whose contribution this study would not have been possible.

Conflicts of Interest: The authors declare no conflict of interest. The funders had no role in the design of the study; in the collection, analyses, or interpretation of data; in the writing of the manuscript; or in the decision to publish the results.

References

1. Croy, I.; Bojanowski, V.; Hummel, T. Men without a sense of smell exhibit a strongly reduced number of sexual relationships, women exhibit reduced partnership security—A reanalysis of previously published data. *Biol. Psychol.* **2013**, *92*, 292–294. [CrossRef]
2. Landolt, P.J.; Heath, R.R.; Chambers, D.L. Oriented flight responses of female Mediterranean fruit flies to calling males, odor of calling males, and a synthetic pheromone blend. *Entomol. Exp. Appl.* **1992**, *65*, 259–266. [CrossRef]
3. Lebreton, S.; Borrero-Echeverry, F.; Gonzalez, F.; Solum, M.; Wallin, E.A.; Hedenström, E.; Hansson, B.S.; Gustavsson, A.-L.; Bengtsson, M.; Birgersson, G.; et al. A Drosophila female pheromone elicits species-specific long-range attraction via an olfactory channel with dual specificity for sex and food. *BMC Biol.* **2017**, *15*, 88. [CrossRef] [PubMed]
4. Li, H.; Wang, P.; Zhang, L.; Xu, X.; Cao, Z.; Zhang, L. Expressions of Olfactory Proteins in Locust Olfactory Organs and a Palp Odorant Receptor Involved in Plant Aldehydes Detection. *Front. Physiol.* **2018**, *9*, 663. [CrossRef]
5. Papadopoulos, N.T.; Shelly, T.E.; Niyazi, N.; Jang, E. Olfactory and Behavioral Mechanisms Underlying Enhanced Mating Competitiveness Following Exposure to Ginger Root Oil and Orange Oil in Males of the Mediterranean Fruit Fly, *Ceratitis capitata* (Diptera: Tephritidae). *J. Insect Behav.* **2006**, *19*, 403–418. [CrossRef]
6. Sollai, G.; Biolchini, M.; Solari, P.; Crnjar, R. Chemosensory basis of larval performance of Papilio hospiton on different host plants. *J. Insect Physiol.* **2017**, *99*, 47–57. [CrossRef] [PubMed]
7. Sollai, G.; Solari, P.; Crnjar, R. Olfactory sensitivity to major, intermediate and trace components of sex pheromone in *Ceratitis capitata* is related to mating and circadian rhythm. *J. Insect Physiol.* **2018**, *110*, 23–33. [CrossRef]
8. Su, C.Y.; Menuz, K.; Carlson, J.R. Olfactory perception: Receptors, cells, and circuits. *Cell* **2009**, *139*, 45–59. [CrossRef]
9. Anton, S.; van Loon, J.J.; Meijerink, J.; Smid, H.M.; Takken, W.; Rospars, J.P. Central projections of olfactory receptor neurons from single antennal and palpal sensilla in mosquitoes. *Arthropod Struct. Dev.* **2003**, *32*, 319–327. [CrossRef] [PubMed]
10. Breer, H. Odor recognition and second messenger signaling in olfactory receptor neurons. *Semin. Cell Biol.* **1994**, *5*, 25–32. [CrossRef]
11. Galizia, C.G.; Rössler, W. Parallel olfactory systems in insects: Anatomy and function. *Annu. Rev. Entomol.* **2010**, *55*, 399–420. [CrossRef] [PubMed]
12. Martin, F.; Boto, T.; Gomez-Diaz, C.; Alcorta, E. Elements of olfactory reception in adult *Drosophila melanogaster*. *Anat. Rec.* **2013**, *296*, 1477–1488. [CrossRef]
13. Mombaerts, P. Odorant receptor gene choice in olfactory sensory neurons: The one receptor-one neuron hypothesis revisited. *Curr. Opin. Neurobiol.* **2004**, *14*, 31–36. [CrossRef] [PubMed]
14. Schilling, B.; Kaiser, R.; Natsch, A.; Gautschi, M. Investigation of odors in the fragrance industry. *Chemoecology* **2010**, *20*, 135–147. [CrossRef]
15. Croy, I.; Mohr, T.; Weidner, K.; Hummel, T.; Junge-Hoffmeister, J. Mother-child bonding is associated with the maternal perception of the child's body odor. *Physiol. Behav.* **2019**, *198*, 151–157. [CrossRef] [PubMed]

16. Croy, I.; Nordin, S.; Hummel, T. Olfactory Disorders and Quality of Life—An Updated Review. *Chem. Senses* **2014**, *39*, 185–194. [CrossRef]
17. Hummel, T.; Nordin, S. Olfactory disorders and their consequences for quality of life. *Acta Oto-Laryngol.* **2005**, *125*, 116–121. [CrossRef] [PubMed]
18. Sollai, G.; Crnjar, R. Age-Related Olfactory Decline Is Associated with Levels of Exercise and Non-exercise Physical Activities. *Front. Aging Neurosci.* **2021**, *13*, 695115. [CrossRef]
19. Stevenson, R.J. An initial evaluation of the functions of human olfaction. *Chem. Senses* **2010**, *35*, 3–20. [CrossRef] [PubMed]
20. Velluzzi, F.; Deledda, A.; Onida, M.; Loviselli, A.; Crnjar, R.; Sollai, G. Relationship between Olfactory Function and BMI in Normal Weight Healthy Subjects and Patients with Overweight or Obesity. *Nutrients* **2022**, *14*, 1262. [CrossRef] [PubMed]
21. Ferreira, V. Revisiting psychophysical work on the quantitative and qualitative odour properties of simple odour mixtures: A flavour chemistry view. Part 2: Qualitative aspects. A review. *Flavour Fragr. J.* **2012**, *27*, 201–215. [CrossRef]
22. Frank, M.E.; Fletcher, D.B.; Hettinger, T.P. Recognition of the Component Odors in Mixtures. *Chem. Senses* **2017**, *42*, 537–546. [CrossRef] [PubMed]
23. Ferreira, V. Revisiting psychophysical work on the quantitative and qualitative odour properties of simple odour mixtures: A flavour chemistry view. Part 1: Intensity and detectability. A review. *Flavour Fragr. J.* **2012**, *27*, 124–140. [CrossRef]
24. d'Acampora Zellner, B.; Dugo, P.; Dugo, G.; Mondello, L. Gas chromatography-olfactometry in food flavour analysis. *J. Chromatogr. A* **2008**, *1186*, 123–143. [CrossRef] [PubMed]
25. Delahunty, C.M.; Eyres, G.; Dufour, J.P. Gas chromatography-olfactometry. *J. Sep. Sci.* **2006**, *29*, 2107–2125. [CrossRef]
26. Jordán, M.J.; Tandon, K.; Shaw, P.E.; Goodner, K.L. Aromatic profile of aqueous banana essence and banana fruit by gas chromatography-mass spectrometry (GC-MS) and gas chromatography-olfactometry (GC-O). *J. Agric. Food Chem.* **2001**, *49*, 4813–4817. [CrossRef]
27. Nuzzi, M.; Scalzo, R.L.; Testoni, A.; Rizzolo, A. Evaluation of Fruit Aroma Quality: Comparison between Gas Chromatography–Olfactometry (GC–O) and Odour Activity Value (OAV) Aroma Patterns of Strawberries. *Food Anal. Methods* **2008**, *1*, 270–282. [CrossRef]
28. Plutowska, B.; Wardencki, W. Application of gas chromatography–olfactometry (GC–O) in analysis and quality assessment of alcoholic beverages—A review. *Food Chem.* **2008**, *107*, 449–463. [CrossRef]
29. El Kazzy, M.; Weerakkody, J.S.; Hurot, C.; Mathey, R.; Buhot, A.; Scaramozzino, N.; Hou, Y. An Overview of Artificial Olfaction Systems with a Focus on Surface Plasmon Resonance for the Analysis of Volatile Organic Compounds. *Biosensors* **2021**, *11*, 244. [CrossRef]
30. Kim, C.; Lee, K.K.; Kang, M.S.; Shin, D.-M.; Oh, J.-W.; Lee, C.-S.; Han, D.-W. Artificial olfactory sensor technology that mimics the olfactory mechanism: A comprehensive review. *Biomater. Res.* **2022**, *26*, 40. [CrossRef]
31. Pelosi, P.; Zhu, J.; Knoll, W. From Gas Sensors to Biomimetic Artificial Noses. *Chemosensors* **2018**, *6*, 32. [CrossRef]
32. Nikolic, M.V.; Milovanovic, V.; Vasiljevic, Z.Z.; Stamenkovic, Z. Semiconductor Gas Sensors: Materials, Technology, Design, and Application. *Sensors* **2020**, *20*, 6694. [CrossRef]
33. Saruhan, B.; Lontio Fomekong, R.; Nahirniak, S. Review: Influences of Semiconductor Metal Oxide Properties on Gas Sensing Characteristics. *Front. Sens.* **2021**, *2*, 657931. [CrossRef]
34. Amoore, J.E.; Venstrom, D.; Davis, A.R. Measurement of specific anosmia. *Percept. Mot. Ski.* **1968**, *26*, 143–164. [CrossRef]
35. O'Connell, R.J.; Stevens, D.A.; Akers, R.P.; Coppola, D.M.; Grant, A.J. Individual differences in the quantitative and qualitative responses of human subjects to various odors. *Chem. Senses* **1989**, *14*, 293–302. [CrossRef]
36. Sollai, G.; Tomassini Barbarossa, I.; Usai, P.; Hummel, T.; Crnjar, R. Association between human olfactory performance and ability to detect single compounds in complex chemical mixtures. *Physiol. Behav.* **2020**, *217*, 112820. [CrossRef]
37. Cain, W.S.; Gent, J.F. Olfactory sensitivity: Reliability, generality, and association with aging. *J. Exp. Psychol. Hum. Percept. Perform.* **1991**, *17*, 382–391. [CrossRef]
38. Calderón-Garcidueñas, L.; Franco-Lira, M.; Henríquez-Roldán, C.; Osnaya, N.; González-Maciel, A.; Reynoso-Robles, R.; Villarreal-Calderon, R.; Herritt, L.; Brooks, D.; Keefe, S.; et al. Urban air pollution: Influences on olfactory function and pathology in exposed children and young adults. *Exp. Toxicol. Pathol. Off. J. Ges. Fur Toxikol. Pathol.* **2010**, *62*, 91–102. [CrossRef] [PubMed]
39. Feldmesser, E.; Bercovich, D.; Avidan, N.; Halbertal, S.; Haim, L.; Gross-Isseroff, R.; Goshen, S.; Lancet, D. Mutations in olfactory signal transduction genes are not a major cause of human congenital general anosmia. *Chem. Senses* **2007**, *32*, 21–30. [CrossRef]
40. Hasin-Brumshtein, Y.; Lancet, D.; Olender, T. Human olfaction: From genomic variation to phenotypic diversity. *Trends Genet. TIG* **2009**, *25*, 178–184. [CrossRef]
41. Jafek, B.W.; Gordon, A.S.; Moran, D.T.; Eller, P.M. Congenital anosmia. *Ear Nose Throat J.* **1990**, *69*, 331–337.
42. Melis, M.; Tomassini Barbarossa, I.; Crnjar, R.; Sollai, G. Olfactory Sensitivity Is Associated with Body Mass Index and Polymorphism in the Voltage-Gated Potassium Channels Kv1.3. *Nutrients* **2022**, *14*, 4986.
43. Melis, M.; Tomassini Barbarossa, I.; Hummel, T.; Crnjar, R.; Sollai, G. Effect of the rs2890498 polymorphism of the OBPIIa gene on the human ability to smell single molecules. *Behav. Brain Res.* **2021**, *402*, 113127. [CrossRef] [PubMed]
44. Seo, H.S.; Guarneros, M.; Hudson, R.; Distel, H.; Min, B.C.; Kang, J.K.; Croy, I.; Vodicka, J.; Hummel, T. Attitudes toward Olfaction: A Cross-regional Study. *Chem. Senses* **2011**, *36*, 177–187. [CrossRef] [PubMed]
45. Silva Teixeira, C.S.; Cerqueira, N.M.; Silva Ferreira, A.C. Unravelling the Olfactory Sense: From the Gene to Odor Perception. *Chem. Senses* **2016**, *41*, 105–121. [CrossRef]

46. Sollai, G.; Melis, M.; Magri, S.; Usai, P.; Hummel, T.; Tomassini Barbarossa, I.; Crnjar, R. Association between the rs2590498 polymorphism of Odorant Binding Protein (OBPIIa) gene and olfactory performance in healthy subjects. *Behav. Brain Res.* **2019**, *372*, 112030. [CrossRef]
47. Sollai, G.; Melis, M.; Mastinu, M.; Paduano, D.; Chicco, F.; Magri, S.; Usai, P.; Hummel, T.; Barbarossa, I.T.; Crnjar, R. Olfactory Function in Patients with Inflammatory Bowel Disease (IBD) Is Associated with Their Body Mass Index and Polymorphism in the Odor Binding-Protein (OBPIIa) Gene. *Nutrients* **2021**, *13*, 703. [CrossRef]
48. Sollai, G.; Melis, M.; Tomassini Barbarossa, I.; Crnjar, R. A polymorphism in the human gene encoding OBPIIa affects the perceived intensity of smelled odors. *Behav. Brain Res.* **2022**, *427*, 113860. [CrossRef]
49. Sorokowska, A.; Sorokowski, P.; Hummel, T. Cross-Cultural Administration of an Odor Discrimination Test. *Chemosens. Percept.* **2014**, *7*, 85–90. [CrossRef]
50. Velluzzi, F.; Deledda, A.; Lombardo, M.; Fosci, M.; Crnjar, R.; Grossi, E.; Sollai, G. Application of Artificial Neural Networks (ANN) to Elucidate the Connections among Smell, Obesity with Related Metabolic Alterations, and Eating Habit in Patients with Weight Excess. *Metabolites* **2023**, *13*, 206. [PubMed]
51. Gerkin, R.C.; Castro, J.B. The number of olfactory stimuli that humans can discriminate is still unknown. *eLife* **2015**, *4*, e08127. [CrossRef] [PubMed]
52. Le Fur, Y.; Mercurio, V.; Moio, L.; Blanquet, J.; Meunier, J.M. A new approach to examine the relationships between sensory and gas chromatography-olfactometry data using Generalized Procrustes analysis applied to six French Chardonnay wines. *J. Agric. Food Chem.* **2003**, *51*, 443–452. [CrossRef]
53. Brattoli, M.; Cisternino, E.; Dambruoso, P.R.; de Gennaro, G.; Giungato, P.; Mazzone, A.; Palmisani, J.; Tutino, M. Gas chromatography analysis with olfactometric detection (GC-O) as a useful methodology for chemical characterization of odorous compounds. *Sensors* **2013**, *13*, 16759–16800. [CrossRef]
54. Dussort, P.; Depretre, N.; Bou-Maroun, E.; Fant, C.; GUICHARD, E.; Brunerie, P.; Le Fur, Y.Y.; Le Quéré, J.-L. An original approach for gas chromatography-olfactometry detection frequency analysis: Application to gin. *Food Res. Int.* **2012**, *49*, 253–262. [CrossRef]
55. Pollien, P.; Ott, A.; Montigon, F.; Baumgartner, M.; Muñoz-Box, R.; Chaintreau, A. Hyphenated Headspace-Gas Chromatography-Sniffing Technique: Screening of Impact Odorants and Quantitative Aromagram Comparisons. *J. Agric. Food Chem.* **1997**, *45*, 2630–2637. [CrossRef]
56. Hummel, T.; Sekinger, B.; Wolf, S.R.; Pauli, E.; Kobal, G. 'Sniffin' sticks': Olfactory performance assessed by the combined testing of odor identification, odor discrimination and olfactory threshold. *Chem. Senses* **1997**, *22*, 39–52. [CrossRef] [PubMed]
57. Hummel, T.; Kobal, G.; Gudziol, H.; Mackay-Sim, A. Normative data for the "Sniffin' Sticks" including tests of odor identification, odor discrimination, and olfactory thresholds: An upgrade based on a group of more than 3,000 subjects. *Eur. Arch. Otorhinolaryngol.* **2007**, *264*, 237–243. [CrossRef]
58. Fischer, M.; Zopf, Y.; Elm, C.; Pechmann, G.; Hahn, E.G.; Schwab, D.; Kornhuber, J.; Thuerauf, N.J. Subjective and objective olfactory abnormalities in Crohn's disease. *Chem. Senses* **2014**, *39*, 529–538. [CrossRef]
59. Rizzolo, A.; Polesello, A.; Polesello, S. Use of headspace capillary GC to study the development of volatile compounds in fresh fruit. *J. High Resolut. Chromatogr.* **1992**, *15*, 472–477. [CrossRef]
60. van Den Dool, H.; Kratz, P.D. A generalization of the retention index system including linear temperature programmed gas—Liquid partition chromatography. *J. Chromatogr. A* **1963**, *11*, 463–471. [CrossRef]
61. Akiyama, M.; Murakami, K.; Ohtani, N.; Iwatsuki, K.; Sotoyama, K.; Wada, A.; Tokuno, K.; Iwabuchi, H.; Tanaka, K. Analysis of volatile compounds released during the grinding of roasted coffee beans using solid-phase microextraction. *J. Agric. Food Chem.* **2003**, *51*, 1961–1969. [CrossRef] [PubMed]
62. Caporaso, N.; Whitworth, M.B.; Cui, C.; Fisk, I.D. Variability of single bean coffee volatile compounds of Arabica and robusta roasted coffees analysed by SPME-GC-MS. *Food Res. Int.* **2018**, *108*, 628–640. [CrossRef]
63. Gloess, A.N.; Yeretzian, C.; Knochenmuss, R.; Groessl, M. On-line analysis of coffee roasting with ion mobility spectrometry–mass spectrometry (IMS–MS). *Int. J. Mass Spectrom.* **2018**, *424*, 49–57. [CrossRef]
64. Lee, S.; Kim, M.; Lee, K.-G. Effect of reversed coffee grinding and roasting process on physicochemical properties including volatile compound profiles. *Innov. Food Sci. Emerg. Technol.* **2017**, *44*, 97–102. [CrossRef]
65. López-Galilea, I.; Fournier, N.; Cid, C.; Guichard, E. Changes in headspace volatile concentrations of coffee brews caused by the roasting process and the brewing procedure. *J. Agric. Food Chem.* **2006**, *54*, 8560–8566. [CrossRef]
66. Majcher, M.A.; Klensporf-Pawlik, D.; Dziadas, M.; Jeleń, H.H. Identification of aroma active compounds of cereal coffee brew and its roasted ingredients. *J. Agric. Food Chem.* **2013**, *61*, 2648–2654. [CrossRef] [PubMed]
67. Sunarharum, W.; Williams, D.; Smyth, H. Complexity of coffee flavor: A compositional and sensory perspective. *Food Res. Int.* **2014**, *62*, 315–325. [CrossRef]
68. Yang, N.; Liu, C.; Liu, X.; Degn, T.K.; Munchow, M.; Fisk, I. Determination of volatile marker compounds of common coffee roast defects. *Food Chem.* **2016**, *211*, 206–214. [CrossRef]
69. Zapata, J.; Londoño, V.; Naranjo, M.; Osorio, J.; Lopez, C.; Quintero, M. Characterization of aroma compounds present in an industrial recovery concentrate of coffee flavour. *CyTA—J. Food* **2018**, *16*, 367–372. [CrossRef]
70. Gonzalez-Kristeller, D.C.; do Nascimento, J.B.; Galante, P.A.; Malnic, B. Identification of agonists for a group of human odorant receptors. *Front. Pharmacol.* **2015**, *6*, 35. [CrossRef]

71. van Ruth, S.M.; O'Connor, C.H. Evaluation of three gas chromatography-olfactometry methods: Comparison of odour intensity-concentration relationships of eight volatile compounds with sensory headspace data. *Food Chem.* **2001**, *74*, 341–347. [CrossRef]
72. Ratner, B. The correlation coefficient: Its values range between +1/−1, or do they? *J. Target. Meas. Anal. Mark.* **2009**, *17*, 139–142. [CrossRef]
73. Wu, S.; Wang, P.; Xie, D.; Jian, F. Correlation analysis of flow parameters in the olfactory cleft and olfactory function. *Sci. Rep.* **2022**, *12*, 20819. [CrossRef] [PubMed]
74. Brattoli, M.; de Gennaro, G.; de Pinto, V.; Loiotile, A.D.; Lovascio, S.; Penza, M. Odour detection methods: Olfactometry and chemical sensors. *Sensors* **2011**, *11*, 5290–5322. [CrossRef]
75. Goyert, H.F.; Frank, M.E.; Gent, J.F.; Hettinger, T.P. Characteristic component odors emerge from mixtures after selective adaptation. *Brain Res. Bull.* **2007**, *72*, 1–9. [CrossRef]
76. Marshall, K.; Laing, D.G.; Jinks, A.; Hutchinson, I. The capacity of humans to identify components in complex odor-taste mixtures. *Chem. Senses* **2006**, *31*, 539–545. [CrossRef] [PubMed]
77. Mayol, A.R.; Acree, T.E. Advances in Gas Chromatography-Olfactometry. In *Gas Chromatography-Olfactometry*; ACS Symposium Series; American Chemical Society: Washington, DC, USA, 2001; Volume 782, pp. 1–10.
78. Jadauji, J.B.; Djordjevic, J.; Lundström, J.N.; Pack, C.C. Modulation of olfactory perception by visual cortex stimulation. *J. Neurosci. Off. J. Soc. Neurosci.* **2012**, *32*, 3095–3100. [CrossRef]

chemosensors

MDPI

Article

Chemoresistive Properties of V_2CT_x MXene and the V_2CT_x/V_3O_7 Nanocomposite Based on It

Artem S. Mokrushin [1,*], Ilya A. Nagornov [1], Aleksey A. Averin [2], Tatiana L. Simonenko [1], Nikolay P. Simonenko [1], Elizaveta P. Simonenko [1] and Nikolay T. Kuznetsov [1]

[1] Kurnakov Institute of General and Inorganic Chemistry of the Russian Academy of Sciences, 31 Leninsky pr., Moscow 119991, Russia
[2] Frumkin Institute of Physical Chemistry and Electrochemistry, Russian Academy of Sciences, 31 Leninsky pr., bldg. 4, Moscow 199071, Russia
* Correspondence: artyom.nano@gmail.com

Abstract: The *in-situ* Raman spectroscopy oxidation of the accordion-like V_2CT_x MXene has been studied. It was found that a nanocomposite of V_2CT_x/V_3O_7 composition was formed as a result. The elemental and phase composition, the microstructure of the synthesized V_2CT_x powder and MXene film as well as the V_2CT_x/V_3O_7 nanocomposite obtained at a minimum oxidation temperature of 250 °C were studied using a variety of physical and chemical analysis methods. It was found that the obtained V_2CT_x and V_2CT_x/V_3O_7 films have an increased sensitivity to ammonia and nitrogen dioxide, respectively, at room temperature and zero humidity. It was shown that the V_2CT_x/V_3O_7 composite material is characterized by an increase in the response value for a number of analytes (including humidity) by more than one order of magnitude, as well as a change in their detection mechanisms compared to the individual V_2CT_x MXene.

Keywords: MXene; chemoresistive gas sensors; V_2C; vanadium oxide; Raman; 2D-nanomaterials

Citation: Mokrushin, A.S.; Nagornov, I.A.; Averin, A.A.; Simonenko, T.L.; Simonenko, N.P.; Simonenko, E.P.; Kuznetsov, N.T. Chemoresistive Properties of V_2CT_x MXene and the V_2CT_x/V_3O_7 Nanocomposite Based on It. *Chemosensors* **2023**, *11*, 142. https://doi.org/10.3390/chemosensors11020142

Academic Editors: Kai Xu and Zhong Li

Received: 5 January 2023
Revised: 6 February 2023
Accepted: 13 February 2023
Published: 15 February 2023
Corrected: 4 December 2023

1. Introduction

The family of two-dimensional (2D) transition d-metal carbides with the general formula $M_{n+1}C_nT_x$ has attracted the attention of the scientific community in recent years due to its unique properties and high variability of the surface chemistry [1,2]. Currently, the most studied MXenes are the Ti_2CT_x and $Ti_3C_2T_x$ carbide compounds to which the vast majority of the work has been devoted [3–6]. However, there are more and more studies devoted to other two-dimensional transition metal carbides, including Mo_2CT_x, Nb_2CT_x, V_2CT_x [7–10] and various nanocomposites based on them [11–14], are also emerging. At the moment, researchers continue to search and find different ways to synthesize and apply MXenes. The vast majority of works on V_2CT_x are devoted to their use in the composition of lithium-ion batteries and supercapacitors [15–17]. V_2CT_x MXene is also used in ferromagnetics [18], antibacterial coatings [19], memristors [20], hydrogen storage devices [21], in catalysis during the oxygen evolution reaction (OER) [22], as well as in chemoresistive gas sensors [23].

The strong interactions of gases with the surface of MXenes corresponding to high negative adsorption energies (as confirmed by DFT calculations) allow this class of compounds to be successfully used as a sensitive materials in chemoresistive gas sensors [24,25]. The mixed and metallic conductivity of individual MXenes allows for obtaining responses at room temperature with a high signal-to-noise ratio (SNR) [26], which distinguishes them from metal oxide semiconductors (MOS), the classical receptor materials for chemoresistive gas sensors [27–30].

For V_2CT_x MXene, there is sporadic work in the literature on its use in the composition of chemoresistive gas sensors. In [31] the responses to various gases capable of hydrogen bonding (NH_3, triethylamine, ethanol, methanol, acetone, and formaldehyde) were studied depending on the type of etching agent. It has been shown that after the synthesis, the surface functionalization of V_2CT_x MXenes changes and different types of responses (so-called *n*- and *p*-type, which correspond to the decrease and increase in electrical resistance when the analyte is injected) to the same gases are observed. In [32], for the obtained V_2CT_x MXene, the authors managed to fix the response to 5–50 ppm NO_2; however, data on the responses to other gases are not given, which makes it difficult to assess the selectivity of the obtained MXene. In [33], Eunji Lee et al. carried out a comprehensive study of the gas-sensitive chemoresistive properties of V_2CT_x MXenes, and found that the greatest response of a large number of analyzed gases was obtained for hydrogen. Xingwei Wang et al. [34] described a synergistic process to fabricate a supercapacitor and simultaneously a gas sensor for ammonia based on a polyaniline/V_2C MXene composite with autonomous power supply from an electromagnetic–triboelectric hybrid generator. The authors were able to obtain responses of 0.3–10 ppm NH_3 at 20 °C with high selectivity. The described approach is extremely promising and state-of-the-art.

The use of MXenes as sensitive materials at room temperatures is due to the fact that they are prone to oxidation at increasing operating temperatures in an air atmosphere. In [35], using TG/DTA and XRD analyses it was shown that V_2CT_x MXene is completely oxidized to V_2O_5 at temperatures above 500 °C. The thermograms presented in the article show an intense exothermic effect with a maximum at 332 °C, accompanied by an increase in mass, when heating the MXenes in an air atmosphere, which is associated with the oxidation process of the MXenes. Unfortunately, the authors did not present data on the phase composition of oxidized MXene in the specified temperature range, which complicates the interpretation of the given data. In [36] the authors studied in situ changes in the phase composition of V_2CT_x MXene in different gas environments using Raman spectroscopy. According to the data obtained, it can be seen that in the air atmosphere the oxidation of MXene begins around 400 °C. Such studies are extremely important for fine control of the phase composition in the context of obtaining V_2CT_x/VO_x nanocomposites for various applications, including chemical gas sensing.

Previously, we have obtained titanium carbide MXenes [37,38], and studied the phase transformation of the Ti_2CT_x MXenes film using in situ Raman spectroscopy and studied the effect of the oxidation process on its chemoresistive properties [39]. The chemoresistive properties of the complex vanadium–titanium carbide $Ti_{0.2}V_{1.8}C$ have also been previously studied [40]. The present work is devoted to the study of the oxidation process of V_2CT_x MXene film using in situ Raman spectroscopy and to the study of the gas-sensitive chemoresistive properties of the obtained nanomaterial to a wide group of analyte gases.

2. Materials and Methods

2.1. Synthesis and Application

Reagents: vanadium metal powders (99.9%, 0.5–100 µm, Ruskhim, Moscow, Russia), aluminum (99.2%, 30 µm, Ruskhim, Moscow, Russia), graphite (MPG-8 grade, Ruskhim, Moscow, Russia), potassium bromide KBr (99%, Ruskhim, Moscow, Russia), sodium fluoride NaF (>99%, Reahim, Moscow, Russia), hydrofluoric acid (50%, Honeywell International Inc, Charlotte, NC, USA), hydrochloric acid (36%, Sigma Tek, Moscow, Russia).

The synthesis of V_2CT_x MXene was carried out by selective etching of aluminum contained in the MAX-phase of V_2AlC under the influence of hydrofluoric and hydrochloric acid solutions. Methods for the synthesis of V_2AlC and V_2CT_x, which are close to those used in this work, are described in detail in [40]. Briefly, to obtain V_2AlC powders of aluminum, vanadium, graphite and potassium bromide were mixed in the ratios n(V):n(Al):n(C) = 2:1.2:0.8 and m(V + Al + C) = m(KBr), co-milled, compacted into tablets and subjected to heat treatment in a muffle furnace at 1000 °C [39,40].

To obtain an accordion-like V_2CT_x MXene, a 1 g MAX-phase V_2AlC sample was usually typically added to a solution containing 12 mL of HF (50%) and 8 mL of HCl (36%). After stirring at room temperature for 30 min, the system was heated to 40 ± 5 °C and kept under these conditions under stirring for 120 h. The resulting powder was separated by centrifugation and washed repeatedly with distilled water until pH ~5–6. The sample was then washed twice with ethanol and centrifuged again. The resulting precipitate was redispersed in ethanol in an ultrasonic bath for 30 min, most of the impurities (primarily MAX-phase particles) were separated by centrifugation of the dispersion at 1000 rpm for 5 min. The dispersion of MXene in ethanol was evaporated, and the resulting phase was dried in a vacuum at 150 °C.

The MXene receptor layer was deposited on a special Al_2O_3 substrate with platinum interdigital electrodes and a heater (on the reverse side) by microextrusion printing (10 μm resolution) using a three-coordinate positioning system equipped with a pneumatic and capillary dispenser in the form of a hollow needle with an internal diameter of 150 μm. A dispersion of V_2CT_x MXene in butanol with a concentration of ~5 mg/mL was used. The coating was dried in a desiccator and then held in a vacuum at 150 °C for 5 min.

2.2. Instrumentation

Micrographs and the chemical composition of the surface of the samples were obtained by scanning (NVision 40 scanning electron microscope, Carl Zeiss, (Oberkochen, Germany), secondary electron detector, accelerating voltage 1–10 kV) and transmission electron microscopy (JEOL, JEM-1011, Akishima, Japan), and by X-ray spectral elemental microanalysis (INCA X-MAX 80 energy dispersive X-ray (EDX) spectrometer, Oxford Instruments (Oxford, UK), accelerating voltage 20 kV). X-ray diffraction was performed on powders and films using a D8 Advance (Bruker, Billerica, MA, USA, CuKα = 1.5418 Å, Ni filter, E = 40 keV, I = 40 mA; 2θ range: 5–45°; resolution: 0.02°; point accumulation time: 0.3 s).

A Renishaw (New Mills Wotton-under-Edge, Gloucestershire, UK, GL12 8JR) inVia Reflex Microscope system equipped with a Peltier-cooled CCD (532-nm Nd:YAG laser line was used) was used to record Raman spectra. The laser light was focused on the sample to ~2 μm through a 50× objective (power on the sample was <0.3 mW). A THMS600 stage (Linkam Scientific Instruments Ltd., Redhill, Unit 9, Perrywood Business Park, Honeycrock Lane, Salfords, Redhill, UK, Surrey RH1 5DZ) was used to record variable temperature Raman scattering. The *in-situ* film heating rate was 5 °C/min. The sample temperature was pre-calibrated using a high-precision Testo 868 thermal imaging camera. When recording Raman spectra, there is local heating of the imaging area, so the spectra were recorded in different areas of the film, but close to each other. All spectra were normalized with respect to the most intense spectrum.

The chemoresistive responses were performed at room temperature using the laboratory setup described in [41]. A gas-air atmosphere was created in a quartz cell (volume -7×10^{-5} m^3) using two Bronkhorst gas flow controllers with a maximum throughput of 100 and 200 mL/min. The electrical properties of the oxide films obtained were measured using a Fluke 8846A Digit Precision Multimeter with an upper detection limit of 1000 MΩ. The sensor temperature was monitored using a pre-calibrated platinum micro-heater. Prior to gas-sensitive measurements, the film was held in a baseline gas atmosphere until a stable signal was obtained.

Different relative humidities (RH) were generated using a special unit with a bubbler flask. The RH of the gas mixture was measured with a digital flow hygrometer "Excis" (EXIS, Russia, Moscow). The temperature value of the relative humidity was set and then measured at 20 °C.

All gas-sensing measurements were performed at room temperature (RT) and 0% RH. The response to H_2, CO, NH_3, benzene (C_6H_6), acetone (C_3H_6O), methane (CH_4) and ethanol (C_2H_5OH) were calculated using the following ratio:

$$S_1 = \frac{|R_{BL} - R_g|}{R_{BL}} \times 100\% \quad (1)$$

where R_{BL}—baseline resistance (synthetic air was used as the baseline gas), R_g—resistance at a given concentration of analyte gas.

The response to humidity was calculated using the following ratio:

$$S_2 = \frac{|R_{BL} - R_{RH}|}{R_{BL}} \times 100\% \tag{2}$$

where R_{BL}—resistance at 0% relative humidity, R_{RH}—at a given relative humidity.

3. Results and Discussion

3.1. The Phase Composition and Microstructure of the Obtained V_2CT_x MXene before Oxidation

Figure 1a shows the X-ray patterns of the V_2AlC MAX-phase powders that were used for the synthesis, as well as an X-ray pattern of the resulting V_2CT_x MXene powder. As can be seen, the set of V_2AlC MAX-phase reflexes correlates well with the data reported in the literature (V_2AlC, ICSD-606283, Space group *P*6₃/*mmc*). In addition to the characteristic intense reflex (002) at 2θ = 7.4° on the X-ray pattern of the obtained V_2CT_x MXene sample, there are reflexes of impurity phases: the V_2AlC MAX-phase and the cubic vanadium carbide phase VC (ICSD-159870, Space group *Fm-3m*), which should not affect the gas sensitive properties of MXene. The position of the observed reflex (002) for V_2CT_x MXene is in good agreement with other experimental data for this compound [42–44]. Figure 1b shows an X-ray pattern of the V_2CT_x MXene film before oxidation on an Al_2O_3 substrate. Because of the low thickness of the obtained film, only the characteristic set of α-Al_2O_3 substrate reflexes (PDF 00-005-0712) is clearly visible in the X-ray pattern shown. The inset on the left shows a region of the X-ray pattern in the 2θ = 4.5–12° interval with a long signal accumulation time at the point where a widened reflex (002) of the V_2CT_x MXene phase is observed. Its position is in good agreement with both the literature data and the data obtained for the powder of the corresponding composition in the present study (Figure 1a).

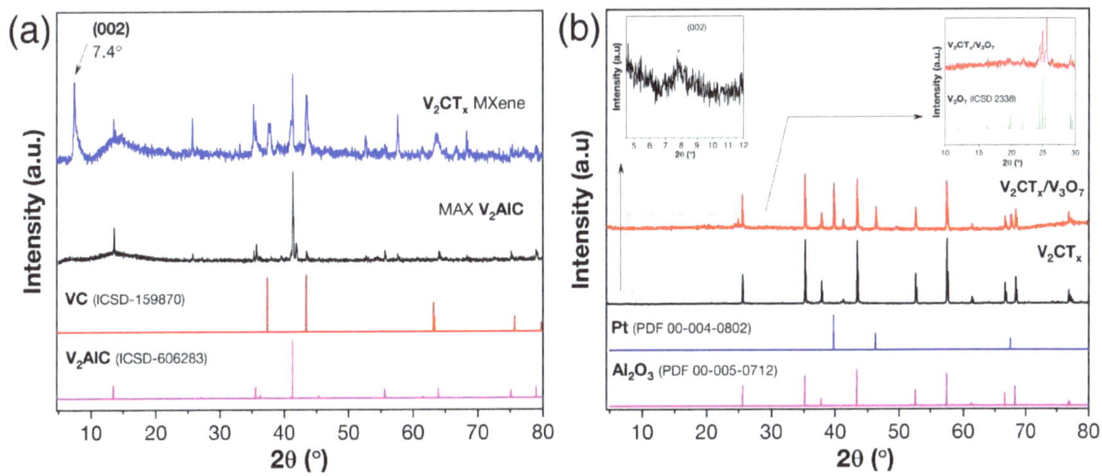

Figure 1. X-ray patterns of the V_2AlC MAX-phase and the obtained accordion-like V_2CT_x MXene powder (**a**), as well as V_2CT_x MXene films on Al_2O_3/Pt substrates before and after in situ oxidation with insets of characteristic bands for the corresponding V_2CT_x MXene (left) and V_3O_7 (right) phases (**b**).

Figure 2 shows SEM and TEM micrographs of the synthesized V_2CT_x MXene powder. The microstructure of V_2CT_x MXene has an accordion-like structure typical of this fabrication technique. The average interlayer distance was calculated from the Figure 2a micrograph, and was found to be 22 ± 5 nm. The value obtained is quite high, indicating a large specific surface area of the obtained multilayer V_2CT_x MXene, which is important for gas sensing. The TEM data (Figure 2c) confirm the formation of the multilayer harmonic, which is in good agreement with the SEM data. Figure 3a–c shows SEM micrographs of the V_2CT_x MXene film on an

Al_2O_3 substrate. As can be observed from the presented data, the microstructure of the films is in complete agreement with the microstructure of the powder: the accordion-like structure of the MXene is maintained, the interlayer spacing values are in the range of the calculated values for the powder. Using EDX analysis, it was found that besides C, V, forming elements V_2CT_x MXene and aluminum (typical for an impurity of the initial MAX-phase V_2AlC or aluminum fluorides), the MXene surface *T*-functional groups contain oxygen, fluorine and chlorine (Figure 4a). Their ratio n(F):n(Cl) = 19:1 is determined by peculiarities of the selective etching process by the chosen system (HF + HCl).

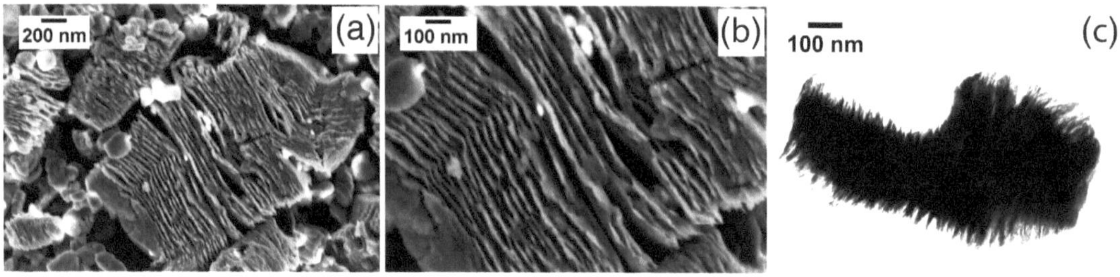

Figure 2. SEM (**a**,**b**) and TEM (**c**) microphotographs of V_2CT_x MXene powder.

Figure 3. SEM micrographs of V_2CT_x MXene film on Al_2O_3/Pt substrates before (**a**–**c**) and after (**d**–**f**) oxidation.

3.2. Thermal Analysis of V_2CT_x Powder

The study of the thermal behavior of the obtained V_2CT_x multilayer powder sample during its heating in an air flow (Figure 4b) allows for a more correct planning of the experiment in the *in-situ* oxidation of the MXene receptor layer during heating during the recording of Raman spectra. Thus, in the initial stages of heating at temperatures <120–230 °C there is a mass loss of ~1.6–1.7%, probably due to the removal of adsorbed water molecules. On heating >240 °C there is a further decrease in mass, which may be due to the removal of MXene functional groups, primarily OH-groups. This process at temperatures above 240–260 °C is overlaid by exothermic processes of the beginning of MXene oxidation: with a maximum at 300 °C for more dispersed MXene aggregates (a stack diameter of 100–300 nm), surface areas of larger V_2CT_x aggregates (~500–1000 nm stack diameter) and internal areas of large accordion-like aggregates, to which diffusion is difficult (with a maximum at 352 °C). In this case, the oxidation process becomes dominant at 350 °C, as the decrease in mass loss due to functional groups removal is replaced by a tendency to increase in mass. A low intensity exothermic effect in the interval 490–570 °C can be attributed to the oxidation of more oxidation-resistant phases VC and V_2AlC. Thus, the onset of the oxidation process of V_2CT_x MXene with simultaneous heating in air and recording of Raman spectra should be expected at temperatures ~240–250 °C.

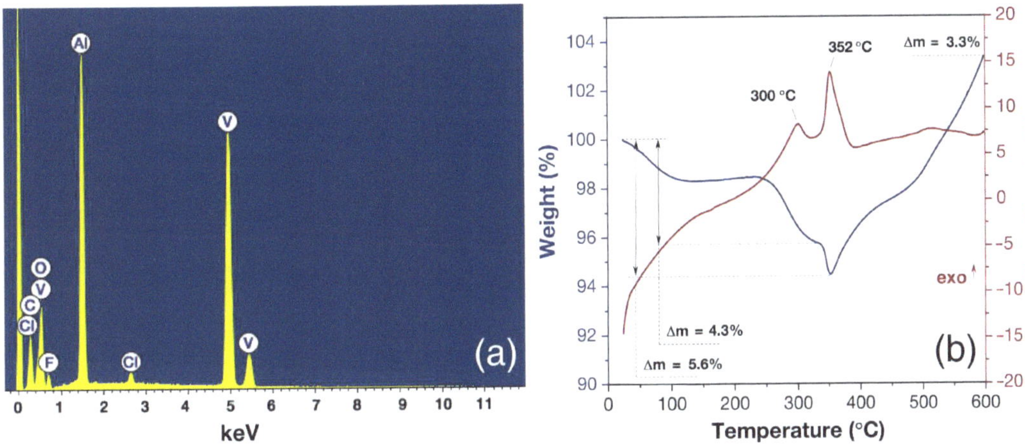

Figure 4. EDX spectrum of V_2CT_x film on Al_2O_3 substrate (**a**) and DSC (red) and TGA (blue) curves of the obtained V_2CT_x powder during heating in an air current (**b**).

3.3. In Situ Raman Spectroscopy during Heating of V_2CT_x MXene Film

The main purpose of using in situ Raman spectroscopy in the present study was to study the oxidation process of the V_2CT_x MXene film as well as the formation of the V_2CT_x/VO_x composite for further study of the gas-sensitive properties.

Figure 5 shows the *in-situ* Raman spectra of the V_2CT_x MXene film when heated in the temperature range RT–250 °C. At room temperature, four MXene characteristic modes ω_1–ω_4 are observed at 269, 431, 689 and 912 cm^{-1}, respectively. These modes can be related to the individual V_2CT_x MXene [36,45–47]. In addition, D-(ω_D) and G-modes (ω_G) are present in the spectra at 1354 and 1582 cm^{-1}, which are common to many carbon systems with sp^2-hybridization of carbon atoms [48]. The presence of D- and G-modes is characteristic of the entire MXene family; it illustrates the formation of layered carbon in their composition or the formation of a graphene-like carbon impurity during excessive etching of the MAX-phase. The described set of characteristic modes ω_1–ω_4 and ω_D, ω_G is maintained up to 200 °C. It is worth noting that all modes except for the D- and G-bands are of low-intensity, which is a consequence of two factors: the nature of the MXenes themselves, which are not characterized by intense Raman bands, and the conditions of the Raman experiment, since the study was performed at low laser power to avoid local overheating of the coating by the laser itself and MXene oxidation. In addition, as we have recently shown for Ti_2CT_x [40], the MXene modes on the Al_2O_3 substrate are much weaker than on the platinum substrate.

The changes in the Raman spectra of the V_2CT_x MXene film start to appear at 250 °C. In addition, new modes are added to the V_2CT_x MXene ω_1–ω_4 bands: intense ω_5 at 144 cm^{-1} as well as ω_6–ω_{11} at 280, 306, 409, 525, 700 and 995 cm^{-1}, respectively. The Raman bands described are characteristic of α-V_2O_5 (orthorhombic crystal lattice, Space group *Pmmn*) [49]: B_{2g}, B_{1g}, A_{1g}, A_{1g}, A_{1g} and (A_{1g} and B_{2g}) modes, respectively [50]. It should be noted that the D- and G-modes are maintained at the heating temperature of 250 °C, which also confirms the fact that the MXene oxidation did not proceed completely and the V_2CT_x/V_2O_5 composite was formed at these temperature conditions. Thus, the temperature of 250 °C was chosen as optimal for in situ heating of the V_2CT_x MXene film using Raman spectroscopy.

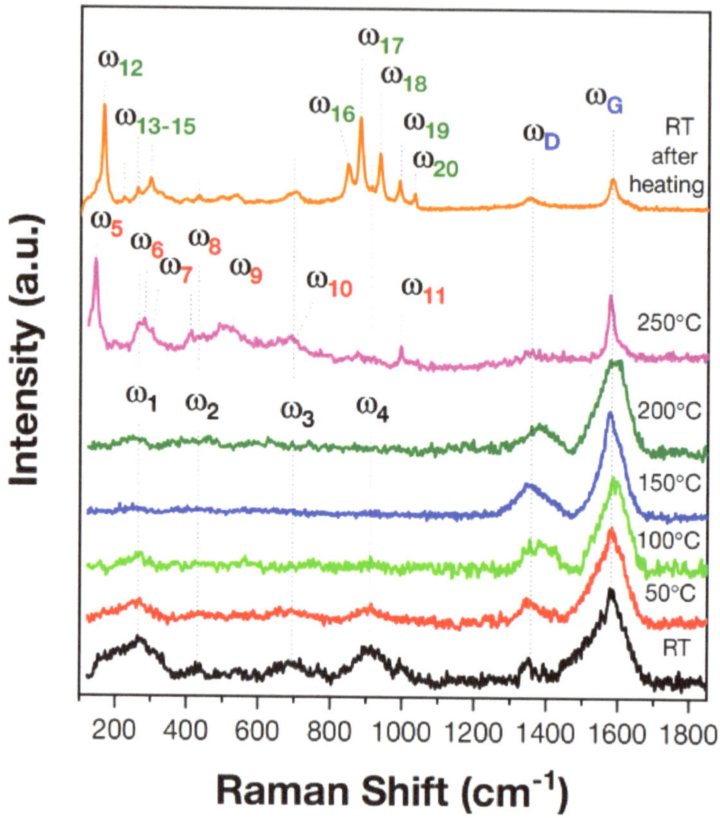

Figure 5. In situ Raman spectra during heating of V_2CT_x MXene film in air in the temperature range RT–250 °C.

After cooling the sample to room temperature, Raman spectra were also recorded for the obtained oxidized V_2CT_x MXene film. It was found that the spectrum obtained is significantly different from that recorded for the sample heated to 250 °C (Figure 5). The spectrum contains a completely different set of Raman modes: ω_{12}–ω_{20} at 168 (intense), 224, 261, 298, 845, 880 (intense), 936, 990 and 1031 cm^{-1}, respectively, which correspond with high accuracy to the characteristic peaks for another vanadium oxide, V_3O_7 (monoclinic crystal lattice, Space group *C2/c*) [50]. It is to be noted that the ω_1–ω_3 and ω_D–ω_D modes characteristic of V_2CT_x MXene are maintained in the spectrum, indicating that the V_2CT_x/V_3O_7 composite has formed during cooling. Thus, after cooling from 250 °C to room temperature, the obtained film undergoes structural changes: the vanadium (V) oxide α-V_2O_5 formed during MXene oxidation is transformed into the mixed oxide V_3O_7 containing vanadium atoms in the +IV and +V oxidation states. It is worth noting that the formation of V_3O_7 is not a common case for vanadium oxides, which are characterized by individual V_2O_5 and VO_2 oxides, as well as composites containing their individual phases [49,51].

3.4. The Phase Composition and Microstructure of the Obtained V_2CT_x MXene after Oxidation

Figure 1b shows an X-ray pattern of the V_2CT_x/V_3O_7 film on an Al_2O_3/Pt substrate after heating to 250 °C using in situ Raman spectroscopy. It shows intense reflexes related to the substrate materials α-Al_2O_3 and Pt (PDF 00-005-0712 and 00-004-0802, respectively), as well as low-intensity reflexes of the V_3O_7 phase. The resulting set of V_3O_7 reflexes correlates well with the available data in the database (ICSD 2338), which is particularly evident in the corresponding inset in Figure 1b, right. The XRD data obtained are in full agreement with the Raman data and confirm the V_3O_7 content in the films obtained after heating to 250 °C in air.

Figure 3d–f shows SEM microphotographs of V_2CT_x/V_3O_7 films after heating to 250 °C using in situ Raman spectroscopy; it is shown that the microstructure of the films undergoes significant changes. The microphotographs show a shape similar to the initial multilayer harmonic, which is characteristic of MXene, but this is no longer observed after oxidation. The formed structures become more closed and porous with a pore size of ~25–40 nm, which is close to the value of the interlayer distances that were calculated for the individual multilayer aggregates of V_2CT_x MXene.

3.5. Gas-Sensing Chemoresistive Properties

In the first step, chemoresistive responses were determined for the obtained V_2CT_x and V_2CT_x/V_3O_7 films when detecting 100 ppm CO, NH_3, NO_2, benzene (C_6H_6), acetone (C_3H_6O), ethanol (C_2H_5OH), and 1000 ppm methane (CH_4) and H_2 at room temperature and 0% RH. Experimental responses (S1) for these gases are shown in Figure 6a,b. Selectivity diagrams (Figure 6c,d) were constructed from the series of responses obtained. The obtained responses for V_2CT_x do not show high numerical values; nevertheless, due to the high SNR value (which is typical for the whole MXene family) it was possible to efficiently detect different gases. For the original V_2CT_x MXene film, the highest response (1.35%) was observed in the detection of NH_3. Notable responses were also recorded for NO_2 (0.96%) and CO (0.94%), while the response for all other analyzed gases did not exceed 0.55%. For the oxidized V_2CT_x/V_3O_7 MXene, a significant increase in the response value for all analyzed gases was observed: the response for NH_3 increased more than 9 times to 12.7%, and for NO_2 more than 15 times to 14.9%. The response to all other analyzed gases did not exceed 10.3%. The numerical values of the responses to all the studied gases of the MXene film before and after oxidation are shown in Figure 6c,d.

For ammonia, for which high responses were observed during detection, the sensitivity to different concentrations was studied. Figure 7a,b shows the responses to 4–100 ppm NH_3 at RT of the MXene film before and after oxidation. As can be seen, there is a consistent increase in the response (S1) with increasing NH_3 concentrations from 4 to 100 ppm in both cases: from 0.18 to 1.35%, and from 2.7 to 12.7% for V_2CT_x and V_2CT_x/V_3O_7, respectively, with a small baseline drift (which can be explained by the high interaction energy of the ammonia molecule with the receptor material). Thus, both V_2CT_x and V_2CT_x/V_3O_7 films demonstrate the ability to precisely detect different analyte content in the gas atmosphere at room temperature.

Figure 7c,d shows the response of the obtained V_2CT_x and V_2CT_x/V_3O_7 films when the relative humidity RH is changed from 0 to 95%. For the original V_2CT_x film, the response (S2) is 10% when the RH is changed to 95%, which is quite high compared to the response for the other gases for which the response does not exceed 1.35%. The oxidized V_2CT_x/V_3O_7 film also shows a significant increase in sensitivity to RH: the response at 70% RH is 85.6%. At the same time, it should be noted that in the range RH = 50–95% the response to humidity changes much less significantly than at humidity <50%. In general, the high sensitivity of MXenes to humidity is typical and well-described in the literature, but we have not found any reports in the available sources of an increase in their sensitivity to humidity during the formation of the V_2CT_x/V_3O_7 composite.

Figure 8 shows a column diagram of the selectivity, taking into account the sign of the response. Positive values correspond to an increase in electrical resistance (*p*-response), and negative values correspond to a decrease (*n*-response). As can be seen from the presented data, for the original V_2CT_x MXene film in the detection of all gases (including an increase in RH) there is a *p*-response, which is typical for MXene, except for NO_2, for which an *n*-response is recorded. However, for the oxidized V_2CT_x/V_3O_7 film, significant changes are observed: for all gases, except NH_3 and NO_2, the response is reversed from *p*- to *n*-response. When NH_3 is detected, the *p*-type response is retained, and when NO_2 is detected, the *n*-type response is retained. The observed processes with the change of direction of resistance change are a consequence of the change of mechanisms of gas detection.

In the detection of various gases, the *p*-response (with an increase of resistance at the analyte injection) is quite typical for various individual MXenes, including V_2CT_x [33]. The resistance increase observed in this case can be explained by the so-called "swelling mechanism", according to which gases can intercalate into the MXenes' interlayer space, which causes an additional increase in the interlayer distance, complicating the charge transfer outside the MXenes plane [33]. In addition, it is believed that the sorption of any gases due to the charge transfer between them and the MXenes metal atoms interferes with the transport of charge carriers and reduces their quantity.

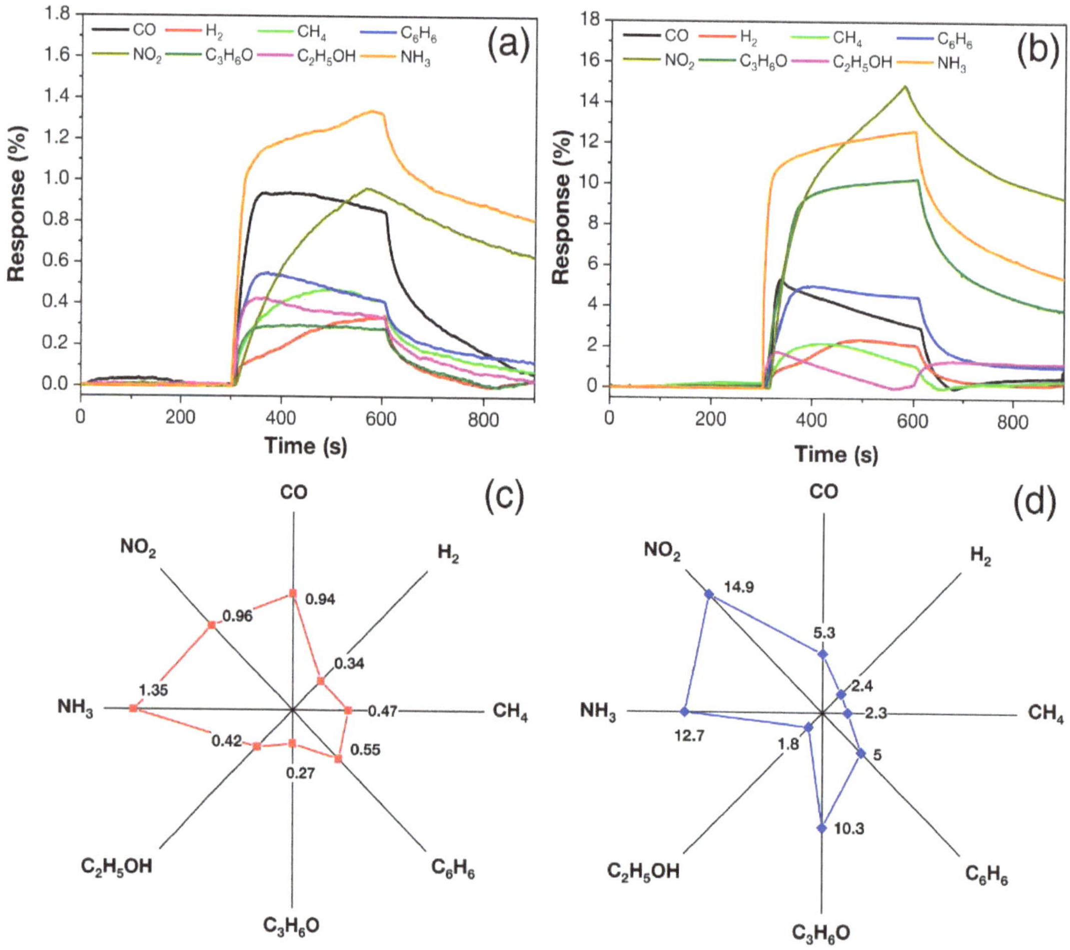

Figure 6. Responses (%) to various gases (100 ppm CO, NH_3, NO_2, C_6H_6, C_3H_6O, C_2H_5OH and 1000 ppm CH_4, H_2) and V_2CT_x MXene film selectivity diagrams before (**a**,**c**) and after oxidation (**b**,**d**); all data obtained at RT and 0%RH.

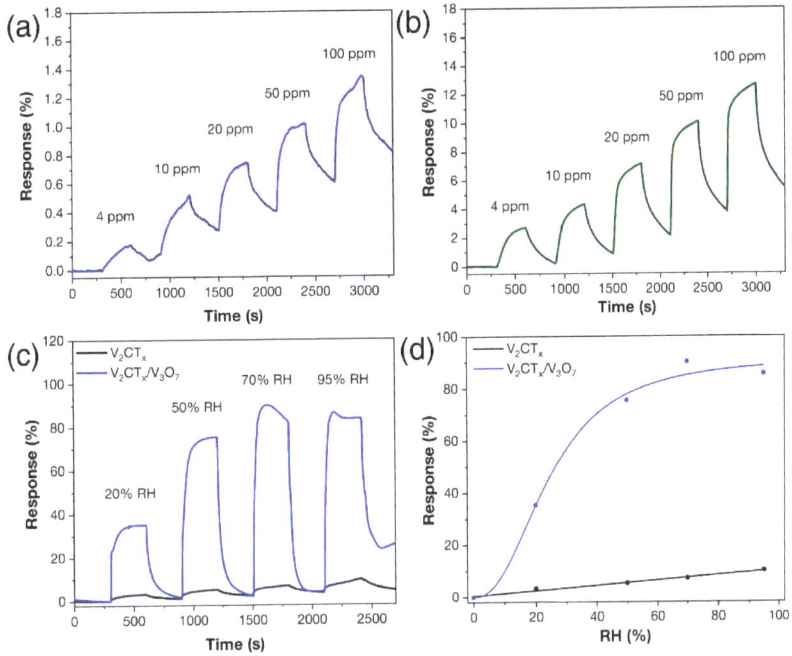

Figure 7. Responses (%) to 4–100 ppm NH_3 of V_2CT_x MXene film before (**a**) and after oxidation (**b**) at RT and 0%RH; responses to 20–95%RH (**c**,**d**) at RT.

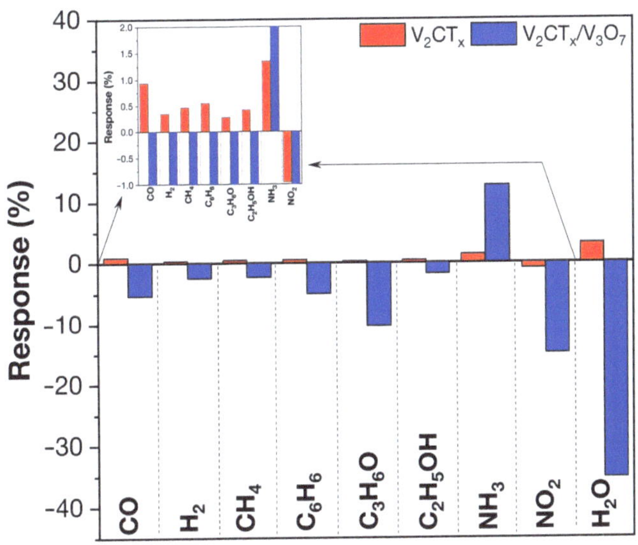

Figure 8. The selectivity diagram of V_2CT_x MXene film and V_2CT_x/V_3O_7 nanocomposite summarizing the response data (taking into account the sign of response, i.e., the direction of resistance change. The "+" sign corresponds to an increase in electrical resistance, and the "−" sign to a decrease) for different gases (100 ppm CO, NH_3, NO_2, C_6H_6, C_3H_6O, C_2H_5OH, 1000 ppm CH_4, H_2 and 20%RH), with the inset with a large increase in response value for some gases.

As in the case described in the present study, the investigation by Yajie Zhang et al. [32] also recorded an *n*-response in the detection of NO_2, which the authors associated with possible reactions between the NO_2 molecule and the functional groups -O and -OH on the surface of the V_2CT_x MXene. As a result, NO_2^--groups may be formed on the MXene surface, which leads to a decrease in electrical resistance. This mechanism is supported by the fact that, according to EDX analysis, the V_2CT_x powder obtained contains a large number of oxygen-containing fragments (O, OH and probably H_2O in the interlayer space) in addition to F- and Cl-groups, which may also be involved in the gas-sensing mechanism and influence the response characteristics [31]. There are a few studies in the literature on the use of an individual V_2CT_x as a sensor material; nevertheless, the available data are in good agreement with the data obtained in the present study [31–34].

Vanadium dioxide (VO_2) and vanadium pentaoxide (V_2O_5) are known to be *n*-type semiconductors [52,53]. The vanadium oxide (V_3O_7) formed in this case is intermediate between VO_2 and V_2O_5 and is also an *n*-type semiconductor. The *n*-type response obtained when the V_2CT_x/V_3O_7 film detected all the gases studied except NH_3 can be explained by the classical detection mechanism for MOS-sensors through the ion adsorbed oxygen on the surface of the V_3O_7 semiconductor nanoparticles [54–56]. CO, C_6H_6, C_3H_6O, C_2H_5OH, CH_4 and H_2 are reducing gases, so when they react with ionized oxygen (O_y^{x-}), they should oxidize to form CO_2 and H_2O, which can lead to the release of additional electrons while the electrical resistance decreases, corresponding to an *n*-response [57–60]. The obtained data on the sensing properties agree well with those reported in the literature for V_2O_5 [30] at elevated sensing temperatures. When NH_3 and NO_2 were detected by the V_2CT_x/V_3O_7 nanocomposite film, *p*- and *n*-responses, respectively, were observed (i.e., the detection character of V_2CT_x MXene was maintained), which is not typical for *n*-type semiconductors. In this regard, it can be assumed that the detection of these gases is mainly due to the characteristic of the MXene mechanism.

The increase in response values for the V_2CT_x/V_3O_7 nanocomposite can be associated with the formation of metal defects in the MXene layers during oxidation and the increase in specific surface area that improved adsorption of gases, as well as with the formation of heterojunctions. This leads to an effective separation of charge carriers at the interface and increases the sensitivity to adsorption of electron-donor and -acceptor gases.

It is to be noted that the oxidation of the V_2CT_x film also changes the direction of the humidity-sensing response from the *n*- to the *p*-type. For the individual V_2CT_x film, the *p*-type response with increasing humidity is also traditionally explained by the swelling mechanism resulting from the adsorption of molecules between the MXene layers. On the other hand, the *n*-type response of the V_2CT_x/V_3O_7 composite film can be explained by the mechanisms of the interaction of water molecules with MOS-materials [54], according to which hydroxylgroups are formed on the surface of the metal oxide semiconductor. As a result, electrons are emitted and a decrease in electrical resistance is observed, which corresponds to the *n*-response [27].

Separately, it should be noted that there are no data in the literature on the use of V_3O_7 as a gas-sensitive chemoresistive material. This fact can be explained by the metastability of this oxide, and the operation of MOS-gas sensors usually occurs at elevated temperatures. The data obtained in this study are probably the first experimental confirmation of the high gas sensitivity of this oxide.

4. Conclusions

In summary, the present study is the first to investigate the effect on the gas-sensitive properties of accordion-like V_2CT_x MXene synthesized by selective etching of V_2AlC MAX-phase by HF-HCl acid mixture, and doping with nanosized V_3O_7 as a result of MXene oxidation controlled in situ by Raman spectroscopy.

The elemental and phase composition, the microstructure of the synthesized V_2CT_x MXene, the coatings based on it, and the V_2CT_x/V_3O_7 nanocomposite obtained at a minimum oxidation temperature of 250 °C were studied by using a variety of physical and chemical analytical methods.

It was found that the obtained V_2CT_x and V_2CT_x/V_3O_7 films are sensitive at room temperature and zero humidity to a wide range of investigated gases (H_2, CO, NH_3, C_6H_6, C_3H_6O, CH_4, C_2H_5OH and NO_2), but the highest responses were observed for ammonia and nitrogen dioxide. At the same time, the partial oxidation of V_2CT_x leads not only to an increase in the response of the V_2CT_x/V_3O_7 material, but also to some change in selectivity. Thus, if for the original V_2CT_x multilayer MXene the responses at 100 ppm (RT, RH = 0%) decreased in the series NH_3 (1.35%) > NO_2 (0.96%) > CO (0.94%), then for the V_2CT_x/V_3O_7 composite the series of highest responses changes: NO_2 (14.9%) > NH_3 (12.7%) > acetone (10.3). Despite the close response values for NO_2 and NH_3, this does not negatively affect the selectivity due to the different nature of the resistance change when these gases are injected.

It is shown that the oxidation of the V_2CT_x film and the formation of the V_2CT_x/V_3O_7 nanocomposite also leads to a significant increase in the humidity response value (RH = 0–95%) and a change in its character from *n*- to *p*-type. At the same time, for V_2CT_x/V_3O_7 there is a saturation of the response value after reaching the relative humidity of 50%, in contrast to the initial V_2CT_x MXene.

Author Contributions: Conceptualization, A.S.M., N.P.S. and E.P.S.; methodology, N.T.K.; validation, E.P.S. and N.T.K.; investigation, A.S.M., I.A.N., A.A.A., T.L.S. and N.P.S.; resources, N.T.K.; writing—original draft preparation, A.S.M., I.A.N., A.A.A. and T.L.S.; writing—review and editing, I.A.N. and E.P.S.; supervision, N.T.K.; project administration, N.P.S.; funding acquisition, N.P.S. All authors have read and agreed to the published version of the manuscript.

Funding: This work was supported by the Russian Science Foundation, project No. 21-73-10251, https://rscf.ru/en/project/21-73-10251/ (accessed on 30 January 2023).

Institutional Review Board Statement: Not applicable.

Informed Consent Statement: Not applicable.

Data Availability Statement: Not applicable.

Conflicts of Interest: The authors declare no conflict of interest.

References

1. Naguib, M.; Mashtalir, O.; Carle, J.; Presser, V.; Lu, J.; Hultman, L.; Gogotsi, Y.; Barsoum, M.W. Two-Dimensional Transition Metal Carbides. *ACS Nano* **2012**, *6*, 1322–1331. [CrossRef] [PubMed]
2. Naguib, M.; Mochalin, V.N.; Barsoum, M.W.; Gogotsi, Y. 25th Anniversary Article: MXenes: A New Family of Two-Dimensional Materials. *Adv. Mater.* **2014**, *26*, 992–1005. [CrossRef] [PubMed]
3. Li, X.; An, Z.; Lu, Y.; Shan, J.; Xing, H.; Liu, G.; Shi, Z.; He, Y.; Chen, Q.; Han, R.P.S.; et al. Room Temperature VOCs Sensing with Termination-Modified $Ti_3C_2T_x$ MXene for Wearable Exhaled Breath Monitoring. *Adv. Mater. Technol.* **2022**, *7*, 2100872. [CrossRef]
4. Wang, J.; Xu, R.; Xia, Y.; Komarneni, S. Ti_2CT_x MXene: A Novel p-Type Sensing Material for Visible Light-Enhanced Room Temperature Methane Detection. *Ceram. Int.* **2021**, *47*, 34437–34442. [CrossRef]
5. Lee, J.; Kang, Y.C.; Koo, C.M.; Kim, S.J. $Ti_3C_2T_x$ MXene Nanolaminates with Ionic Additives for Enhanced Gas-Sensing Performance. *ACS Appl. Nano Mater.* **2022**, *5*, 11997–12005. [CrossRef]
6. Majhi, S.M.; Ali, A.; Greish, Y.E.; El-Maghraby, H.F.; Qamhieh, N.N.; Hajamohideen, A.R.; Mahmoud, S.T. Accordion-like-Ti_3C_2 MXene-Based Gas Sensors with Sub-ppm Level Detection of Acetone at Room Temperature. *ACS Appl. Electron. Mater.* **2022**, *4*, 4094–4103. [CrossRef]
7. Ho, D.H.; Choi, Y.Y.; Jo, S.B.; Myoung, J.M.; Cho, J.H. Sensing with MXenes: Progress and Prospects. *Adv. Mater.* **2021**, *33*, 2005846. [CrossRef]
8. Alwarappan, S.; Nesakumar, N.; Sun, D.; Hu, T.Y.; Li, C.Z. 2D Metal Carbides and Nitrides (MXenes) for Sensors and Biosensors. *Biosens. Bioelectron.* **2022**, *205*, 113943. [CrossRef]
9. Ganesh, P.S.; Kim, S.Y. Electrochemical Sensing Interfaces Based on Novel 2D-MXenes for Monitoring Environmental Hazardous Toxic Compounds: A Concise Review. *J. Ind. Eng. Chem.* **2022**, *109*, 52–67. [CrossRef]

10. SiSivasankarapillai, V.S.; Sharma, T.S.K.; Wabaidur, K.Y.H.S.M.; Angaiah, S.; Dhanusuraman, R. MXene Based Sensing Materials: Current Status and Future Perspectives. *ES Energy Environ.* **2022**, *5*, 4–14. [CrossRef]
11. Zhao, Q.; Zhou, W.; Zhang, M.; Wang, Y.; Duan, Z.; Tan, C.; Liu, B.; Ouyang, F.; Yuan, Z.; Tai, H.; et al. Edge-Enriched $Mo_2TiC_2T_x/MoS_2$ Heterostructure with Coupling Interface for Selective NO_2 Monitoring. *Adv. Funct. Mater.* **2022**, *32*, 2203528. [CrossRef]
12. Nahirniak, S.; Saruhan, B. MXene Heterostructures as Perspective Materials for Gas Sensing Applications. *Sensors* **2022**, *22*, 972. [CrossRef] [PubMed]
13. Devaraj, M.; Rajendran, S.; Hoang, T.K.A.; Soto-Moscoso, M. A Review on MXene and Its Nanocomposites for the Detection of Toxic Inorganic Gases. *Chemosphere* **2022**, *302*, 134933. [CrossRef]
14. Sett, A.; Rana, T.; Rajaji, U.; Sha, R.; Liu, T.Y.; Bhattacharyya, T.K. Emergence of Two-Dimensional Nanomaterials-Based Breath Sensors for Non-Invasive Detection of Diseases. *Sens. Actuators A Phys.* **2022**, *338*, 113507. [CrossRef]
15. Wang, Z.; Yu, K.; Feng, Y.; Qi, R.; Ren, J.; Zhu, Z. VO_2(p)-V_2C(MXene) Grid Structure as a Lithium Polysulfide Catalytic Host for High-Performance Li-S Battery. *ACS Appl. Mater. Interfaces* **2019**, *11*, 44282–44292. [CrossRef]
16. Shan, Q.; Mu, X.; Alhabeb, M.; Shuck, C.E.; Pang, D.; Zhao, X.; Chu, X.F.; Wei, Y.; Du, F.; Chen, G.; et al. Two-Dimensional Vanadium Carbide (V_2C) MXene as Electrode for Supercapacitors with Aqueous Electrolytes. *Electrochem. Commun.* **2018**, *96*, 103–107. [CrossRef]
17. Zhou, J.; Gao, S.; Guo, Z.; Sun, Z. Ti-Enhanced Exfoliation of V_2AlC into V_2C MXene for Lithium-Ion Battery Anodes. *Ceram. Int.* **2017**, *43*, 11450–11454. [CrossRef]
18. Tan, H.; Wang, C.; Duan, H.; Tian, J.; Ji, Q.; Lu, Y.; Hu, F.; Hu, W.; Li, G.; Li, N.; et al. Intrinsic Room-Temperature Ferromagnetism in V_2C MXene Nanosheets. *ACS Appl. Mater. Interfaces* **2021**, *13*, 33363–33370. [CrossRef]
19. Zada, S.; Lu, H.; Yang, F.; Zhang, Y.; Cheng, Y.; Tang, S.; Wei, W.; Qiao, Y.; Fu, P.; Dong, H.; et al. V_2C Nanosheets as Dual-Functional Antibacterial Agents. *ACS Appl. Bio Mater.* **2021**, *4*, 4215–4223. [CrossRef]
20. He, N.; Zhang, Q.; Tao, L.; Chen, X.; Qin, Q.; Liu, X.; Lian, X.; Wan, X.; Hu, E.; Xu, J.; et al. VC-Based Memristor for Applications of Low Power Electronic Synapse. *IEEE Electron. Device Lett.* **2021**, *42*, 319–322. [CrossRef]
21. Liu, H.; Lu, C.; Wang, X.; Xu, L.; Huang, X.; Wang, X.; Ning, H.; Lan, Z.; Guo, J. Combinations of V2C and Ti_3C_2 MXenes for Boosting the Hydrogen Storage Performances of MgH_2. *ACS Appl. Mater. Interfaces* **2021**, *13*, 13235–13247. [CrossRef] [PubMed]
22. Chen, Y.; Yao, H.; Kong, F.; Tian, H.; Meng, G.; Wang, S.; Mao, X.; Cui, X.; Hou, X.; Shi, J. V_2C MXene Synergistically Coupling FeNi LDH Nanosheets for Boosting Oxygen Evolution Reaction. *Appl. Catal. B Environ.* **2021**, *297*, 120474. [CrossRef]
23. Junkaew, A.; Arróyave, R. Enhancement of the Selectivity of MXenes (M_2C, M = Ti, V, Nb, Mo) via Oxygen-Functionalization: Promising Materials for Gas-Sensing and -Separation. *Phys. Chem. Chem. Phys.* **2018**, *20*, 6073–6082. [CrossRef] [PubMed]
24. Choi, S.J.; Kim, I.D. *Recent Developments in 2D Nanomaterials for Chemiresistive-Type Gas Sensors*; The Korean Institute of Metals and Materials: Seoul, Republic of Korea, 2018; Volume 14, ISBN 0123456789.
25. Lee, E.; Kim, D.-J. Review—Recent Exploration of Two-Dimensional MXenes for Gas Sensing: From a Theoretical to an Experimental View. *J. Electrochem. Soc.* **2020**, *167*, 037515. [CrossRef]
26. Kim, S.J.; Koh, H.J.; Ren, C.E.; Kwon, O.; Maleski, K.; Cho, S.Y.; Anasori, B.; Kim, C.K.; Choi, Y.K.; Kim, J.; et al. Metallic $Ti_3C_2T_x$ MXene Gas Sensors with Ultrahigh Signal-to-Noise Ratio. *ACS Nano* **2018**, *12*, 986–993. [CrossRef]
27. Mokrushin, A.S.; Nagornov, I.A.; Simonenko, T.L.; Simonenko, N.P.; Gorobtsov, P.Y.; Arkhipushkin, I.A.; Simonenko, E.P.; Sevastyanov, V.G.; Kuznetsov, N.T. Gas-Sensitive Nanostructured ZnO Films Praseodymium and Europium Doped: Electrical Conductivity, Selectivity, Influence of UV Irradiation and Humidity. *Appl. Surf. Sci.* **2022**, *589*, 152974. [CrossRef]
28. Nagornov, I.A.; Mokrushin, A.S.; Simonenko, E.P.; Simonenko, N.P.; Sevastyanov, V.G.; Kuznetsov, N.T. Liquid-Phase Growth of Nanocrystalline ZnO Thin Films and Their Gas-Sensitive Properties. *Russ. J. Inorg. Chem.* **2022**, *67*, 539–546. [CrossRef]
29. Mokrushin, A.S.; Simonenko, T.L.; Simonenko, N.P.; Yu, P.; Bocharova, V.A.; Kozodaev, M.G.; Markeev, A.M.; Lizunova, A.A.; Volkov, I.A.; Simonenko, E.P.; et al. Microextrusion Printing of Gas-Sensitive Planar Anisotropic NiO Nanostructures and Their Surface Modification in an H2S Atmosphere. *Appl. Surf. Sci.* **2022**, *578*, 151984. [CrossRef]
30. Gorobtsov, P.Y.; Mokrushin, A.S.; Simonenko, T.L.; Simonenko, N.P.; Simonenko, E.P.; Kuznetsov, N.T. Microextrusion Printing of Hierarchically Structured Thick V_2O_5 Film with Independent from Humidity Sensing Response to Benzene. *Materials* **2022**, *15*, 7837. [CrossRef]
31. Wu, M.; An, Y.; Yang, R.; Tao, Z.; Xia, Q.; Hu, Q.; Li, M.; Chen, K.; Zhang, Z.; Huang, Q.; et al. V_2CT_x and $Ti_3C_2T_x$ MXenes Nanosheets for Gas Sensing. *ACS Appl. Nano Mater.* **2021**, *4*, 6257–6268. [CrossRef]
32. Zhang, Y.; Jiang, Y.; Duan, Z.; Huang, Q.; Wu, Y.; Liu, B.; Zhao, Q.; Wang, S.; Yuan, Z.; Tai, H. Highly Sensitive and Selective NO_2 Sensor of Alkalized V_2CT_x MXene Driven by Interlayer Swelling. *Sens. Actuators B Chem.* **2021**, *344*, 130150. [CrossRef]
33. Lee, E.; Vahidmohammadi, A.; Yoon, Y.S.; Beidaghi, M.; Kim, D.J. Two-Dimensional Vanadium Carbide MXene for Gas Sensors with Ultrahigh Sensitivity Toward Nonpolar Gases. *ACS Sens.* **2019**, *4*, 1603–1611. [CrossRef] [PubMed]
34. Wang, X.; Zhang, D.; Zhang, H.; Gong, L.; Yang, Y.; Zhao, W.; Yu, S.; Yin, Y.; Sun, D. In Situ Polymerized Polyaniline/MXene (V_2C) as Building Blocks of Supercapacitor and Ammonia Sensor Self-Powered by Electromagnetic-Triboelectric Hybrid Generator. *Nano Energy* **2021**, *88*, 106242. [CrossRef]

35. Wu, M.; Wang, B.; Hu, Q.; Wang, L.; Zhou, A. The Synthesis Process and Thermal Stability of V_2C MXene. *Materials* **2018**, *11*, 2112. [CrossRef]
36. Thakur, R.; Vahidmohammadi, A.; Moncada, J.; Adams, W.R.; Chi, M.; Tatarchuk, B.; Beidaghi, M.; Carrero, C.A. Insights into the Thermal and Chemical Stability of Multilayered V_2CT:X MXene. *Nanoscale* **2019**, *11*, 10716–10726. [CrossRef] [PubMed]
37. Simonenko, E.P.; Simonenko, N.P.; Nagornov, I.A.; Simonenko, T.L.; Mokrushin, A.S.; Sevastyanov, V.G.; Kuznetsov, N.T. Synthesis of MAX Phases in the Ti_2AlC–V_2AlC System as Precursors of Heterometallic MXenes $Ti_{2-X}V_XC$. *Russ. J. Inorg. Chem.* **2022**, *67*, 705–714. [CrossRef]
38. Simonenko, E.P.; Simonenko, N.P.; Nagornov, I.A.; Simonenko, T.L.; Gorobtsov, P.Y.; Mokrushin, A.S.; Kuznetsov, N.T. Synthesis and Chemoresistive Properties of Single-Layer MXene Ti_2CT_X. *Russ. J. Inorg. Chem.* **2022**, *67*, 1838–1847. [CrossRef]
39. Mokrushin, A.S.; Nagornov, I.A.; Gorobtsov, P.Y.; Averin, A.A.; Simonenko, T.L.; Simonenko, N.P.; Simonenko, E.P.; Kuznetsov, N.T. Effect of Ti_2CT_x MXene Oxidation on Its Gas-Sensitive Properties. *Chemosensors* **2023**, *11*, 13. [CrossRef]
40. Simonenko, N.P.; Glukhova, O.E.; Plugin, I.A.; Kolosov, D.A.; Nagornov, I.A.; Simonenko, T.L.; Varezhnikov, A.S.; Simonenko, E.P.; Sysoev, V.V.; Kuznetsov, N.T. The $Ti_{0.2}V_{1.8}C$ MXene Ink-Prepared Chemiresistor: From Theory to Tests with Humidity versus VOCs. *Chemosensors* **2023**, *11*, 7. [CrossRef]
41. Mokrushin, A.S.; Nagornov, I.A.; Simonenko, T.L.; Simonenko, N.P.; Yu, P.; Khamova, T.V.; Kopitsa, G.P.; Evzrezov, A.N.; Simonenko, E.P.; Sevastyanov, V.G.; et al. Chemoresistive Gas-Sensitive ZnO/Pt Nanocomposites Films Applied by Microplotter Printing with Increased Sensitivity to Benzene and Hydrogen. *Mater. Sci. Eng. B* **2021**, *271*, 115233. [CrossRef]
42. Xu, M.; Zhou, D.; Wu, T.; Qi, J.; Du, Q.; Xiao, Z. Self-Regulation of Spin Polarization Density Propelling the Ion Diffusion Kinetics for Flexible Potassium-Ion Batteries. *Adv. Funct. Mater.* **2022**, 32, 2203263. [CrossRef]
43. Wang, J.; Guan, Y.; Zhang, Q.; Zhu, H.; Li, X.; Li, Y.; Dong, Z.; Yuan, G.; Cong, Y. Well-Dispersed Ultrafine Pt Nanoparticles Anchored on Oxygen-Rich Surface of V_2CT_x (MXene) for Boosting Hydrogen Evolution Reaction. *Appl. Surf. Sci.* **2022**, *582*, 152481. [CrossRef]
44. Chen, G.; Lv, J.; Han, Y.; Zhang, Q.; Liu, Y.; Lang, J.; Wu, X.; Wang, J.; Lu, M.; Zhang, J. Electron and Ion Transport Behavior of Vanadium Based MXene Induced by Pressure for Lithium Ion Intercalated Electrodes. *J. Colloid Interface Sci.* **2023**, *633*, 207–217. [CrossRef] [PubMed]
45. Wu, X.; Wang, H.; Zhao, Z.; Huang, B. Interstratification-Assembled 2D Black Phosphorene and V_2CT:X MXene as Superior Anodes for Boosting Potassium-Ion Storage. *J. Mater. Chem. A* **2020**, *8*, 12705–12715. [CrossRef]
46. Kim, Y.; Gkountaras, A.; Chaix-Pluchery, O.; Gélard, I.; Coraux, J.; Chapelier, C.; Barsoum, M.W.; Ouisse, T. Elementary Processes Governing V_2AlC Chemical Etching in HF. *RSC Adv.* **2020**, *10*, 25266–25274. [CrossRef]
47. Champagne, A.; Shi, L.; Ouisse, T.; Hackens, B.; Charlier, J.C. Electronic and Vibrational Properties of V_2C-Based MXenes: From Experiments to First-Principles Modeling. *Phys. Rev. B* **2018**, *97*, 115439. [CrossRef]
48. Melchior, S.A.; Raju, K.; Ike, I.S.; Erasmus, R.M.; Kabongo, G.; Sigalas, I.; Iyuke, S.E.; Ozoemena, K.I. High-Voltage Symmetric Supercapacitor Based on 2D Titanium Carbide (MXene, Ti_2CT_x)/Carbon Nanosphere Composites in a Neutral Aqueous Electrolyte. *J. Electrochem. Soc.* **2018**, *165*, A501–A511. [CrossRef]
49. Ureña-Begara, F.; Crunteanu, A.; Raskin, J.P. Raman and XPS Characterization of Vanadium Oxide Thin Films with Temperature. *Appl. Surf. Sci.* **2017**, *403*, 717–727. [CrossRef]
50. Shvets, P.; Dikaya, O.; Maksimova, K.; Goikhman, A. A Review of Raman Spectroscopy of Vanadium Oxides. *J. Raman Spectrosc.* **2019**, *50*, 1226–1244. [CrossRef]
51. Tolosa, A.; Fleischmann, S.; Grobelsek, I.; Presser, V. Electrospun Hybrid Vanadium Oxide/Carbon Fiber Mats for Lithium- and Sodium-Ion Battery Electrodes. *ACS Appl. Energy Mater.* **2018**, *1*, 3790–3801. [CrossRef]
52. Zhou, Y.; Ramanathan, S. GaN/VO_2 Heteroepitaxial p-n Junctions: Band Offset and Minority Carrier Dynamics. *J. Appl. Phys.* **2013**, *113*, 213703. [CrossRef]
53. Armer, C.F.; Lübke, M.; Johnson, I.; McColl, K.; Cora, F.; Yeoh, J.S.; Reddy, M.V.; Darr, J.A.; Li, X.; Lowe, A. Enhanced Electrochemical Performance of Electrospun V_2O_5 Fibres Doped with Redox-Inactive Metals. *J. Solid State Electrochem.* **2018**, *22*, 3703–3716. [CrossRef]
54. Mokrushin, A.S.; Fisenko, N.A.; Gorobtsov, P.Y.; Simonenko, T.L.; Simonenko, E.P.; Simonenko, N.P.; Glumov, O.V.; Melnikova, N.A.; Bukunov, K.A.; Sevastyanov, V.G.; et al. Pen Plotter Printing of ITO Thin Film as a Highly CO Sensitive Component of a Resistive Gas Sensor. *Talanta* **2021**, *221*, 121455. [CrossRef] [PubMed]
55. Simonenko, T.L.; Simonenko, N.P.; Gorobtsov, P.Y.; Mokrushin, A.S.; Solovey, V.R.; Pozharnitskaya, V.M.; Simonenko, E.P.; Glumov, O.V.; Melnikova, N.A.; Lizunova, A.A.; et al. Pen Plotter Printing of Co_3O_4 Thin Films: Features of the Microstructure, Optical, Electrophysical and Gas-Sensing Properties. *J. Alloys Compd.* **2020**, *832*, 154957. [CrossRef]
56. Nagornov, I.A.; Mokrushin, A.S.; Simonenko, E.P.; Simonenko, N.P.; Gorobtsov, P.Y.; Sevastyanov, V.G.; Kuznetsov, N.T. Zinc Oxide Obtained by the Solvothermal Method with High Sensitivity and Selectivity to Nitrogen Dioxide. *Ceram. Int.* **2020**, *46*, 7756–7766. [CrossRef]
57. Ji, H.; Zeng, W.; Li, Y. Gas Sensing Mechanisms of Metal Oxide Semiconductors: A Focus Review. *Nanoscale* **2019**, *11*, 22664–22684. [CrossRef]
58. Deng, Y. *Semiconducting Metal Oxides for Gas Sensing*; Elsivier: Amsterdam, The Netherlands, 2019; ISBN 9789811358524.

59. Degler, D.; Weimar, U.; Barsan, N. Current Understanding of the Fundamental Mechanisms of Doped and Loaded Semiconducting Metal-Oxide-Based Gas Sensing Materials. *ACS Sens.* **2019**, *4*, 2228–2249. [CrossRef]
60. Jeong, S.Y.; Kim, J.S.; Lee, J.H. Rational Design of Semiconductor-Based Chemiresistors and Their Libraries for Next-Generation Artificial Olfaction. *Adv. Mater.* **2020**, *32*, 2002075. [CrossRef]

chemosensors

Article

Humidity and Temperature Sensing of Mixed Nickel–Magnesium Spinel Ferrites

Milena P. Dojcinovic [1], Zorka Z. Vasiljevic [1], Lazar Rakocevic [2], Vera P. Pavlovic [3], Souad Ammar-Merah [4], Jelena D. Vujancevic [5] and Maria Vesna Nikolic [1,*]

1 Institute for Multidisciplinary Research, University of Belgrade, 11030 Belgrade, Serbia
2 Institute of Nuclear Sciences of Vinca, University of Belgrade, 11351 Belgrade, Serbia
3 Faculty of Mechanical Engineering, University of Belgrade, 11120 Belgrade, Serbia
4 ITODYS Laboratory, Université Paris Cité, CNRS UMR-7086, 75013 Paris, France
5 Institute of Technical Sciences of SASA, 11000 Belgrade, Serbia
* Correspondence: mariavesna@imsi.rs

Abstract: Temperature- and humidity-sensing properties were evaluated of Ni_xMg_{1-x} spinel ferrites ($0 \leq x \leq 1$) synthesized by a sol-gel combustion method using citric acid as fuel and nitrate ions as oxidizing agents. After the exothermic reaction, amorphous powders were calcined at 700 °C followed by characterization with XRD, FTIR, XPS, EDS and Raman spectroscopy and FESEM microscopy. Synthesized powders were tested as humidity- and temperature-sensing materials in the form of thick films on interdigitated electrodes on alumina substrate in a climatic chamber. The physicochemical investigation of synthesized materials revealed a cubic spinel $Fd\bar{3}m$ phase, nanosized but agglomerated particles with a partially to completely inverse spinel structure with increasing Ni content. $Ni_{0.1}Mg_{0.9}Fe_2O_4$ showed the highest material constant ($B_{30,90}$) value of 3747 K and temperature sensitivity (α) of $-4.08\%/K$ compared to pure magnesium ferrite ($B_{30,90}$ value of 3426 K and α of $-3.73\%/K$) and the highest average sensitivity towards humidity of 922 kΩ/%RH in the relative humidity (RH) range of 40–90% at the working temperature of 25 °C.

Keywords: nickel; magnesium; spinel; ferrite; humidity; temperature; sensing; structure; morphology

Citation: Dojcinovic, M.P.; Vasiljevic, Z.Z.; Rakocevic, L.; Pavlovic, V.P.; Ammar-Merah, S.; Vujancevic, J.D.; Nikolic, M.V. Humidity and Temperature Sensing of Mixed Nickel–Magnesium Spinel Ferrites. *Chemosensors* **2023**, *11*, 34. https://doi.org/10.3390/chemosensors11010034

Academic Editor: Pi-Guey Su

Received: 4 December 2022
Revised: 23 December 2022
Accepted: 28 December 2022
Published: 2 January 2023

1. Introduction

Spinel ferrites are metal oxides with the general formula MFe_2O_4, where M could be Cu, Mg, Ca, Co, Ni, Zn, Mn, Sr, Ba, etc., or may be a combination of different metal cations. Ferrites have a spinel structure, which is defined by 16 d octahedral positions, called B-sites, and 8 tetrahedral positions, called A-sites, occupied by metal cations, while 32 e positions at polyhedron vertexes are occupied by O ions [1]. Spinels can be normal, where all B-sites are occupied by Fe ions and all A-sites are occupied by other metal cations, wherein the inversion degree λ is zero and the formula is $\left(M^{2+}\right)\left(Fe_2^{3+}\right)O_4$, or inverse, when iron ions occupy A-sites and more Fe ions and other cations occupy B-sites. Parameter λ is then 1, and the formula is $\left(Fe^{3+}\right)\left(M^{2+}Fe^{3+}\right)O_4$ [1]. Spinels can be partially inverse, with the inversion parameter ranging between 0 and 1; the resulting formula is $\underbrace{\left(M^{2+}_{1-\lambda}Fe^{3+}{}_{\lambda}\right)}_{\text{tetrahedral sites}}\underbrace{\left(M^{2+}_{\lambda}Fe^{3+}_{2-\lambda}\right)}_{\text{octahedral sites}}O_4$, with λ representing the inversion degree.

Due to their magnetic, electric, dielectric and optical properties, natural abundance and also high biocompatibility, spinel ferrites, especially in nano form, have been the subject of much research for application in gas sensing [2], water and wastewater treatment [3], adsorption [4], catalysis [5] and photocatalysis [6], as magnetic nanocarriers for medical applications [7] and more recently for energy storage [8].

Many different and versatile methods have been applied for the synthesis of ferrites and include high-temperature solid-state synthesis from oxides as precursors [5,9], thermal decomposition of metal nitrates [1], pulsed laser deposition [1], sol-gel synthesis [10], co-precipitation [11], chemical vapor deposition [1] and others [5], depending on the cost, simplicity and also the desired morphology and particle size. The electric, electromagnetic and optical characteristics of the resulting ferrites are dependent on their synthesis conditions, cation substitution, inversion degree and elemental composition [12]. Citrate combustion synthesis is a popular synthesis method because of the short reaction time, cheap chemicals and citrate serving as a chelating agent and fuel at the same time [13]. Citrate combustion synthesis, as a sol-gel synthesis, ensures the obtaining of highly crystalline nanoparticle powders [8].

Monitoring humidity is a widespread necessity for the sake of human health and safety, and also for the quality of products and efficacy of industrial processes. Most of the commercially available humidity sensors measure relative humidity, and they can be categorized based on material types or operating principle. As for material types, those could be ceramic/semiconductor, polymer or hybrid materials. All of the mentioned detect and measure humidity by absorbing and desorbing water molecules, which change physical and electrical properties such as impedance, capacitance, electrical resistance and others [14]. Ceramic-type humidity sensors exhibit advantages over polymer films such as physical stability and chemical and thermal resistance [14].

Temperature sensing is another scientific and engineering discipline employed in almost every aspect of human life. The main types of temperature sensors are thermocouples, RTDs, thermistors and semiconductor-based ICs. Thermistors respond to temperature with impedance and resistance changes. Metal oxide spinels are mainly NTC (negative temperature coefficient) thermistors, which means that with an increase in temperature, their electric impedance decreases. The change can be described by the Arrhenius equation, with the B-value determining the steepness of the temperature change. Ceramic NTC thermistors are low-cost, simple to synthesize and chemically and physically stable. "They offer the best sensitivity and accuracy at the lowest price" [15].

We incorporated citrate synthesis into this work to synthesize mixed Ni-Mg ferrites. While both magnesium ferrite and nickel ferrite have been tested as humidity sensors [16,17], neither has been considered as a temperature sensor, so detailed research has been conducted in this work to comprehend the impact of combining nickel and magnesium in a ferrite material with regard to its sensing properties.

2. Materials and Methods

The materials used in this synthesis are citric acid (Sigma Aldrich, ACS, St. Louis, MI, USA) and the metal nitrates $Mg(NO_3)_2 \cdot 6\,H_2O$ (Sigma Aldrich, puriss, p. a.), $Ni(NO_3)_2 \cdot 6\,H_2O$ (Sigma Aldrich, purum, p. a.), and $Fe(NO_3)_3 \cdot 9\,H_2O$ (Sigma Aldrich, puriss, p. a.).

The sol-gel combustion method with citric acid as a reducing agent and nitrate ions as oxidizers was used to synthesize nickel–magnesium ferrites $Ni_xMg_{1-x}Fe_2O_4$, with x = 0, 0.1, 0.3, 0.5, 0.7, 0.9 and 1. The precursory nickel, magnesium and iron nitrate 1 M aqueous solutions were mixed and put on a heated magnetic stirrer. Water was evaporated from the mixture at 80 °C until a gel was formed. Then, the temperature was set to 250 °C until the combustion reaction occurred. The resulting powders were amorphous, brown-black in color, very light and flaky. Powders were calcined in a chamber furnace for 3 h at 700 °C with a heating rate of 10 °C/min.

To study the powder structure, X-ray diffraction (XRD) data were acquired using a PANalytical X'Pert PRO diffractometer in Bragg–Bretano geometry with a scattering angle from 10 to 120°, a step of 0.017 s, a hold time of 24.76 s and Co K_α radiation (wavelength of 1.78901 Å). Room temperature Raman spectra (Raman shift region of 150–1000 cm^{-1}, 2.5 mW power at sample) were taken with an XploRA (Horiba Jobin Yvon) spectrometer with a 532 nm laser. An FTIR Nicolet 6700 ATR device (range 400–2000 cm^{-1}, resolution 4 cm^{-1}) was used to collect Fourier transform infrared (FTIR) spectra. To study the powder morphology and microstructure, field emission scanning electron microscopy (FESEM) was applied on a TESCAN MIRA3 XM and ZEISS GeminiSEM 360 with an Oxford Instruments EDX. An FEI Talos F200X microscope operated at 200 kV was used to collect transmission electron microscopy (TEM) micrographs. To study the elemental composition, X-ray photoelectron spectroscopy (XPS) was performed by a SPECS System with an XP50M X-ray source for a Focus 500 and PHOIBOS 100/150 analyzer (AlKα source (1486.74 eV) at 12.5 kV and 32 mA). A constant pass energy of 40 eV, step size of 0.5 eV and dwell time of 0.2 s in the FAT mode were applied to record survey spectra with a 0–1000 eV binding energy and at a pressure of 9×10^{-9} mbar using the SpecsLab data analysis software. A constant pass energy of 20 eV, step size of 0.1 eV and dwell time of 2 s in the FAT mode were applied to record detailed spectra of Mg 1s, Ni 2p, and Fe 2p peaks. All the peak positions were referenced against C 1s at 284.5 eV. The commercial CasaXPS software package was used to analyze all recorded spectra.

Humidity and temperature sensing measurements were performed in a JEIO TECH TH-KE-025 humidity and temperature climatic chamber with the relative humidity (RH) ranging from 40 to 90% (at T = 25 °C) and the temperature ranging from 25 to 90 °C (at RH = 40%). The spinel ferrite powders were mixed with water and ultrasonically dispersed to prepare a paste that was drop-cast on interdigitated Au electrodes on a ceramic substrate (Drop Sense IDEAU200). After drying at 50 °C, the paste formed a thick film (as shown in Figure 1). The thick film layer was measured with a laboratory micrometer, and the average thickness was estimated as 100 µm. Impedance was measured with a Hioki LCR 3536 analyzer in the frequency range of 8 Hz–1 MHz at a voltage of 1 V.

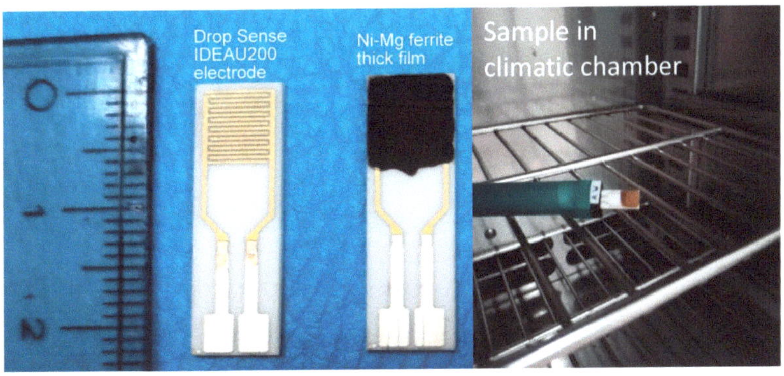

Figure 1. Ni-Mg ferrite thick film sample, used electrodes and setup in climatic chamber.

3. Results and Discussion

3.1. Structural Characterization

Analysis of the measured XRD diffractograms of synthesized nickel–magnesium ferrite powders (Figure 2) using the HighScorePlus software showed well-resolved, highly crystalline peaks that can be indexed as a cubic spinel structure (space group $Fd\bar{3}m$), taking into account ICSD cards 98-016-7491 and 98-024-6894 for $MgFe_2O_4$ and $NiFe_2O_4$, respectively. Traces of hematite (ICSD 98-002-2505) were noted also, and the hematite content varied depending on the composition (Table 1), which is in accordance with previous results [10]. The Scherrer equation was used to calculate the crystallite size, as shown in Table 1. It varied in the range 30–39 nm, with 39 nm obtained for pure nickel ferrite and 37 nm obtained for pure magnesium ferrite. If we observe the (311) peak of the cubic spinel structure more closely, as shown in Figure 2b, we can note that the substitution of magnesium for nickel in magnesium ferrite led to a slight shift of the diffraction peak to higher values. The shift of diffraction peaks can be attributed to the fact that Ni^{2+} has a smaller ionic radius of 0.69 Å compared to the Mg^{2+} ionic radius of 0.72 Å, so when nickel is exchanged for magnesium, the lattice volume shrinks slightly. This has been noted before for Mg- or Zn (A)-substituted $Ni_{1-x}A_xFe_2O_4$ [11] and cobalt-substituted nickel ferrites [18]. The introduction of Mg or Zn with a larger ionic radius (0.72 and 0.74 Å) instead of Ni, with a smaller ionic radius (0.69 Å), led to a small distortion of the lattice and a shift of diffraction peaks [11]. Rietveld refinement enabled determination of the lattice constant and inversion degree, as shown in Table 1. Good agreement between measured and fitted data was observed, as shown in the example given in Figure 2c for $Ni_{0.3}Mg_{0.7}Fe_2O_4$.

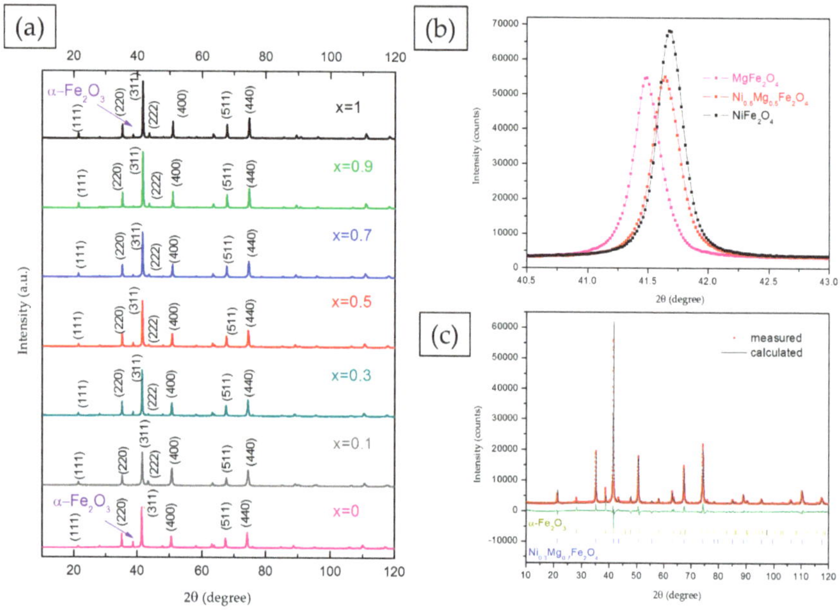

Figure 2. XRD patterns of $Ni_xMg_{1-x}Fe_2O_4$ ($0 \leq x \leq 1$) (**a**); (311) peak shift with magnesium substitution with nickel (**b**); and refined XRD diffractogram of $Ni_{0.3}Mg_{0.7}Fe_2O_4$ (**c**).

Table 1. Lattice parameters, inversion degree, crystallite size and hematite content determined for $Ni_xMg_{1-x}Fe_2O_4$ ($0 \leq x \leq 1$).

Sample	Lattice Constant (Å)	Inversion Degree	Crystallite Size (nm)	Hematite (wt.%)
$MgFe_2O_4$	8.37854(10)	0.8	37	17.7
$Ni_{0.1}Mg_{0.9}Fe_2O_4$	8.36136(2)	0.8	30	0
$Ni_{0.3}Mg_{0.7}Fe_2O_4$	8.36820(8)	0.8	38	6.2
$Ni_{0.5}Mg_{0.5}Fe_2O_4$	8.34918(11)	0.8	37	6.9
$Ni_{0.7}Mg_{0.3}Fe_2O_4$	8.34304(10)	1	34	6.9
$Ni_{0.9}Mg_{0.1}Fe_2O_4$	8.33014(10)	1	39	4.2
$NiFe_2O_4$	8.33677(10)	1	39	0.4

In cubic spinel ferrites the lattice constant, peak position, peak intensity ratio and inversion degree of the spinel structure also depend greatly on the cation distribution on A and B sites [19]. In the case of magnesium ferrite, there are varied opinions on the preference of magnesium ions for tetrahedral and octahedral sites, and magnesium ferrite has most often been described as a "partially inverse spinel" ferrite with magnesium ions present at both tetrahedral and octahedral sites [20,21]. In our previous work magnesium ferrite synthesized using the sol-gel combustion method and citric acid as fuel showed a partially inverse spinel structure, $[Mg_{0.21}Fe_{0.79}]_A[Mg_{0.79}Fe_{1.21}]_BO_4$, with Mg^{2+} ions at both A and B sites [22]. The substitution of magnesium with nickel leads to changes in the lattice parameters, peak position, peak intensity ratio and other parameters of the cubic spinel structure [23,24]. Nickel ions show a strong preference for octahedral (B) sites both in pure nickel ferrite—where often all nickel ions are located at octahedral sites, with Fe^{3+} ions at both tetrahedral and octahedral sites [25]—and in mixed metal spinel structures such as magnesium–nickel ferrites [26–28]. This was the case in this work, as Rietveld analysis showed that when nickel substituted Mg, it was located at octahedral sites, while Mg^{2+} and Fe^{3+} were distributed between octahedral and tetrahedral sites. The inversion degree was 0.8 for compositions with x = 0.1, 0.3 and 0.5, with the cation distribution as follows: $[Mg_{0.2}Fe_{0.8}]_A[Mg_{0.7}Ni_{0.1}Fe_{1.2}]_BO_4$, $[Mg_{0.2}Fe_{0.8}]_A[Mg_{0.5}Ni_{0.3}Fe_{1.2}]_BO_4$ and $[Mg_{0.2}Fe_{0.8}]_A[Mg_{0.3}Ni_{0.5}Fe_{1.2}]_BO_4$, respectively. For x = 0.7, 0.9 and 1 (pure nickel ferrite), the inversion degree was determined as 1, making the cation distribution $[Fe_1]_A[Mg_{0.3}Ni_{0.7}Fe_1]_BO_4$, $[Fe_1]_A[Mg_{0.1}Ni_{0.9}Fe_{0.5}]_BO_4$ and $[Fe_1]_A[Ni_1Fe_1]_BO_4$, respectively. The determined lattice parameter values for $Ni_xMg_{1-x}Fe_2O_4$ powders also varied with the cation distribution and were higher for samples with higher Mg content, as noted before [19,29].

Raman spectra of synthesized samples are shown in Figure 3a. The spectra of $MgFe_2O_4$ and $NiFe_2O_4$ samples are similar to those already reported in the literature, when considering completely or partially inverse spinel structures [25,30–33]. It is known that in the case of completely inverse or normal spinel structure, five first-order Raman-active modes are expected (A_{1g} + E_g + $3T_{2g}$) [25,34,35], and all of those five peaks are observed in spectra in Figure 3a, even though the $T_{2g}(3)$ mode becomes very weak in $MgFe_2O_4$.

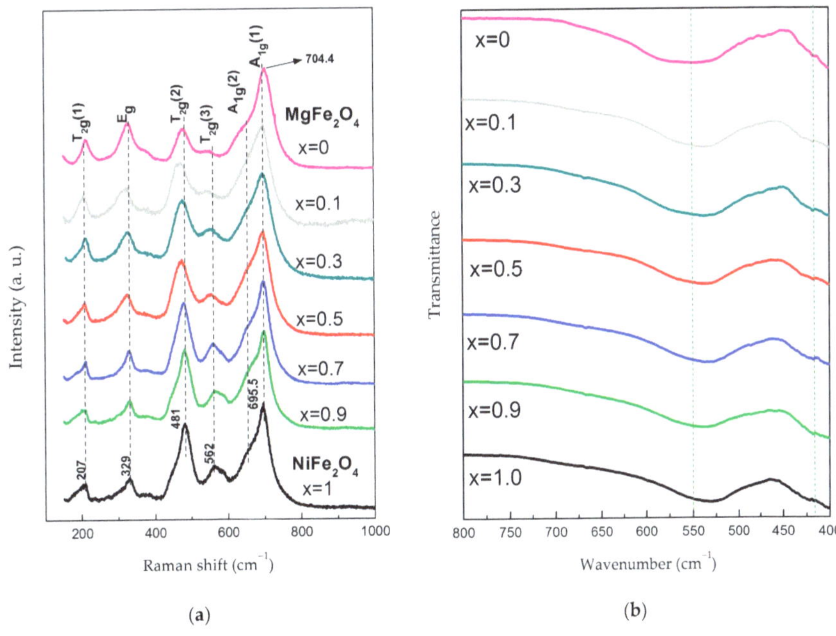

Figure 3. (**a**) Raman spectra and (**b**) FTIR spectra of synthesized $Ni_xMg_{1-x}Fe_2O_4$ ($0 \leq x \leq 1$).

According to the literature, the A_{1g} peak originates from a symmetric stretching vibration of oxygen within the tetrahedral AO_4 groups, i.e., symmetric stretching of (Fe/M)–O bonds for metal ions at the tetrahedral sites (tetrahedral breath mode) [27,36,37]. Although some authors report that modes below 600 cm^{-1} originate from the vibrations of ions in the octahedral group [27,35,38], others claim that this refers only to the $T_{2g}(2)$ and $T_{2g}(3)$ modes, which originate from asymmetric stretching and asymmetric bending of (Fe/M)–O bonds, respectively, while T_{2g} (1) comes from translational vibration of the whole AO_4 tetrahedron. The E_g mode is mostly attributed to symmetric bending vibrations of oxygen with respect to cations in tetrahedral surroundings [27,37,39]. The spectra in Figure 3a show the asymmetry of Raman peaks, indicating the occurrence of the additional modes, which is common for spinel ferrites, especially nanocrystalline ferrites. Even though a certain asymmetry of Raman peaks for nanocrystalline samples may be caused by the confinement effect [40,41], the redistribution of cations (M^{2+} and Fe^{3+}) between octahedral and tetrahedral sites has been recognized as the main reason for the asymmetry of Raman modes in spinel ferrites. In this regard, some authors assign the occurrence of additional modes to the non-equivalent bonds caused by different metal ions in octahedral positions (in partially or completely inverse spinels) and to the short-range ordering of M^{2+} and Fe^{3+} ions at the B site, which induces the lowering of the symmetry [30,42–45]. On the other hand, the distribution of M^{2+} cations between A and B positions is also often reported as the cause of the appearance of additional modes [11,27,46–49]. Having in mind the latter, it is important to outline that the spectra of the ferrites $NiFe_2O_4$ and $MgFe_2O_4$ generally have a pronounced asymmetry of the A_{1g} peak, indicating the splitting of this peak into the $A_{1g}(1)$ and $A_{1g}(2)$ modes, and such an effect is also perceived in our spectra. Considering the existence of different cations at tetrahedral positions as the possible reason for A_{1g} peak splitting, in the case of partially inverse spinels, some authors attribute the Mg–O bond to the $A_{1g}(1)$ mode in $MgFe_2O_4$ [48,50], while others ascribe the same bond to the $A_{1g}(2)$ mode [32]. However, some research indicates that in completely inverse spinels, both the A_{1g} (1) and A_{1g} (2) modes can originate only from the contribution of Fe-O bonds in tetrahedral coordination [18].

When considering the spectra of the $Ni_xMg_{1-x}Fe_2O_4$ system in Figure 3a, it can be concluded that the addition of Ni to $MgFe_2O_4$ causes an alteration in the relative intensity, shape and position of Raman peaks. Namely, with increase in the x value from 0 to 0.5, an increase in relative intensity as well as a certain change of the shape of the $T_{2g}(3)$ and $T_{2g}(2)$ modes are observed, probably due to more pronounced incorporation of Ni^{2+} ions into octahedral positions, i.e., due to the occurrence of both Ni^{2+} and Mg^{2+} ions as M ions at these sites. Along with that, changes in the E_g mode are also noticed. When x exceeds the value of 0.5, a certain modification of the shape and width of the A_{1g} peak becomes evident compared to the lower x values, which together with the results of Rietveld analysis may indicate that the incorporation of Mg^{2+} ions into the tetrahedral positions becomes negligible. There is also an additional difference in the relative intensity, shape and position of the $T_{2g}(3)$ and $T_{2g}(2)$ modes, with respect to the samples with $x \leq 0.5$, which may be a consequence of the dominant presence of Ni^{2+} and Fe^{3+} ions at octahedral positions for the higher values of x. Taking into account the modifications of the shape and intensity of the $T_{2g}(3)$ mode with a change of x, it could be assumed that the shoulder effect detected at 585–590 cm^{-1} in $NiFe_2O_4$ originates from Ni^{2+} ions at B sites, while the mode at 562 cm^{-1} comes from Fe^{3+} ions at B sites. It is also possible that those modes are influenced by the creation of Fe^{2+} ions, as well as by the creation of pairs of Fe^{2+} and Ni^{3+} ions in $NiFe_2O_4$, because of the charge transfer between Fe^{3+} and Ni^{2+} at the B sites. Namely, the mentioned mechanism is a leading cause of conductivity in $NiFe_2O_4$ nanoparticles [36]. A slight shift of the $A_{1g}(1)$ mode towards lower values with an increase of x was also detected, which is in accordance with investigations by D. Varshney and K. Verma [11]. The shift may be a result of the higher atomic mass of Ni^{2+} as compared to the Mg^{2+}ion. An analogous shift is also noted for the $T_{2g}(1)$ peak.

The measured FTIR spectra of calcined $Ni_xMg_{1-x}Fe_2O_4$ powders are displayed in Figure 3b in the range of 400–800 cm^{-1}. No other bands were detected because there are no organic phases present in the synthesized powders, but only peaks originating from the cubic spinel structure, as shown in Figure 3b. The band in the range of 400–450 cm^{-1} originates from metal–oxygen vibrations at the octahedral sites, while the band in the range of 510–550 cm^{-1} is due to stretching vibration of the metal–oxygen bond at tetrahedral sites [22,28]. Analysis of the band position shows that there is a peak shift to lower wavenumbers of the band showing Mg, Ni, Fe-O bonds at the tetrahedral sites with an increase in x starting from about 550 cm^{-1} for $MgFe_2O_4$ and ending at about 530 cm^{-1}

for $NiFe_2O_4$. The lowering of the vibration frequency can be explained if we consider the higher atomic mass of Fe, which is 55.845 a.u. compared to the 24.305 a.u. of magnesium. Heavier iron ions replace magnesium ions at the tetrahedral sites (while nickel ions have a preference for octahedral sites and replace magnesium and iron ions), causing the vibration energy to decline and the band to move to lower wavenumbers. The bands showing metal–oxygen bonds at the octahedral sites are in similar positions, from 414 cm^{-1} for $MgFe_2O_4$ to 422 cm^{-1} for $NiFe_2O_4$. In $MgFe_2O_4$, the partially inverse structure means that most of the octahedral sites are occupied by Fe ions. With the addition of nickel, for x = 0.7, 0.9 and 1, the structure is completely inverse, and octahedral sites are evenly occupied by Ni and Fe ions. There is no significant shift due to the interchanging of iron ions with nickel ions at the octahedral sites, as they have similar ionic radii and atomic masses, 58.7 a.u. compared to 55.8 a.u., respectively.

3.2. Morphology

Obtained FESEM micrographs are displayed in Figure 4a–c and Figure S1. The synthesized $Ni_xMg_{1-x}Fe_2O_4$ powders are composed of nanocrystalline particles with a spherical shape and noticeable agglomeration, similar to published research on magnesium and nickel spinel ferrites [26,51]. Calcination caused coarsening of the agglomerates in the synthesized powders [52]. As the nickel content was higher, the structure was more compact. A similar phenomenon was observed for mixed Mg-Co spinel ferrites when the cobalt amount increased [22]. Strong agglomeration in synthesized powders, especially in the ones with high nickel content, can be explained by the magnetic nature of the material [53]. A similar phenomenon was noticed in TEM images, where also agglomeration was the lowest for $MgFe_2O_4$ and the highest for $NiFe_2O_4$.

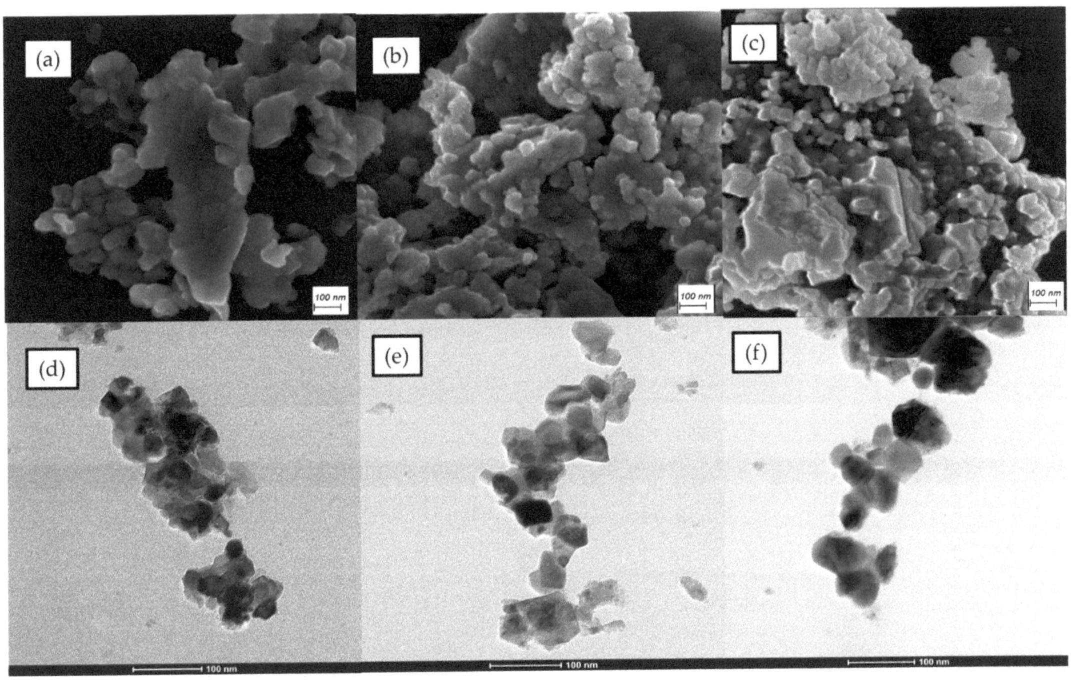

Figure 4. FESEM and TEM images of $MgFe_2O_4$ (**a**,**d**); $Ni_{0.5}Mg_{0.5}Fe_2O_4$ (**b**,**e**); and $NiFe_2O_4$ (**c**,**f**).

High-resolution TEM (HRTEM) images of selected areas of $MgFe_2O_4$, $Ni_{0.5}Mg_{0.5}Fe_2O_4$ and $NiFe_2O_4$ individual particles are shown in Figure 5. Analysis of periodic lattice fringes was performed using fast Fourier transform (FFT) analysis (insets in Figure 5) and revealed crystal lattice spacings of the (311), (220) and (111) planes of the cubic spinel phase, measured as 0.25, 0.29 and 0.48 nm, respectively. Though analysis of XRD diffractograms showed that the cation (Mg, Ni and Fe) distribution and composition influenced the peak position, inversion degree and lattice parameter values (Figure 2 and Table 1), it was not possible to determine noticeable differences in crystal lattice spacing values, but the high crystallinity degree of the powder samples was confirmed.

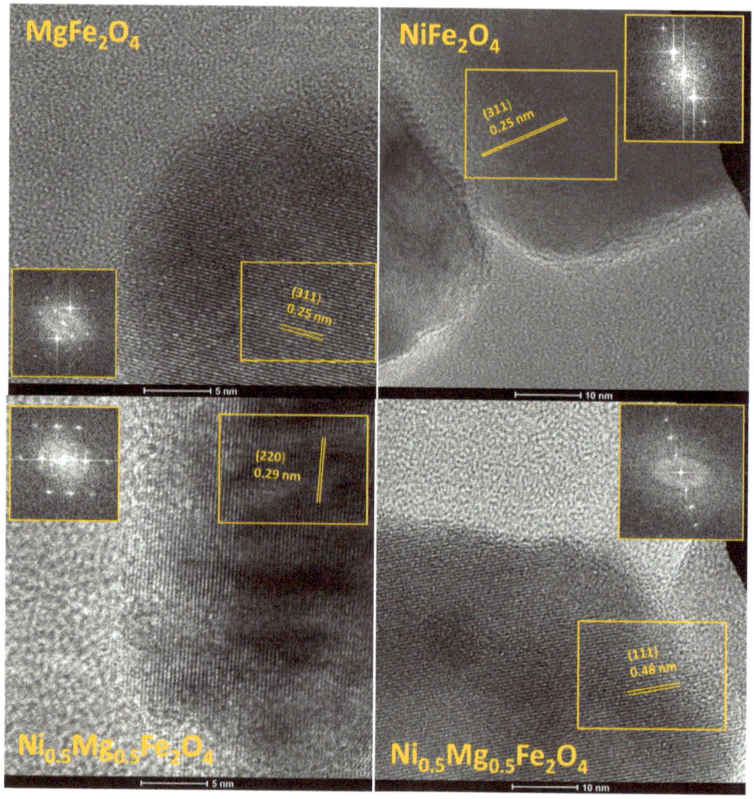

Figure 5. HRTEM micrographs of selected $MgFe_2O_4$, $Ni_{0.5}Mg_{0.5}Fe_2O_4$ and $NiFe_2O_4$ particles showing crystal lattice planes with d-spacing of 0.25, 0.29 and 0.48 nm, confirming the presence of (311), (220) and (111) crystal faces of the cubic spinel lattice phase.

3.3. Elemental Composition and Surface Analysis

The elemental compositions of $MgFe_2O_4$, $Ni_{0.5}Mg_{0.5}Fe_2O_4$ and $NiFe_2O_4$ were determined by EDX analysis. The results showed that magnesium, nickel and iron were homogeneously distributed in the $Ni_5Mg_{0.5}Fe_2O_4$ powder samples (Figure 6a,b). The carbon present in the sample comes from the process of fuming the sample with carbon powder to enable the measurement. The silicon in the spectra comes from the supporting material for the measured samples and as shown in Figure 6b is surrounding the powder samples. Quantitative elemental analysis was conducted by measuring spectra on several points of each material and calculating the average elemental percentage, such as in Figure 6c for $Ni_{0.5}Mg_{0.5}Fe_2O_4$. Elemental quantitative analysis of all three investigated materials corresponded well to the desired stoichiometry, as presented in Figure 6d. A slight oxygen deficiency has been noted before in magnesium ferrite and nickel ferrite, and it may be explained by the abundance of oxygen vacancies which occur throughout the sol-gel combustion synthesis process and subsequent sintering treatment [54,55].

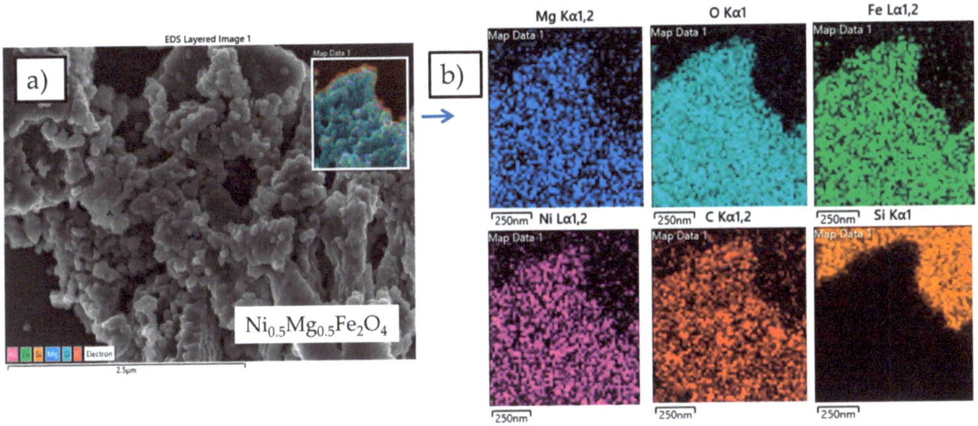

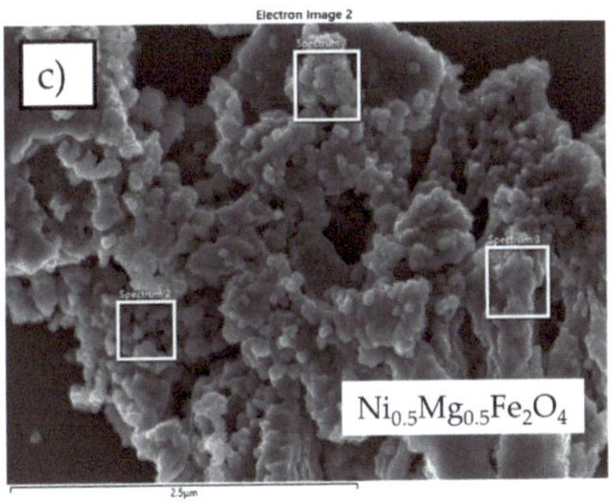

d)	$MgFe_2O_4$	$Ni_{0.5}Mg_{0.5}Fe_2O_4$	$NiFe_2O_4$
El.	Atomic %		
Mg	9.73	8.22	/
Ni	/	8.10	14.85
O	21.53	52.82	51.78
Fe	44.18	30.02	32.54
Si	24.56	0.85	0.82
	$Mg_1Fe_{2.25}O_{3.59}$	$Ni_{0.5}Mg_{0.51}Fe_{1.875}O_{3.3}$	$Ni_1Fe_{2.2}O_{3.5}$

Figure 6. (**a**) SEM micrograph collected on $Ni_{0.5}Mg_{0.5}Fe_2O_4$; (**b**) Elemental mapping of $Mg_{0.5}Ni_{0.5}Fe_2O_4$ inferred from Energy dispersive spectrometry (EDX) coupled to SEM; (**c**) Selected area for EDX analysis and (**d**) Table with quantitative elemental analysis and resulting stoichiometry.

The chemical states of the compositional elements contained in nickel–magnesium ferrites were analyzed by XPS, as illustrated in the survey scan spectrum in Figure 7. X-ray photoelectron spectroscopy can examine the elemental composition, oxidation states and chemical states of the elements in a sample by determining binding energies that correspond to the specific chemical bonds of present elements [56]. When considering spinel ferrite systems, XPS peak positions are dependent on cation distribution because the same cations, depending on their octahedral or tetrahedral coordination (A and B sites), have different cation–oxygen bond lengths and therefore different bond energies [56]. Because bonds in an octahedral coordination are longer and weaker than bonds in a tetrahedral coordination, we expect octahedral bonds to occur at lower binding energy values [57–59].

The Mg 1s peak (Figure 7a) is found at 1306.1 eV for both $MgFe_2O_4$ and $Ni_{0.5}Mg_{0.5}Fe_2O_4$. The positions of the peaks indicate that all magnesium ions are in a 2+ oxidation state. The peak positions are in accordance with the results that were obtained by Mittal et al. [60]. Dhanyaprabha et al. [56] attributed this peak to a purely tetrahedral Mg^{2+}. On the other hand, Mittal et al. [60] deconvoluted the Mg 1s peak into two components to obtain the distribution of Mg^{2+} ions at tetrahedral and octahedral sites. Similarly, we deconvoluted the Mg 1s peak into 1306.3 and 1305.6 eV (the left, tetrahedral component) and 1304.6 and 1304.2 eV (the right, octahedral component) for $MgFe_2O_4$ and $Ni_{0.5}Mg_{0.5}Fe_2O_4$, respectively. The results show that magnesium ions are coordinated at both tetrahedral and octahedral sites.

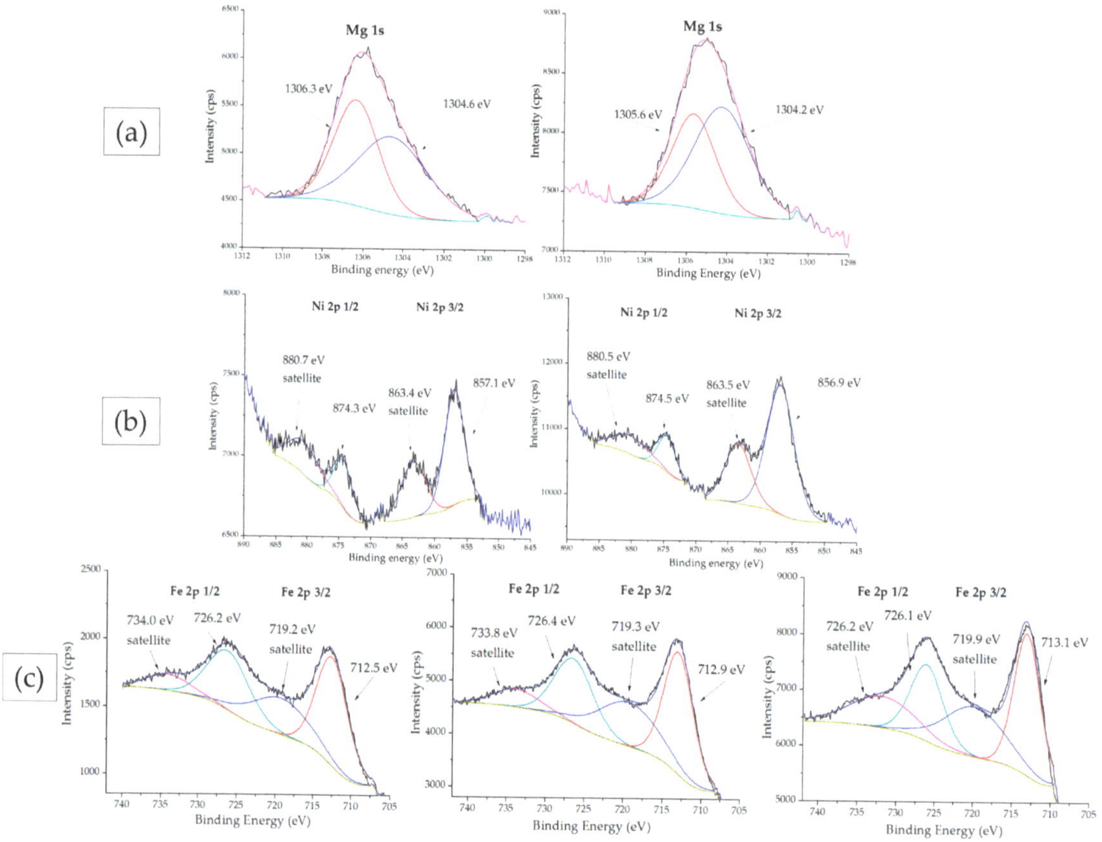

Figure 7. XPS spectra of: (**a**) Mg 1s, (**b**) Ni 2p and (**c**) Fe 2p peaks for the materials $MgFe_2O_4$, $Ni_{0.5}Mg_{0.5}Fe_2O_4$ and $NiFe_2O_4$, from top to bottom.

The nickel 2p peak is shown in Figure 7b for $Mg_{0.5}Ni_{0.5}Fe_2O_4$ and $NiFe_2O_4$. Ni 2p 1/2 and 3/2 peaks were identified at 857.1 eV (with its satellite peak at 863.4 eV) and 874.3 eV (with its satellite peak at 880.7 eV), respectively, for $Ni_{0.5}$ $Mg_{0.5}Fe_2O_4$ and at 856.9 eV (satellite 863.5 eV) and 874.5 eV (satellite 880.5 eV), respectively, for $NiFe_2O_4$. All of the peak positions are in accordance with the already published results and originate from the Ni^{2+} ion [56,61]. The distance between the main Ni 2p 3/2 peak and its satellite peak is 6.3 eV for $Ni_{0.5}Mg_{0.5}Fe_2O_4$ and 6.6 eV for $NiFe_2O_4$. Töpfer et al. [62] attributed the increase in this distance to more octahedrally coordinated nickel ions and the increase in inversion degree, which is in accordance with our findings that all additional nickel ions prefer octahedral sites.

The Fe 3p spectra in Figure 7c show two characteristic doublet peaks, 2p 1/2 and 2p 3/2, along with their satellite peaks. The peaks were fitted, and their positions are located at 726.2 eV (satellite at 734.0 eV) and 712.5 eV (satellite 719.2 eV), respectively, for $MgFe_2O_4$; 726.4 eV (satellite 733.8 eV) and 712.9 eV (satellite 719.3 eV), respectively, for $Mg_{0.5}Ni_{0.5}Fe_2O_4$; and 726.1 eV (satellite 731.8 eV) and 713.1 eV (satellite 719.9 eV), respectively, for $NiFe_2O_4$. These peaks are attributed to Fe in a 3+ oxidation state [61]. The positions of the peaks are in good agreement with already published results for Fe in spinels [57].

XPS enables quantitative elemental surface analysis. The results are summarized in Table 2. Both $MgFe_2O_4$ and $Mg_{0.5}Ni_{0.5}Fe_2O_4$ show higher Mg/Fe and Mg/Ni ratios than expected. The Mg/Fe ratio is 2.147 in $MgFe_2O_4$ and 0.6 in $Mg_{0.5}Ni_{0.5}Fe_2O_4$. The Mg/Ni ratio is 7.6 in $Mg_{0.5}Fe_{0.5}Fe_2O_4$, while the Ni/Fe ratio in $NiFe_2O_4$ is 0.58, which is close to the expected stoichiometric ratio of 0.5. The differences between values obtained by XPS and the expected stoichiometric values can be attributed to the fact that only the first 10 nm of surface are analyzed by XPS, which shows different values than the bulk material. The magnesium-rich surface has already been noted in the literature [56,60], and it is explained by the higher Mg^{2+} ion mobility [9].

Table 2. Quantitative surface elemental analysis obtained by X-ray photoelectron spectroscopy.

El.	$MgFe_2O_4$ (%)	$Ni_{0.5}Mg_{0.5}Fe_2O_4$ (%)	$NiFe_2O_4$ (%)
Mg	23.4	9.9	/
Ni	/	1.3	13.1
Fe	10.9	72.4	22.3
C	3.3	/	8.9
O	62.4	16.4	55.7

3.4. Temperature Sensing

In the measured temperature range (30–90 °C) at a set RH of 40% for all analyzed $Ni_xMg_{1-x}Fe_2O_4$ thick film samples except pure $NiFe_2O_4$, we noted a noticeable decrease of DC resistance with an increase in temperature, indicating that nickel–magnesium ferrites show NTC thermistor properties. An example of the change of normalized electrical resistance for samples of $Ni_{0.7}Mg_{0.3}Fe_2O_4$, $Ni_{0.5}Mg_{0.5}Fe_2O_4$ and $Ni_{0.1}Mg_{0.9}Fe_2O_4$ is shown in Figure 8. The results show that the degree of change depended on the composition (Ni content). For pure $NiFe_2O_4$ the change of resistance with temperature was small. This is probably due to the fact that we obtained p-type nickel ferrite rather than n-type, which influences the materials' conductivity. According to Sutka et al. [63] nickel ferrite with an inverse spinel structure is commonly p-type. With an increase in environmental temperature, the resistance of NTC thermistors decreases in accordance with the Arrhenius equation [15]:

$$R = R_{\infty} e^{\frac{B}{T}} \quad (1)$$

where B is the B-value, a material constant that describes the resistance change; R_{∞} is the resistance at infinite temperature; and T is the measured temperature. The slope of the linear fit of the graph of ln R = f(1/T) represents the B-value (as shown in the example given in the inset in Figure 8), which should be 2000–5000 K in order to be adequate for use in temperature sensors [15]. The calculated $B_{30/90}$ values for the synthesized nickel–magnesium ferrites are shown in Table 3.

Table 3. Material constant ($B_{30/90}$), temperature sensitivity (α), activation energy for conduction (E_a) and activation energy for the relaxation process (E_r).

Sample	$B_{30/90}$ (K)	α (%/K)	E_a (eV)	E_r (eV)
$MgFe_2O_4$	3426	−3.73	0.343	0.306
$Ni_{0.1}Mg_{0.9}Fe_2O_4$	3747	−4.08	0.347	0.319
$Ni_{0.3}Mg_{0.7}Fe_2O_4$	3177	−3.46	0.308	0.286
$Ni_{0.5}Mg_{0.5}Fe_2O_4$	2849	−3.10	0.294	0.263
$Ni_{0.7}Mg_{0.3}Fe_2O_4$	2218	−2.41	0.246	0.211
$Ni_{0.9}Mg_{0.1}Fe_2O_4$	1348	−1.47	0.119	0.039
$NiFe_2O_4$	/	/	/	/

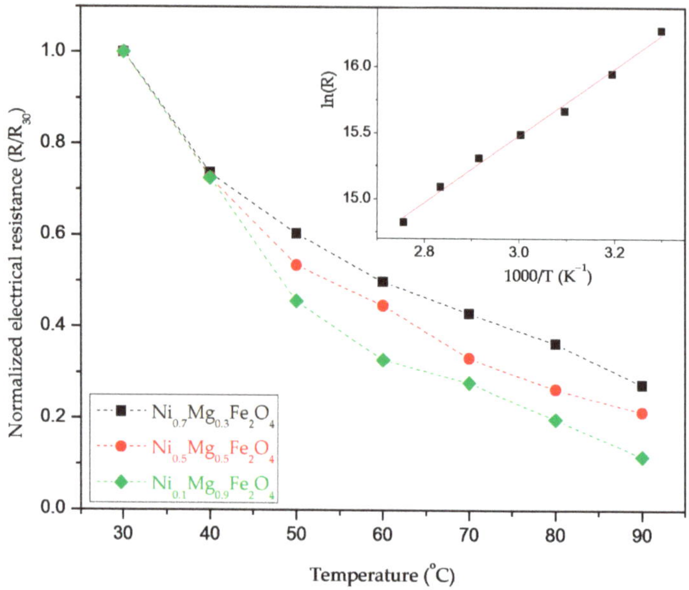

Figure 8. Change of DC resistance of $Ni_{0.7}Mg_{0.3}Fe_2O_4$, $Ni_{0.5}Mg_{0.5}Fe_2O_4$ and $Ni_{0.1}Mg_{0.9}Fe_2O_4$ in the temperature range of 30–90 °C; inset represents estimation of the material constant for $Ni_{0.7}Mg_{0.3}Fe_2O_4$.

The obtained values varied with the nickel content, and except for $Ni_{0.9}Mg_{0.1}Fe_2O_4$ and pure nickel ferrite, these values were within the range used in commercial NTC bulk ceramics (2000–5000 K), showing that magnesium ferrite and nickel–magnesium ferrites can be applied in temperature sensing as NTC materials [15]. The highest material constant $B_{30/90}$ was obtained for $Ni_{0.1}Mg_{0.9}Fe_2O_4$ as 3747 K, followed by pure magnesium ferrite with a $B_{30/90}$ of 3426 K. The temperature sensitivity at room temperature (30 °C) was determined as $\alpha = 1/R \cdot dR/dT = -B/T^2$, and the obtained values are shown in Table 3. For $Ni_{0.1}Mg_{0.9}Fe_2O_4$ we obtained $\alpha = -4.08\%/K$, and this value is comparable with commercial NTC devices ($-4\%/K$). This confirms the potential for applying $Ni_{0.1}Mg_{0.9}Fe_2O_4$ in temperature sensing, especially as these values were obtained for thick film samples composed of synthesized powder with no high-temperature treatment, thus enabling future application in flexible electronics where only low-temperature treatment of the sensing layer is possible [64].

Further research will involve investigating the aging and resistivity of these materials, as these parameters are also of great significance when selecting NTC thermistor materials for temperature sensing [65].

Analysis of the measured impedance for all samples showed a decrease of impedance with an increase in frequency and also with an increase in temperature, as shown in Figure 9a for $Ni_{0.7}Mg_{0.3}Fe_2O_4$. The change of impedance with temperature corresponded to the change of DC electrical resistance, so it reduced with increasing nickel content in nickel–magnesium ferrites. The measured complex impedance spectra (Figure 9b) were analyzed using an equivalent circuit (shown as inset in Figure 9b) consisting of a parallel resistance and constant phase element (CPE), reflecting the influence of both grain and grain boundary components and non-ideal Debye capacitance behavior [66,67]. The temperature dependence of the determined resistance was analyzed using the small-polaron hopping (SPH) model as described in [68]:

$$\frac{R}{T} = A_0 e^{\frac{E_A}{kT}} \quad (2)$$

where A_0 is the pre-exponential factor, k is the Boltzmann constant, and E_A is the activation energy for conduction. The obtained values for the activation energy for conduction are given in Table 3. We can see that the activation energy values vary in relation to the nickel content and decrease with an increase of nickel in nickel–magnesium ferrite samples. The highest value was obtained for $Ni_{0.1}Mg_{0.9}Fe_2O_4$, again confirming that a small amount of added Ni improved the temperature-sensing properties of magnesium ferrite. The determined relaxation time also has an Arrhenius dependence on temperature [68]:

$$\tau = \tau_0 e^{\frac{E_r}{kT}} \tag{3}$$

where τ_0 is the pre-exponential factor, k is the Boltzmann constant, and E_r is the activation energy for the relaxation process at grains and grain boundaries. The values obtained for the activation energy for the relaxation process are given also in Table 3. The highest value was obtained for $Ni_{0.1}Mg_{0.9}Fe_2O_4$.

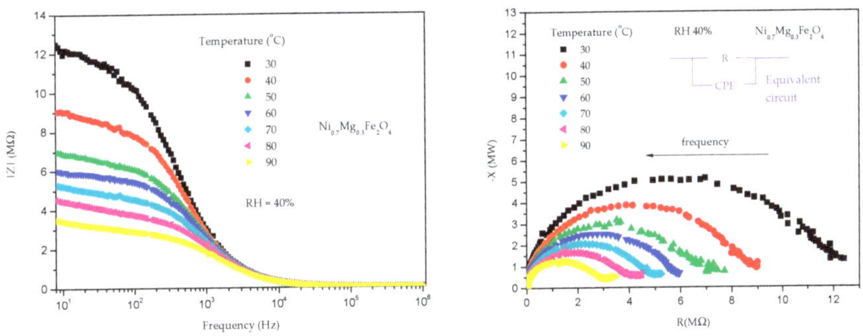

(a) (b)

Figure 9. Change of impedance component (**a**) and complex impedance (**b**) with the change in temperature at a constant RH of 40% for $Ni_{0.7}Mg_{0.3}Fe_2O_4$.

3.5. Humidity Sensing

Porous semiconducting materials can be used for sensing changes in ambient relative humidity (RH). The principle lying behind the sensing properties is the adsorption of water molecules on the surface of the material. Active sites on the metal oxide surface retain water molecules during humidity exposure. A detailed scheme of the humidity detection mechanism is shown in Figure 10. When the RH is low, water molecules interact with the porous sample surface by forming a chemisorbed layer on the sample surface. Water first dissociates to a hydroxyl that is firmly attached to the surface. The interaction energy is high, and significant energy is required for proton hopping. As the RH increases, a physisorbed layer of double hydrogen-bonded water molecules is formed, followed by further single hydrogen-bonded physisorbed layers. In the layers of the adsorbed water, the high mobility of water molecules and the Grotthus chain reaction cause low energy for proton hopping, increased conductivity and a decrease in the impedance value when the RH is high [66,69].

In the case of nickel–magnesium ferrites, all of the synthesized samples were tested at a constant temperature of 25 °C (room temperature). With an increase in relative humidity from 40% to 90%, there is a noticeable decrease in the impedance and complex impedance magnitude in each of the synthesized materials, as shown in Figure 11 for $Ni_{07}Mg_{0.3}Fe_2O_4$. The impedance of pure $MgFe_2O_4$ decreased 277 times with an RH increase from 40% to 90%, which corresponds well to and is higher than in the research of Jeseentharani et al. [70], in whose case the impedance of $MgFe_2O_4$ in the form of pellets decreased 230 times from 5% to 98% RH.

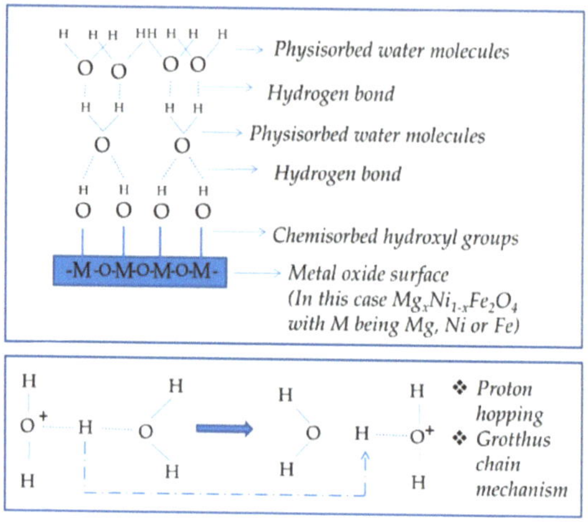

Figure 10. Scheme of the humidity detection mechanism.

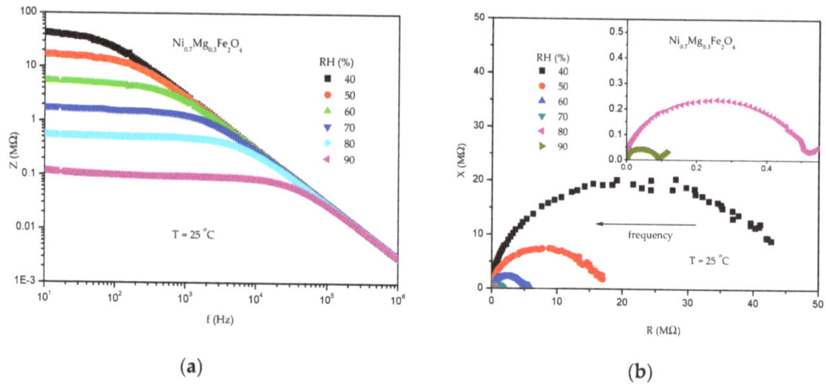

Figure 11. Change of impedance with frequency (**a**) and complex impedance (**b**) in the RH range of 40–90% for $Ni_{0.7}Mg_{0.3}Fe_2O_4$.

The impedance decreased also with an increase in frequency (Figure 11a), and the change of the impedance with frequency is larger in the lower frequency range. We selected 100 Hz as the working frequency, and this is commonly the case for humidity sensing with metal oxides [71]. The complex impedance consisted of one semicircle, which can be noted in Figure 11b, and this can be attributed to overlapping dielectric relaxation processes due to highly conducting grains and the resistive nature of grain boundaries [72]. The semicircle magnitude decreases with an increase in RH due to the increase in ionic conductivity in accordance with the humidity-sensing mechanism of metal oxides [66]. The small tail at high RH of 80 and 90% (seen in the inset in Figure 11b) has been noted before for spinel metal oxides and can be attributed to a charge diffusion process [67].

The change of impedance with RH at the frequency of 100 Hz is shown in Figure 12a. Magnesium ferrite and nickel–magnesium ferrites with varying nickel contents show a similar trend of impedance decrease with an increase in RH, with a rapid decrease in impedance as soon as the RH starts to increase from RH 40–70%, while the curve shape for nickel ferrite is different, and the impedance decreases only slightly until RH 70% and then more noticeably only in the high RH range of 70–90%. This difference is due to the different conducting mechanisms of p-type nickel ferrite and n-type magnesium ferrite and nickel–magnesium ferrites with varying nickel contents [63]. A lower impedance in $MgFe_2O_4$ compared to $NiFe_2O_4$ is also seen in [73].

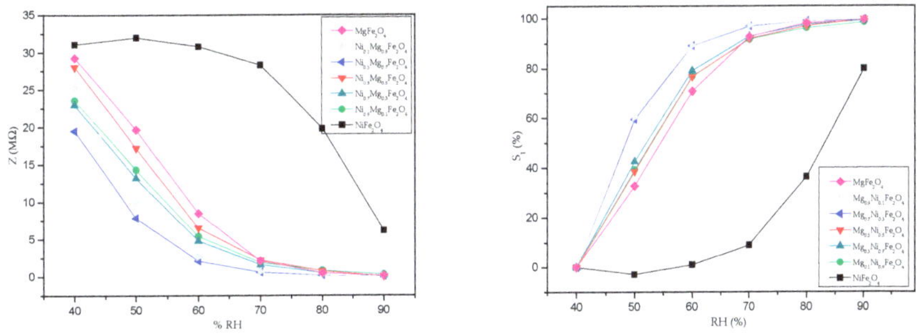

(**a**) (**b**)

Figure 12. Impedance change with the relative humidity at 100 Hz for $Ni_xMg_{1\text{-}x}Fe_2O_4$ (**a**); sensitivity change with relative humidity at 100 Hz (**b**).

Sensitivity S_1, calculated as:

$$S_1 = \frac{Z_{max} - Z}{Z_{max}} \tag{4}$$

is the parameter that shows how the impedance is changing in comparison to the starting value (Z_{max}). Calculated values are shown in Figure 12b. The greatest change in absolute impedance value can be attributed to $MgFe_2O_4$ (where it changes from 29.3 MΩ at 40% to 110 kΩ at 90%). All of the synthesized materials except nickel ferrite showed sensitivity S_1 values of nearly 100% and a similar trend, while the highest sensitivity of 99.4% is noted for $Ni_{0.1}Mg_{0.9}Fe_2O_4$ (where it changes from 25.3 MΩ at 40% to 40 kΩ at 90%); a value of 99.82% is noted for $Ni_{0.5}Mg_{0.5}Fe_2O_4$ (where it changes from 28 MΩ at 40% to 50 kΩ at 90%). The most rapid changes of sensitivity for RH in the range of 40–60% were noted for $Ni_{0.1}Mg_{0.9}Fe_2O_4$ and $Ni_{0.3}Mg_{0.7}Fe_2O_4$.The dissimilar and lower sensitivity of nickel ferrite can be explained by the differences due to p-type conductivity when cation vacancies are present due to oxygen attraction during the calcination/heating process [63,74].

Another sensitivity/sensor response parameter that is used to characterize the humidity response is $S_2 = \frac{\Delta Z}{\Delta RH}$, and it presents the ratio between the change of sensor impedance and the RH at 100 Hz [16,66]. The sensitivity value changed depending on the RH humidity region, in accordance with the humidity-sensing mechanism, as shown in Table 4. For magnesium ferrite and nickel–magnesium ferrites with varying nickel contents, the sensitivity was higher in the lower RH range, and this can be linked to the porous surface of the spinel ferrite thick film. The highest sensitivity towards change in the RH was shown by $Ni_{0.1}Mg_{0.9}Fe_2O_4$, with an average sensitivity of 922.6 kΩ/%RH.

Table 4. Sensitivity of $Ni_xMg_{1\text{-}x}Fe_2O_4$ samples in the RH range of 40–90%, calculated as $\Delta Z/\Delta RH$.

ΔRH (%)	ΔZ/ΔRH (kΩ/%RH)					
	$Ni_{0.9}Mg_{0.1}Fe_2O_4$	$Ni_{0.7}Mg_{0.3}Fe_2O_4$	$Ni_{0.5}Mg_{0.5}Fe_2O_4$	$Ni_{0.5}Mg_{0.5}Fe_2O_4$	$Ni_{0.1}Mg_{0.9}Fe_2O_4$	$MgFe_2O_4$
10	925	980	1083	1163	1530	958
20	908	910	1078	870	1131	1038
30	541	712	860	630	820	905
40	568	562	682	483	627	719
50	465	457	560	389	505	583
average	681.4	724.2	852.6	707	922.6	840.6

Table 5 shows a comparison between the resistance/impedance values in the measured RH range and the sensitivity values determined for different metal oxide humidity sensors in the available literature. The sensing material developed in this work shows good sensitivity and impedance reduction in the analyzed RH range, comparable with other humidity-sensing metal oxides.

Table 5. Literature comparison of humidity sensing metal oxides.

Sensing Material	Impedance/Resistance Change	Sensitivity	Ref.
$ZnFe_2O_4$, solid-state synthesis, pellets	43935 MΩ (RH 5%)–15 MΩ (RH 98%), R	2895 ($R_{5\%}/R_{98\%}$)	[70]
$CuFe_2O_4$ solid-state synthesis, pellets	1930.9 MΩ (RH 5%)–7.22 MΩ (RH 98%), R	267 ($R_{5\%}/R_{98\%}$)	[70]
$CoFe_2O_4$ solid-state synthesis, pellets	1506.8 MΩ (RH 5%)–5.8 MΩ (RH 98%), R	260 ($R_{5\%}/R_{98\%}$)	[70]
$NiFe_2O_4$ solid-state synthesis, pellets	2907.5 MΩ (RH 5%)–11.6 MΩ (RH 98%), R	249 ($R_{5\%}/R_{98\%}$)	[70]
$MgFe_2O_4$ solid-state synthesis, pellets	26452 MΩ (RH 5%)–114.8 MΩ (RH 98%), R	230 ($R_{5\%}/R_{98\%}$)	[70]
$MgFe_2O_4$, RF sputtered thin film, calcined at 800 °C	10^{12} Ω (RH 10%)–10^9 Ω (RH 90%), R	20.888 ($R_{10\%}/R_{90\%}$)	[75]
$NiFe_2O_4$, solid-state synthesis, pellets	4.07 MΩ (RH 15%)–32.5 kΩ (RH 85%) Z at 2.5 kHz	57.6 kΩ/% RH	[76]
$MgFe_2O_4$-Fe_2O_3-SnO_2 composite, solid-state synthesis, pellet	26.1 MΩ (RH 30%)–1.77 MΩ (RH 90%), Z at 105 Hz	391 kΩ/% RH (RH 30–90%)	[77]
MnZn ferrite, thin film	−83 kOm RH (30%)–53 kOm (RH 90%), R	1.54%/% RH	[78]
$Mg_{0.2}Zn_{0.8}Fe_2O_4$, coprecipitation synthesis, thick film	3100 MΩ (RH 30%)–600 MΩ (RH 95%), R	60 MΩ/% RH (RH 30–90%)	[79]
$NiMn_2O_4$ synthesized by electrospinning, thick film	31 MΩ (40% RH)–8.8 MΩ (90% RH), R	327.36 kΩ/% RH (RH 40–90%)	[66]
$MgFe_2O_4$, sol-gel synthesis, thick film	29.3 MΩ (RH 40%)–110 kΩ (RH 90%), Z at 100 Hz	840.6 kΩ/% RH (RH 40–90%)	This work
$Ni_{0.1}Mg_{0.9}Fe_2O_4$, sol-gel synthesis, thick film	25.3 MΩ (RH 40%)–40 kΩ (RH 90%) Z at 100 Hz	922.6 kΩ/% RH (RH 40–90%)	This work

The response time of a sensor can be measured as the time the sensor needs to reach 90% of the total response when subjected to a specific relative humidity value, while the time required to go back to 90% of the starting signal can be defined as the sensor recovery time. We measured this at 100 Hz with the thick film sensor at room temperature and ambient humidity (estimated at RH 45%) and exposed it to an RH of 90% in the humidity chamber, as shown in Figure 13 for $Ni_{0.1}MgFe_2O_4$. The average response time was about 20 s, while the average recovery time was about 45 s. There was no noticeable drift in the signal. Similar results were obtained for the other nickel–magnesium ferrite samples and for magnesium ferrite.

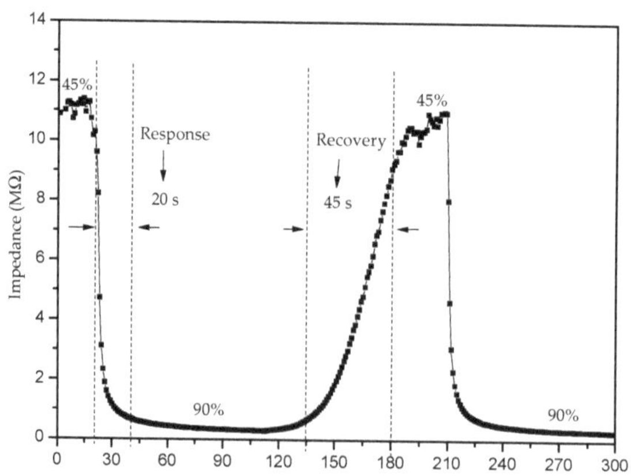

Figure 13. Response and recovery times obtained for $Ni_{0.1}Mg_{0.9}Fe_2O_4$ in the RH range of 45–90% at the working temperature of 25 °C.

The results obtained both for temperature and humidity sensing of $Ni_{0.1}Mg_{0.9}Fe_2O_4$ show that this sensing material has potential as a multifunctional material for both temperature and humidity sensing.

4. Conclusions

Through sol-gel combustion synthesis with citric acid as fuel and subsequent calcination (annealing) at 700 °C, we successfully synthesized magnesium–nickel spinel ferrites, $Ni_xMg_{1-x}Fe_2O_4$ ($0 \leq x \leq 1$). The obtained materials showed a partially or completely inverse cubic spinel structure, nanocrystalline but agglomerated particles and a magnesium-rich surface. All of the synthesized materials showed a response to changes in ambient temperature and humidity, with resistive properties decreasing with increases in relative humidity and temperature. Good sensitivity values were obtained for magnesium ferrite and nickel–magnesium ferrites, while nickel ferrite showed a different kind of conduction mechanism and therefore lower sensing performances. The best response toward temperature and relative humidity changes was shown by $Ni_{0.1}Mg_{0.9}Fe_2O_4$. The obtained results comprehensively indicate that doping magnesium ferrite with nickel in small amounts, x = 0.1, increases the activation energy for conduction and enhances the conductivity, which enables a better temperature- and humidity-sensing performance of $Ni_{0.1}Mg_{0.9}Fe_2O_4$.

Supplementary Materials: The following supporting information can be downloaded at: https://www.mdpi.com/article/10.3390/chemosensors11010034/s1, Figure S1: FESEM images of $Ni_xMg_{1-x}Fe_2O_4$ ($0 \leq x \leq 1$).

Author Contributions: M.P.D.: conceptualization, data curation, investigation and writing—original draft; Z.Z.V.: investigation and writing—review and editing; L.R.: investigation and writing—review and editing; V.P.P.: investigation and writing—original draft; S.A.-M.: investigation and writing—review and editing; J.D.V.: investigation and writing—review and editing; M.V.N.: supervision, conceptualization, data curation, investigation, writing—original draft and writing—review and editing. All authors have read and agreed to the published version of the manuscript.

Funding: This research was funded by the Ministry for Education, Science and Technology Development of the Republic of Serbia under contracts 451-03-68/2022-14/200053 (M.P.D, Z.Z.V, M.V.N.), 451-03-68/2022-14/200105 (V.P.P.) and 451-03-68/2022-14/200175 (J.D.V.).

Institutional Review Board Statement: Not applicable.

Informed Consent Statement: Not applicable.

Data Availability Statement: The data presented in this study are available on request from the corresponding author. The data are not publicly available due to ongoing research.

Acknowledgments: We are grateful to Sarra Gam-Derouich and Alexandre Chevillot-Birraud at ITODYS, Université Paris Cité, for their help with SEM, EDS and FTIR measurements.

Conflicts of Interest: The authors declare no conflict of interest. The funders had no role in the design of the study; in the collection, analyses or interpretation of data; in the writing of the manuscript; or in the decision to publish the results.

References

1. Zhao, Q.; Yan, Z.; Chen, C.; Chen, J. Spinels: Controlled Preparation, Oxygen Reduction/Evolution Reaction Application, and Beyond. *Chem. Rev.* **2017**, *117*, 10121–10211. [CrossRef]
2. Šutka, A.; Gross, K.A. Spinel ferrite oxide semiconductor gas sensors. *Sens. Actuators B Chem.* **2016**, *222*, 95–105. [CrossRef]
3. Kefeni, K.K.; Mamba, B.B.; Msagati, T.A. Application of spinel ferrite nanoparticles in water and wastewater treatment: A review. *Sep. Purif. Technol.* **2017**, *188*, 399–422. [CrossRef]
4. Li, Y.; Wang, T.; Zhang, S.; Zhang, Y.; Yu, L.; Liu, R. Adsorption and electrochemical behavior investigation of methyl blue onto magnetic nickel-magnesium ferrites prepared via the rapid combustion process. *J. Alloys Compd.* **2021**, *885*, 160969. [CrossRef]
5. Qin, H.; He, Y.; Xu, P.; Huang, D.; Wang, Z.; Wang, H.; Wang, Z.; Zhao, Y.; Tian, Q.; Wang, C. Spinel ferrites (MFe_2O_4): Synthesis, improvement and catalytic application in environment and energy field. *Adv. Colloid Interface Sci.* **2021**, *294*, 102486. [CrossRef] [PubMed]
6. Kefeni, K.K.; Mamba, B.B. Photocatalytic application of spinel ferrite nanoparticles and nanocomposites in wastewater treatment: Review. *Sustain. Mater. Technol.* **2019**, *23*, e00140. [CrossRef]
7. Amiri, M.; Salavati-Niasari, M.; Akbari, A. Magnetic nanocarriers: Evolution of spinel ferrites for medical applications. *Adv. Colloid Interface Sci.* **2019**, *265*, 29–44. [CrossRef] [PubMed]
8. Malaie, K.; Ganjali, M.R. Spinel nano-ferrites for aqueous supercapacitors; linking abundant resources and low-cost processes for sustainable energy storage. *J. Energy Storage* **2020**, *33*, 102097. [CrossRef]
9. Mittal, V.; Bera, S.; Nithya, R.; Srinivasan, M.; Velmurugan, S.; Narasimhan, S. Solid state synthesis of Mg–Ni ferrite and characterization by XRD and XPS. *J. Nucl. Mater.* **2004**, *335*, 302–310. [CrossRef]

10. Anumol, C.N.; Chithra, M.; Rout, S.; Sahoo, S.C. Effect of Magnesium Substitution on Structural and Magnetic Properties of Nickel Ferrite Nanoparticles. *J. Supercond. Nov. Magn.* **2019**, *33*, 1611–1617. [CrossRef]
11. Varshney, D.; Verma, K. Substitutional effect on structural and dielectric properties of Ni1−xAxFe$_2$O$_4$ (A = Mg, Zn) mixed spinel ferrites. *Mater. Chem. Phys.* **2013**, *140*, 412–418. [CrossRef]
12. Almessiere, M.A.; Slimani, Y.A.; Hassan, M.; Gondal, M.A.; Cevik, E.; Baykal, A. Investigation of hard/soft $CoFe_2O_4/NiSc_{0.03}Fe_{1.97}O_4$ nanocomposite for energy storage applications. *Int. J. Energy Res.* **2021**, *45*, 16691–16708. [CrossRef]
13. Nair, V.; Jose, R.; Raju, K.; Wariar, P. Optimization of citrate complex combustion for synthesis of transition metal oxide nanostructures. *J. Alloys Compd.* **2013**, *552*, 180–185. [CrossRef]
14. Farahani, H.; Wagiran, R.; Hamidon, M.N. Humidity Sensors Principle, Mechanism, and Fabrication Technologies: A Comprehensive Review. *Sensors* **2014**, *14*, 7881–7939. [CrossRef] [PubMed]
15. Feteira, A. Negative Temperature Coefficient Resistance (NTCR) Ceramic Thermistors: An Industrial Perspective. *J. Am. Ceram. Soc.* **2009**, *92*, 967–983. [CrossRef]
16. Priya, R.S.; Chaudhary, P.; Kumar, E.R.; Balamurugan, A.; Srinivas, C.; Prasad, G.; Yadav, B.; Sastry, D. Evaluation of structural, dielectric and electrical humidity sensor behaviour of $MgFe_2O_4$ ferrite nanoparticles. *Ceram. Int.* **2021**, *47*, 15995–16008. [CrossRef]
17. Dumitrescu, A.; Lisa, G.; Iordan, A.; Tudorache, F.; Petrila, I.; Borhan, A.; Palamaru, M.; Mihailescu, C.; Leontie, L.; Munteanu, C. Ni ferrite highly organized as humidity sensors. *Mater. Chem. Phys.* **2015**, *156*, 170–179. [CrossRef]
18. Nandan, B.; Bhatnagar, M.; Kashyap, S.C. Cation distribution in nanocrystalline cobalt substituted nickel ferrites: X-ray diffraction and Raman spectroscopic investigations. *J. Phys. Chem. Solids* **2019**, *129*, 298–306. [CrossRef]
19. Henderson, C.M.B.; Charnock, J.M.; Plant, D.A. Cation occupancies in Mg, Co, Ni, Zn, Al ferrite spinels: A multi-element EXAFS study. *J. Physics Condens. Matter* **2007**, *19*, 076214. [CrossRef]
20. Akbari, S.; Masoudpanah, S.; Mirkazemi, S.; Aliyan, N. PVA assisted coprecipitation synthesis and characterization of $MgFe_2O_4$ nanoparticles. *Ceram. Int.* **2017**, *43*, 6263–6267. [CrossRef]
21. Gateshki, M.; Petkov, V.; Pradhan, S.K.; Vogt, T. Structure of nanocrystalline $MgFe_2O_4$from X-ray diffraction, Rietveld and atomic pair distribution function analysis. *J. Appl. Crystallogr.* **2005**, *38*, 772–779. [CrossRef]
22. Dojcinovic, M.P.; Vasiljevic, Z.Z.; Pavlovic, V.P.; Barisic, D.; Pajic, D.; Tadic, N.B.; Nikolic, M.V. Mixed Mg–Co spinel ferrites: Structure, morphology, magnetic and photocatalytic properties. *J. Alloys Compd.* **2021**, *855*, 157429. [CrossRef]
23. Deraz, N. Effects of magnesia addition on structural, morphological and magnetic properties of nano-crystalline nickel ferrite system. *Ceram. Int.* **2012**, *38*, 511–516. [CrossRef]
24. Joshi, H.; Gowreesan, S.; Kumar, A.R. Influences of Ni2+ on magnetic property and dielectric property in spinel structure of Mg ferrite (Mg1−xNixFe2O4). *J. Mater. Sci. Mater. Electron.* **2017**, *29*, 3449–3457. [CrossRef]
25. Ahlawat, A.; Sathe, V.G. Raman study of $NiFe_2\ O_4$ nanoparticles, bulk and films: Effect of laser power. *J. Raman Spectrosc.* **2010**, *42*, 1087–1094. [CrossRef]
26. Moradmard, H.; Shayesteh, S.F.; Tohidi, P.; Abbas, Z.; Khaleghi, M. Structural, magnetic and dielectric properties of magnesium doped nickel ferrite nanoparticles. *J. Alloys Compd.* **2015**, *650*, 116–122. [CrossRef]
27. Chavan, P.; Naik, L.R. Investigation of energy band gap and conduction mechanism of magnesium substituted nickel ferrite nanoparticles. *Phys. Status Solidi (A)* **2017**, *214*, 1700077. [CrossRef]
28. Shobana, M.; Kim, K.; Kim, J.-H. Impact of magnesium substitution in nickel ferrite: Optical and electrochemical studies. *Phys. E Low-Dimens. Syst. Nanostructures* **2018**, *108*, 100–104. [CrossRef]
29. Chavan, P.; Naik, L.R.; Belavi, P.B.; Chavan, G.; Ramesha, C.K.; Kotnala, R.K. Studies on Electrical and Magnetic Properties of Mg-Substituted Nickel Ferrites. *J. Electron. Mater.* **2016**, *46*, 188–198. [CrossRef]
30. Ugendar, K.; Samanta, S.; Rayaprol, S.; Siruguri, V.; Markandeyulu, G.; Nanda, B.R.K. Effect of frustrated exchange interactions and spin-half-impurity on the electronic structure of strongly correlated $NiFe_2O_4$. *Phys. Rev. B* **2017**, *96*, 035138. [CrossRef]
31. Ortiz-Quiñonez, J.L.; Pal, U.; Villanueva, M.S. Structural, Magnetic, and Catalytic Evaluation of Spinel Co, Ni, and Co–Ni Ferrite Nanoparticles Fabricated by Low-Temperature Solution Combustion Process. *ACS Omega* **2018**, *3*, 14986–15001. [CrossRef] [PubMed]
32. Yan, Z.; Gao, J.; Li, Y.; Zhang, M.; Guo, M. Hydrothermal synthesis and structure evolution of metal-doped magnesium ferrite from saprolite laterite. *RSC Adv.* **2015**, *5*, 92778–92787. [CrossRef]
33. Mund, H.; Ahuja, B. Structural and magnetic properties of Mg doped cobalt ferrite nano particles prepared by sol-gel method. *Mater. Res. Bull.* **2016**, *85*, 228–233. [CrossRef]
34. Yu, T.; Shen, Z.X.; Shi, Y.; Ding, J. Cation migration and magnetic ordering in spinel $CoFe_2O_4$powder: Micro-Raman scattering study. *J. Physics Condens. Matter* **2002**, *14*, L613–L618. [CrossRef]
35. Singh, J.P.; Srivastava, R.C.; Agrawal, H.M.; Kumar, R. Micro-Raman investigation of nanosized zinc ferrite: Effect of crystallite size and fluence of irradiation. *J. Raman Spectrosc.* **2011**, *42*, 1510–1517. [CrossRef]
36. Yadav, R.S.; Kuřitka, I.; Vilcakova, J.; Havlica, J.; Masilko, J.; Kalina, L.; Tkacz, J.; Enev, V.; Hajdúchová, M. Structural, magnetic, dielectric, and electrical properties of $NiFe_2O_4$ spinel ferrite nanoparticles prepared by honey-mediated sol-gel combustion. *J. Phys. Chem. Solids* **2017**, *107*, 150–161. [CrossRef]
37. Lazarević, Z.Ž; Jovalekić, Č.; Milutinović, A.; Sekulić, D.; Ivanovski, V.N.; Rečnik, A.; Cekić, B.; Romčević, N. Nanodimensional spinel $NiFe_2O_4$ and $ZnFe_2O_4$ ferrites prepared by soft mechanochemical synthesis. *J. Appl. Phys.* **2013**, *113*, 187221. [CrossRef]

38. Wang, T.; Zhu, T.; Brunet, M.; Deshayes, C.; Sciau, P. Raman study of Yuan Qinghua porcelain: The highlighting of dendritic $CoFe_2O_4$crystals in blue decorations. *J. Raman Spectrosc.* **2016**, *48*, 267–270. [CrossRef]
39. Kumar, K.A.; Bhowmik, R. Micro-structural characterization and magnetic study of Ni1.5Fe1.5O4 ferrite synthesized through coprecipitation route at different pH values. *Mater. Chem. Phys.* **2014**, *146*, 159–169. [CrossRef]
40. Gao, Y.; Yin, P. Origin of asymmetric broadening of Raman peak profiles in Si nanocrystals. *Sci. Rep.* **2017**, *7*, srep43602. [CrossRef]
41. Tanwar, M.; Yogi, P.; Lambora, S.; Mishra, S.; Saxena, S.K.; Sagdeo, P.R.; Krylov, A.S.; Kumar, R. Generalisation of phonon confinement model for interpretation of Raman line-shape from nano-silicon. *Adv. Mater. Process. Technol.* **2018**, *4*, 227–233. [CrossRef]
42. Himcinschi, C.; Vrejoiu, I.; Salvan, G.; Fronk, M.; Talkenberger, A.; Zahn, D.R.T.; Rafaja, D.; Kortus, J. Optical and magneto-optical study of nickel and cobalt ferrite epitaxial thin films and submicron structures. *J. Appl. Phys.* **2013**, *113*, 084101. [CrossRef]
43. Ivanov, V.G.; Abrashev, M.V.; Iliev, M.N.; Gospodinov, M.M.; Meen, J.; Aroyo, M.I. Short-rangeB-site ordering in the inverse spinel ferrite$NiFe_2O_4$. *Phys. Rev. B* **2010**, *82*, 024104. [CrossRef]
44. Iliev, M.N.; Mazumdar, D.; Ma, J.X.; Gupta, A.; Rigato, F.; Fontcuberta, J. MonitoringB-site ordering and strain relaxation in NiFe2O4epitaxial films by polarized Raman spectroscopy. *Phys. Rev. B* **2011**, *83*, 014108. [CrossRef]
45. Fritsch, D.; Ederer, C. Effect of epitaxial strain on the cation distribution in spinel ferrites $CoFe_2O_4$ and $NiFe_2O_4$: A density functional theory study. *Appl. Phys. Lett.* **2011**, *99*, 081916. [CrossRef]
46. Kirchberg, K.; Becker, A.; Bloesser, A.; Weller, T.; Timm, J.; Suchomski, C.; Marschall, R. Stabilization of Monodisperse, Phase-Pure $MgFe_2O_4$ Nanoparticles in Aqueous and Nonaqueous Media and Their Photocatalytic Behavior. *J. Phys. Chem. C* **2017**, *121*, 27126–27138. [CrossRef]
47. Puli, V.S.; Adireddy, S.; Ramana, C. Chemical bonding and magnetic properties of gadolinium (Gd) substituted cobalt ferrite. *J. Alloys Compd.* **2015**, *644*, 470–475. [CrossRef]
48. Da Silva, S.W.; Nakagomi, F.; Silva, M.S.; Franco, A.; Garg, V.K.; Oliveira, A.C.; Morais, P.C.; Jr., A.F. Raman study of cations' distribution in Zn x Mg1−x Fe2O4 nanoparticles. *J. Nanoparticle Res.* **2012**, *14*, 798. [CrossRef]
49. Wang, Z.; Lazor, P.; Saxena, S.; O'Neill, H.S. High pressure Raman spectroscopy of ferrite $MgFe_2O_4$. *Mater. Res. Bull.* **2002**, *37*, 1589–1602. [CrossRef]
50. Gawas, S.G.; Meena, S.S.; Bhatt, P.; Verenkar, V.M.S. Nanoscale-driven structural changes and associated superparamagnetism in magnetically diluted Ni–Zn ferrites. *Mater. Chem. Front.* **2018**, *2*, 300–312. [CrossRef]
51. Surya, R.M.; Yulizar, Y.; Cahyana, A.H.; Apriandanu, D.O.B. One-pot *Cajanus cajan* (L.) Millsp. leaf extract-mediated preparation of MgFe2O4 nanoparticles: Optical, structural, morphological and particle size analyses. *Solid State Commun.* **2020**, *326*, 114170. [CrossRef]
52. Varma, A.; Mukasyan, A.S.; Rogachev, A.S.; Manukyan, K.V. Solution Combustion Synthesis of Nanoscale Materials. *Chem. Rev.* **2016**, *116*, 14493–14586. [CrossRef] [PubMed]
53. Prabhakaran, T.; Hemalatha, J. Combustion synthesis and characterization of cobalt ferrite nanoparticles. *Ceram. Int.* **2016**, *42*, 14113–14120. [CrossRef]
54. Nadargi, D.; Umar, A.; Nadargi, J.; Patil, J.; Mulla, I.; Akbar, S.; Suryavanshi, S. Spinel Magnesium Ferrite ($MgFe_2O_4$): A Glycine-Assisted Colloidal Combustion and Its Potentiality in Gas-Sensing Application. *Chemosensors* **2022**, *10*, 361. [CrossRef]
55. He, Y.; Zhang, L.; Xiong, H.-W.; Kang, X. Evolution of lattice defects in nickel ferrite spinel: Oxygen vacancy and cation substitution. *J. Alloys Compd.* **2022**, *917*, 165494. [CrossRef]
56. Dhanyaprabha, K.C.; Jacob, B.; Mohan, M.; Al-Omari, I.A.; Al-Harthi, S.H.; Myint, M.T.; Thomas, H. Structural, Magnetic, and Optical Studies of Ni–Mg Ferrites Synthesized by Polyol Method. *Phys. Status Solidi (A)* **2021**, *218*, 2100193. [CrossRef]
57. Thota, S.; Kashyap, S.C.; Sharma, S.K.; Reddy, V. Cation distribution in Ni-substituted Mn0.5Zn0.5Fe2O4 nanoparticles: A Raman, Mössbauer, X-ray diffraction and electron spectroscopy study. *Mater. Sci. Eng. B* **2016**, *206*, 69–78. [CrossRef]
58. Li, F.; Liu, X.; Yang, Q.; Liu, J.; Evans, D.G.; Duan, X. Synthesis and characterization of Ni1−xZnxFe2O4 spinel ferrites from tailored layered double hydroxide precursors. *Mater. Res. Bull.* **2005**, *40*, 1244–1255. [CrossRef]
59. Arillo, M.; López, M.; Pico, C.; Veiga, M.; Jiménez-López, A.; Rodríguez-Castellón, E. Surface characterisation of spinels with Ti(IV) distributed in tetrahedral and octahedral sites. *J. Alloys Compd.* **2001**, *317–318*, 160–163. [CrossRef]
60. Mittal, V.K.; Chandramohan, P.; Bera, S.; Srinivasan, M.P.; Velmurugan, S.; Narasimhan, S.V. Cation distribution in NixMg1−xFe2O4 studied by XPS and Mössbauer spectroscopy. *Solid State Commun.* **2006**, *137*, 6–10. [CrossRef]
61. Li, W.; Li, F.; Yang, H.; Wu, X.; Zhang, P.; Shan, Y.; Sun, L. A bio-inspired coordination polymer as outstanding water oxidation catalyst via second coordination sphere engineering. *Nat. Commun.* **2019**, *10*, 5074. [CrossRef] [PubMed]
62. Töpfer, J.; Feltz, A.; Gräf, D.; Hackl, B.; Raupach, L.; Weissbrodt, P. Cation Valencies and Distribution in the Spinels $NiMn_2O_4$ and $MzNiMn_2{-}zO_4$ (M = Li, Cu) Studied by XPS. *Phys. Status Solidi (A)* **1992**, *134*, 405–415. [CrossRef]
63. Šutka, A.; Pärna, R.; Käämbre, T.; Kisand, V. Synthesis of p-type and n-type nickel ferrites and associated electrical properties. *Phys. B Condens. Matter* **2015**, *456*, 232–236. [CrossRef]
64. Shin, J.; Jeong, B.; Kim, J.; Nam, V.B.; Yoon, Y.; Jung, J.; Hong, S.; Lee, H.; Eom, H.; Yeo, J.; et al. Sensitive Wearable Temperature Sensor with Seamless Monolithic Integration. *Adv. Mater.* **2020**, *32*, e1905527. [CrossRef] [PubMed]
65. Liu, T.; Zhang, H.; Ma, P.; Chang, A.; Jiang, H. Core–shell NTC materials with low thermal constant and high resistivity for wide-temperature thermistor ceramics. *J. Am. Ceram. Soc.* **2019**, *102*, 4393–4398. [CrossRef]

66. Dojcinovic, M.P.; Vasiljevic, Z.Z.; Krstic, J.B.; Vujancevic, J.D.; Markovic, S.; Tadic, N.B.; Nikolic, M.V. Electrospun Nickel Manganite ($NiMn_2O_4$) Nanocrystalline Fibers for Humidity and Temperature Sensing. *Sensors* **2021**, *21*, 4357. [CrossRef]
67. Nikolic, M.V.; Labus, N.J.; Pavlovic, V.P.; Markovic, S.; Lukovic, M.D.; Tadic, N.B.; Vujancevic, J.D.; Vlahovic, B.; Pavlovic, V.B. Nanocrystalline Zn_2SnO_4/SnO_2: Crystal structure and humidity influence on complex impedance. *J. Electroceramics* **2020**, *45*, 135–147. [CrossRef]
68. Nikolic, M.V.; Sekulic, D.L.; Vasiljevic, Z.Z.; Lukovic, M.D.; Pavlović, V.B.; Aleksic, O.S. Dielectric properties, complex impedance and electrical conductivity of Fe2TiO5 nanopowder compacts and bulk samples at elevated temperatures. *J. Mater. Sci. Mater. Electron.* **2016**, *28*, 4796–4806. [CrossRef]
69. Nikolic, M.V.; Lukovic, M.D.; Vasiljevic, Z.Z.; Labus, N.J.; Aleksic, O.S. Humidity sensing potential of Fe_2TiO_5—Pseudobrookite. *J. Mater. Sci. Mater. Electron.* **2018**, *29*, 9227–9238. [CrossRef]
70. Jeseentharani, V.; George, M.; Jeyaraj, B.; Dayalan, A.; Nagaraja, K.S. Synthesis of metal ferrite (MFe_2O_4, M = Co, Cu, Mg, Ni, Zn) nanoparticles as humidity sensor materials. *J. Exp. Nanosci.* **2013**, *8*, 358–370. [CrossRef]
71. Nikolic, M.V.; Dojcinovic, M.P.; Vasiljevic, Z.Z.; Lukovic, M.D.; Labus, N.J. Nanocomposite Zn_2SnO_4/SnO_2 Thick Films as a Humidity Sensing Material. *IEEE Sensors J.* **2020**, *20*, 7509–7516. [CrossRef]
72. Selmi, M.; Smida, A.; El Kossi, S. Effect of Polaron formation in conduction and dielectric behavior in $La_{0.7}Sr_{0.25}K_{0.05}MnO_3$ oxide. *J. Mater. Sci. Mater. Electron.* **2021**, *32*, 6014–6027. [CrossRef]
73. Mocanu, Z.V.; Airimioaei, M.; Ciomaga, C.E.; Curecheriu, L.; Tudorache, F.; Tascu, S.; Iordan, A.R.; Palamaru, N.M.; Mitoseriu, L. Investigation of the functional properties of Mg x Ni1−x Fe_2O_4 ceramics. *J. Mater. Sci.* **2014**, *49*, 3276–3286. [CrossRef]
74. Sutka, A. The Role of Stoichiometry on Gas Response of Nanostructured Sol–Gel Auto Combustion Derived Nickel Ferrite. *Sens. Lett.* **2013**, *11*, 2010–2013. [CrossRef]
75. Kotnala, R.K.; Shah, J.; Mathpal, M.C.; Verma, K.C.; Singh, S.; Lovkush. Influence of annealing on humidity response of RF sputtered nanocrystalline $MgFe_2O_4$ thin films. *Thin Solid Film.* **2011**, *519*, 6135–6139. [CrossRef]
76. Sekulic, D.L.; Lazarevic, Z.Z.; Romcevic, N.Z. Nanocrystalline Porous Nickel Ferrite Ceramics for Humidity Sensing Applications. In Proceedings of the IEEE 31st International Conference on Microelectronics (MIEL), Nis, Serbia, 16–18 September 2019; pp. 95–98. [CrossRef]
77. Nikolic, M.V.; Lukovic, M.D. Influence of SnO_2 Content on the Humidity Dependent Impedance of the $MgFe_2O_4$-Fe_2O_3-SnO_2 Compound. *Chemosensors* **2020**, *8*, 39. [CrossRef]
78. Arshaka, K.; Twomey, K.; Egan, D. A Ceramic Thick Film Humidity Sensor Based on MnZn Ferrite. *Sensors* **2002**, *2*, 50–61. [CrossRef]
79. Patil, S.; Pawar, A.; Tilekar, S.; Ladgaonkar, B. Investigation of magnesium substituted nano particle zinc ferrites for relative humidity sensors. *Sensors Actuators A Phys.* **2016**, *244*, 35–43. [CrossRef]

chemosensors

MDPI

Article

Hydrogen Sensors Based on In_2O_3 Thin Films with Bimetallic Pt/Pd Catalysts on the Surface and Tin and Dysprosium Impurities in the Bulk

Nadezhda Maksimova [1,*], Tatyana Malinovskaya [1], Valentina Zhek [1], Nadezhda Sergeychenko [2], Evgenii Chernikov [3], Ivan Lapin [4] and Valery Svetlichnyi [4,*]

1 Laboratory of Optical Materials and Coatings, Siberian Physical Technical Institute, Tomsk State University, Tomsk 634050, Russia
2 Laboratory of Semiconductor Devices, Siberian Physical Technical Institute, Tomsk State University, Tomsk 634050, Russia
3 Laboratory of Metal Oxide Semiconductors, Research and Development Center "Perspective Technologies in Microelectronics", Tomsk State University, Tomsk 634050, Russia
4 Laboratory of Advanced Materials and Technology, Siberian Physical Technical Institute, Tomsk State University, Tomsk 634050, Russia
* Correspondence: nkmax3@yandex.ru (N.M.); v_svetlichnyi@bk.ru (V.S.)

Abstract: This paper presents the results of studying the characteristics of hydrogen sensors based on thin In_2O_3 films modified with tin and dysprosium with dispersed double Pt/Pd catalysts deposited on the surface. To control the content of Sn and Dy in the films, an original technology was developed, and ceramic targets were fabricated from powders of the In–Dy–O, Dy–Sn–O, and In–Dy–Sn–O systems synthesized by the sol–gel method. Films of complex composition were obtained by RF magnetron sputtering of the corresponding targets. Structural features of the obtained thin films were studied by Raman spectroscopy. It is shown that various combinations of tin and dysprosium concentrations, as well as the presence of Pt/Pd catalysts on the surface, have a significant effect on the defectiveness of the films and the density of oxygen adsorption centers. As a result, the resistance of sensors in pure air (R_0), the activation energies of the temperature dependences of R_0, the bending of the energy bands at the grain boundaries of the semiconductor, and the responses to the action of hydrogen in the concentration range of 20–25,000 ppm change. A unique feature of Pt/Pd/ In_2O_3: Sn (0.5 at%), Dy (4.95 at%) films is their high sensitivity at 20–100 ppm and the absence of signal saturation in the region of high hydrogen concentrations of 5000–25,000 ppm, allowing them to be used to detect H_2 in a wide range of concentrations.

Keywords: thin films; indium oxide; tin; dysprosium; chemoresistive gas sensors; Raman spectroscopy; ceramic targets; hydrogen sensors

Citation: Maksimova, N.; Malinovskaya, T.; Zhek, V.; Sergeychenko, N.; Chernikov, E.; Lapin, I.; Svetlichnyi, V. Hydrogen Sensors Based on In_2O_3 Thin Films with Bimetallic Pt/Pd Catalysts on the Surface and Tin and Dysprosium Impurities in the Bulk. *Chemosensors* **2023**, *11*, 23. https://doi.org/10.3390/chemosensors11010023

Academic Editors: Kai Xu and Zhong Li

Received: 18 November 2022
Revised: 21 December 2022
Accepted: 23 December 2022
Published: 27 December 2022

1. Introduction

Since the 1960s, transparent conducting oxides have been widely used for optoelectronic applications [1–3]. Tin-doped indium oxide (ITO) is one of themost important transparent conductive oxides for such applications. To date, ITO has the best combination of characteristics in terms of electrical conductivity and optical transmission. Low resistance (about 1–2 × 10^{-4} Ω·cm) and a high concentration of charge carriers (about 1 × 10^{20} cm^{-3}), together with high transparency in the visible range (about 90%) [4], make it possible to widely use ITO in photovoltaics and solar cells, spintronics, electrochromic devices, OLED and LSD displays, touch screens, biosensors, transparent heaters, and chemoresistive gas sensors [5–7].

The technology for producing ITO films is well developed. The films are stable, reproducible, and have good surface morphology. Therefore, ITO is an attractive material for creating new materials on its basis and expanding the field of application. Films based on indium oxide are promising for resistive gas sensors. To create high-speed, highly sensitive sensors with low power consumption and high stability of parameters during operation, it is necessary to introduce catalytic additives into the volume and deposit them onto the surface. The effectiveness of such an approach was demonstrated experimentally in a number of studies conducted on films with thicknesses >500–1000 nm obtained by thick-film technologies modified with oxides of transition metals [8–13] and Dy [14]. In [14], 5% Dy^{3+} doped In_2O_3 nanoparticles were prepared by the hydrolysis-assisted co-precipitation method, using dysprosium nitrate. X-ray diffraction (XRD) revealed that the synthesized nanoparticles have a cubic bixbyite phase and that the addition of Dy^{3+} as a dopant noticeably enhanced the sensing response of In_2O_3 to ethanol.

The published information on the properties of structures based on SnO_2 with rare earth elements (PEEs) is also limited and usually refers to samples obtained by using thick-film technology [15–19]. The introduction of the rare earth elements Nb, Ce, La, Y, and Dy or their oxides during the synthesis of SnO_2 films was used to control the crystallite size and gas-sensitive properties of sensors. As a rule, no phases of oxides of REEs are detected in the films. During thermal annealing, additives segregate on the surface of microcrystals in the form of ions, such as Y^{3+}, Dy^{3+}, etc.

The industrial production of gas sensors requires the development of a technology that includes the magnetron sputtering of targets of complex composition to obtain thin-film sensors with additives of noble, 3d, and rare earth metals. In combination with microelectronic technology in one technological cycle, it is possible to obtain a large number of miniature sensitive elements with identical characteristics. In the patent [20], Kazuyoshi Inoue et al. state that the introduction of one of the lanthanides (La, Nb, Sm, Eu, Gd, Dy, Ho, Er, Tm, Yb, and Lu) into the In_2O_3 film, which is part of the TFT field-effect transistors, allows us to control the characteristics of the device through a smooth change in the film conductivity and free charge carriers' mobility and also to stabilize the device parameters. The oxide semiconductor film obtained in [20] by the DC sputtering method contained indium and at least one of the following elements: gadolinium, dysprosium, holmium, erbium, and ytterbium. The X-ray diffraction of the film showed only a peak obtained from a compound of bixbyite structure.

Previously, we carried out detailed research aimed at the developing an original technology for producing gas sensors by DC sputtering of mosaic metal targets in oxygen-argon plasma in an A 500 (Edwards) magnetron. The doping additives were applied to the surface of the targets. The theory and mechanisms of operation of sensors based on SnO_2 and WO_3 thin films with various catalysts in the bulk (Pt, Au, Ni, Co, Y, and Sc) and on the surface (Pt, Pd, and Au) were developed [21–25]. The paper [21] presents the results of an investigation of the nanostructure, element, and phase composition of thin (100–140 nm) tin dioxide films containing Ag, Y, Sc, Ag + Y, and Ag + Sc additives in the bulk and catalytic Pt/Pd clusters on the surface. The methods of atomic-force microscopy (AFM), scanning Auger microprobe equipped with an Ar^+ ions sputtering system (AES), and X-ray photoelectron spectroscopy (XPS) were used. Sensors based on Pt/Pd/SnO_2: Sb, Ag, Y films were characterized by ultrahigh response values at 50–1000 ppm of H_2 and the stabilization of parameters during long-term tests. The mechanisms of the influence of PEE impurities on the characteristics of thin films of tin dioxide are considered.

In this work, the structural, electrophysical, and gas-sensing properties of thin (about 100 nm) In_2O_3 films were studied. The films were obtained via the RF magnetron sputtering of specially designed original oxide targets with various contents of tin and dysprosium. The dispersed Pt and Pd catalysts were deposited on the surface of the targets. Raman spectroscopy was used to characterize the nanosized films. This is the only method to analyze thin films directly in the miniature sensors. It provides important information about film defects. The Raman spectra of bulk materials differ greatly from those of polycrystalline samples, where the size of particles (grains, agglomerates, clusters, etc.) lies in the nanometer range and the classical modes shift and expand. Doping with impurities and the deposition of dispersed catalyst clusters on the film surface can lead to the appearance of additional peaks and a change in the shape of the spectrum. Of particular interest is the prospect of creating semiconductor sensors for high pre-explosive concentrations of hydrogen that are not available on the market.

2. Materials and Methods

Ceramic targets (CTs) were fabricated by uniaxial static pressing of In_2O_3, SnO_2, and Dy_2O_3 powder compacts on a PLG 12 hydraulic press (LabTools, Saint Petersburg, Russia) at a pressure of 1500 atmospheres. The targets were sintered in a muffle furnace in an air atmosphere at 1000 °C for 1 h. When obtaining powders for compaction, a chemical method was used based on the co-deposition of metal hydroxides by neutralizing solutions of their salts with an alkaline agent, washing of the obtained precipitation of hydroxides, filtering them, drying, and thermal destruction. All reagents were of special-purity grade. The application of this method to obtain powders of mixed oxides of indium, tin, and dysprosium is described in [26].

The targets were 75 mm in diameter and 6 mm thick. The 150 μm thick sapphire plates were used as the substrates. The deposition of thin films of modified In_2O_3 was carried out by RF magnetron sputtering in an A-500 magnetron (Edwards, Burgess Hill, UK) at a discharge power of 100 W. The working gas contained 56 ± 0.5 vol.% of oxygen, and the rest was argon. The pressure in the chamber was within (6.5 – 7) × 10^{-3} mBar. The target-substrate distance was 90 mm. For 100 nm thick films, the sputtering time was 30 min. Two-layer Pt/Pd catalysts were deposited on the surface of the films by DC magnetron sputtering. First palladium was deposited and then platinum was deposited (each layer was deposited for 15 s).

Contacts to the sensitive In_2O_3 layers (Figure 1a) and a heater on the reverse side of the substrate (Figure 1b) were formed via the sputtering of platinum, followed by photolithographic engraving prior to the deposition of semiconductor films. After the formation of the films, the finished samples were subjected to technological annealing in air at a temperature of 720 K for 24 h. Up to 500 sensors were obtained on one substrate with a diameter of 30 mm. Then the samples were cut into individual elements, 0.7 × 0.7 mm^2 in size (sensitive area of 0.3 × 0.3 mm^2), and gold electrical wires measuring 50 μm in diameter were welded to the contact pads of the sensors by thermal compression. Next, the sensors were assembled into TO-8 cases (Figure 1c).

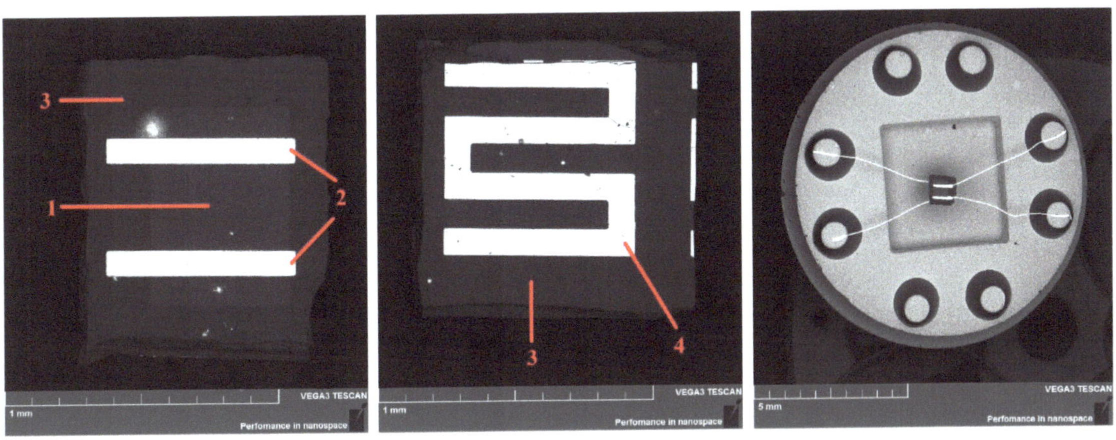

(a) (b) (c)

Figure 1. SEM images obtained in the backscattering (BSE) mode: the sensitive element from the side of (**a**) semiconductor SnO_2 layer and (**b**) heater. Sensors assembled into TO-8 cases (**c**): 1—sensitive layer; 2—Pt electrodes; 3—sapphire substrate; and 4—Pt heater.

Let us introduce the following numbering and notations for films of various compositions:
(I) In_2O_3 (90 at%): Sn (10 at%).
(II) Pt/Pd/In_2O_3 (90 at%): Sn (10 at%).
(III) Pt/Pd/In_2O_3 (86 at%): Sn (10 at%), Dy (4 at%).
(IV) Pt/Pd/In_2O_3 (94.05 at%): Sn (1 at%), Dy (4.95 at%).
(V) Pt/Pd/In_2O_3 (94.55 at%): Sn (0.5 at%), Dy (4.95 at%).
(VI) Pt/Pd/In_2O_3 (97 at%): Sn (1 at%), Dy (2 at%).

The Raman spectra of the films were measured on an inVia Basic confocal Raman spectrometer (Renishaw, Wotton-under-Edge, UK), operating in the backscattering mode at room temperature. The laser excitation wavelength was 785 nm (details of the research technique are presented in [23,24]).

To determine the sensor characteristics, the resistance (R_0) and conductivity (G_0) of the films were measured in pure air. The values of $R_1(G_1)$ were also measured under various hydrogen concentrations, depending on the operating temperature in the constant and pulsed heating modes. The relative humidity in the chamber was maintained at the level of RH = 35%. The ratio of conductivities, G_1/G_0, was taken as the adsorption response to hydrogen. The settling time of 0.9 G_{st}, where G_{st} is the stationary value of the conductivity, was taken as the response time, t_r. Based on the time dependences of the conductivity of sensors in the thermal cycling mode in pure air, the values of the energy band bending, $e\varphi_s$, at the grain boundaries in the studied nanocrystalline films were measured by using the method developed in [25]. The measuring stand and measurement techniques are described in [22]. The principle of operation and the theory of thin-film gas sensors are presented in [22–25].

3. Results and Discussion

3.1. *Raman Spectra of Pt/Pd/In_2O_3: Sn, Dy Films of Various Compositions*

The crystallographic structure of In_2O_3 belongs to the cubic system, the structure of bixbyite [27,28]. A complete description and identification of the Raman spectrum, even for undoped In_2O_3, is still being researched. At room temperature, no more than 10 modes can be observed, with all peaks in the range of 100–650 cm^{-1} [27–30]. In most cases, the following peaks are present in the Raman spectra of In_2O_3: 109, 132, 307, 366, 497, 517, and 630 cm^{-1}. These are approximate values since their positions vary depending on the methods of obtaining the metal oxide and the measurement technique.

Figure 2 shows the Raman spectra of all compositions of the obtained films, as well as the spectra of SnO_2 and In_2O_3 powders. The spectra of the powders, in principle, only slightly differ from those given in the literature for coarse-grained samples (bixbyite structure) [27–30]. In the case of In_2O_3 modified with Sn and Dy, broad bands with a number of maxima are observed in the range of 200–800 cm^{-1}. Similar bands observed for many nanostructured metal oxide films are attributed to dimensional effects [30–33]. The intensities of all modes are higher for films of Composition (V) containing 4.95 at% dysprosium. The low-frequency A_g modes appear as broadened bands at 100–170 cm^{-1}, with a maximum-intensity peak at 119 cm^{-1}, as well as a peak at 299 cm^{-1} shifted by 8 cm^{-1} relative to 307 cm^{-1}. Two modes, F_{2g} at 213 and 366 cm^{-1}, are most noticeable for the samples of Composition (VI). The intense peak at 630 cm^{-1} inherent in indium and tin oxides (visible in the spectra of the corresponding powders) is weakly manifested only in the spectrum of samples from Series (VI) with a reduced content of Dy (2 at%) and a higher Sn content (1 at%).

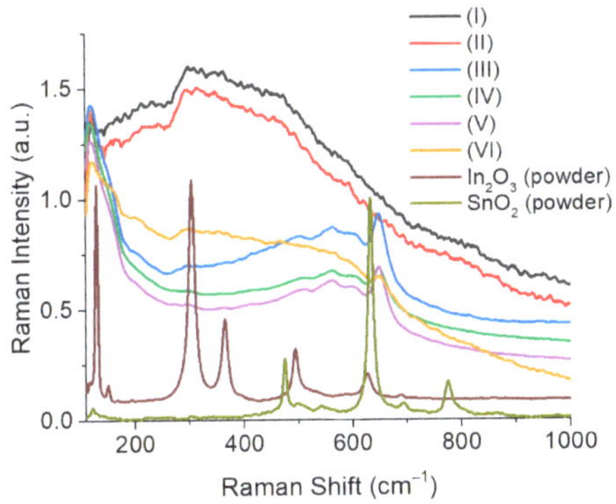

Figure 2. Raman spectra of obtained films and powders.

A feature of the Raman spectra of the studied films is the presence of high-frequency, low-intensity peaks at 565 and 590 cm^{-1}, as well as a strongly pronounced peak at 648 cm^{-1} having the highest intensity for films from Series (III) to (V). These three peaks are not typical of In_2O_3. Previously [23,24], we showed that the Raman spectra of thin films of tin dioxide with dispersed Pt/Pd catalytic layers deposited on the surface also exhibit maxima at 565 and 590 cm^{-1}. Upon the introduction of 13–14 at% Pt into the bulk of SnO_2, the Raman spectra show an intense peak with a maximum at 590 cm^{-1} [23]. A detailed analysis of the experimental data showed that two states of platinum are present in tin dioxide: three-dimensional particles of metallic Pt^0, which is not active in the Raman spectra, and an intermediate oxide PtO. This oxide is identified as two-dimensional dispersed platinum in the Pt^{2+} state responsible for the maximum at 590 cm^{-1}. The maximum at 565 cm^{-1} corresponds to the similar state of Pd^{2+}. The intense peak at 648 cm^{-1}, the broadening of the low-frequency band, and the shift of the maximum to 119 cm^{-1} are apparently due to the presence of dysprosium.

According to [34], REEs are characterized by higher energies of breaking the bonds with oxygen compared to tin and indium: ΔHo_{298} = 171 kcal/mol for Y–O, ΔHo_{298} = 161 kcal/mol for Sc–O, ΔHo_{298} = 144 kcal/mol for Dy–O, ΔHo_{298} = 76 kcal/mol for In–O, and ΔHo_{298} = 127 kcal/mol for Sn–O. It was suggested [21] that during heat treatment of the tin dioxide thin films obtained by magnetron sputtering, yttrium and scandium atoms segregate on the surface of microcrystals, form strong bonds with the lattice oxygen, and are present in the form of Y^{+3} and Sc^{+3} ions. Similar phenomena can also take place under the modification of indium oxide with dysprosium. As a result of formation of a bond of Dy with lattice oxygen, the surface density of superstoichiometric indium and, possibly, tin atoms, which are centers of oxygen chemisorption, increases. The REE impurity ions can serve as catalysts on the surface of microcrystals to create additional active centers and increase sensitivity to hydrogen [21], ethanol [14], methanol [15], and acetone [18].

3.2. Electrical and Gas-Sensitive Properties of Sensors

Oxygen chemisorption occurs on the surface of the indium oxide thin films in the atmosphere. In the near-surface layer of the semiconductor, a space charge region (SCR) of width d_0 depleted by electrons is formed, and the film conductivity decreases. During the dissociative adsorption of hydrogen on the surface of a thin-film sensor based on In_2O_3, atomic hydrogen interacts with a pre-chemisorbed oxygen ion, O^-. The reaction product is water, which is desorbed from the surface, and the electron returns to the conduction band of the semiconductor. The conductivity of the sensor increases. The response value, G_1/G_0, is determined by the density of oxygen adsorption centers, N_i. The centers of adsorption are superstoichiometric atoms of indium and other metals present on the surface [22]. An important role is played by the nanocrystalline structure and the ratio between the doubled SCR width, $2d_0$, and the thickness of the films (or conduction bridges, d_B).

An analysis of the experimental data shows that the characteristics of the studied films depend significantly on the ratio of the Sn and Dy concentrations in the bulk and the presence of Pt/Pd catalysts on the surface. Sn^{4+} ions substitute In^{3+} ones in the bulk and are donor impurities. As a result, the resistance of films of Composition (I) containing 10 at% Sn in pure air is R_0 = 0.11 kΩ (Table 1). The deposition of dispersed Pt/Pd catalysts in the case of samples of Composition (II) promotes the active adsorption of oxygen in the form of O^- on the surface and some increase in the SCR width and R_0. The introduction of a Dy impurity (Composition (III)) into the bulk of In_2O_3 leads to a further increase in R_0. The maximum values of R_0 are observed for the sensors of Compositions (IV) and (V), in which the Sn concentration is reduced to 0.5–1 at% and the Dy content is 4.95 at%. A decrease in the concentration of dysprosium to 2 at% (Composition (VI)) contributes to the decrease in R_0. The effect of the dysprosium impurity on the resistance of the films agrees with the features of the Raman spectra and confirms the assumption about the formation of strong bonds between Dy^{+3} ions and lattice oxygen. In this case, the density of chemisorbed oxygen on the film surface increases, and this contributes to the increase of the SCR width and resistance, R_0.

Table 1. Sensor resistances in pure air (R_0) measured at 300 K; activation energies ΔE_1 and ΔE_2 of temperature dependences of R_0; operating temperatures, T_{max}, at which maximum responses to 100 ppm of H_2 are observed; and response time, t_r, at 670 K.

Sample	R_0, kΩ	ΔE_1, eV	ΔE_2, eV	T_{max}, K	t_r, s
(I)	0.11	0.007–0.009	0.030–0.048	820	16
(II)	0.46	0.013–0.014	0.071–0.075	740	18
(III)	2.13	0.019–0.022	0.18–0.19	720	15
(IV)	694	0.09–0.18	0.38–0.46	640	24
(V)	803	0.044–0.15	0.58–0.61	540	30
(VI)	9.9	0.02–0.03	0.35–0.42	640	24

The temperature dependences of the film resistance in pure air are *N*-shaped (Figure 3a), which is characteristic of thin tin dioxide films modified with the various impurities studied earlier [22]. These dependences are determined by three independent parameters: the concentration and mobility of electrons in the bulk of the film and the negative charge density on the surface. Upon heating from room temperature to 470–500 K, the resistances of all the studied samples decrease mainly due to the ionization of shallow and deep centers in the bulk of the films, and the dependences $\ln R_0$ on $1000/T$ can be approximated by Arrhenius curves. It can be seen that the Arrhenius curves contain two linear sections, the activation energies of which ΔE_1 and ΔE_2 depend on the concentration of impurities and deposited catalysts (Table 1). In the samples of Compositions (I)–(III), characterized by low d_0 and R_0, mainly centers with low ionization energies are found. As the SCR width increases in case of films of Compositions (IV)–(VI), the Fermi level decreases with increasing temperature, and the ionization of deeper centers occurs.

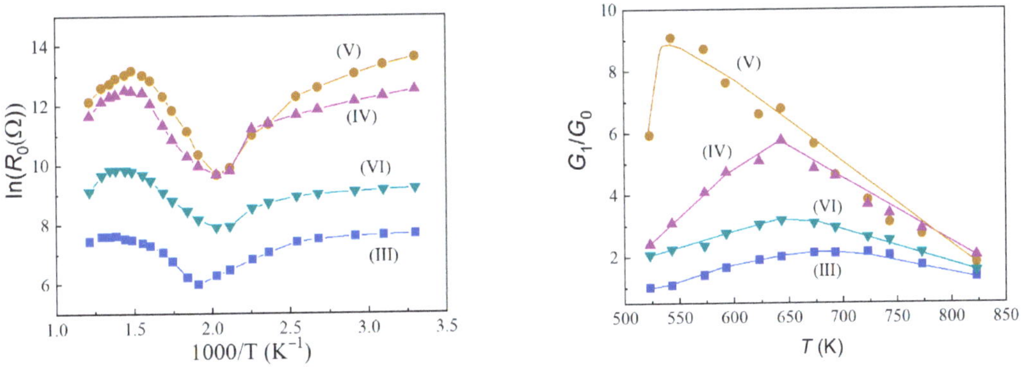

(a) (b)

Figure 3. Temperature dependences of sensors of various compositions: resistance, R_0, in pure air (**a**); and sensor response to 100 ppm H_2 (**b**).

An increase of the resistance (R_0) at T > 470 K and its decrease at ≥700 K (Figure 4a) are determined by an increase and subsequent decrease in the density of the negative charge on the surface of the semiconductor. The dependences of the adsorption response on the operating temperature for all the studied films have the form of curves with a maximum at T_{max} (Figure 3b and Table 1). With increasing temperature, the response increases when atomic hydrogen interacts with chemisorbed oxygen in the form of O^-, the density of which increases in the temperature range 400–700 K. The decrease in response at T > 540–700 K is due to the predominant desorption of gas from the overheated surface of the sensitive element. As a rule, higher values of T_{max} = 720–820 K are observed for films of Compositions (I)–(III). Subsequently, the measurements of the characteristics of sensors were carried out at 670 K, since at this temperature, there is an optimal ratio between the sufficiently high response values and the performance: the response time, t_r, does not exceed 30 s.

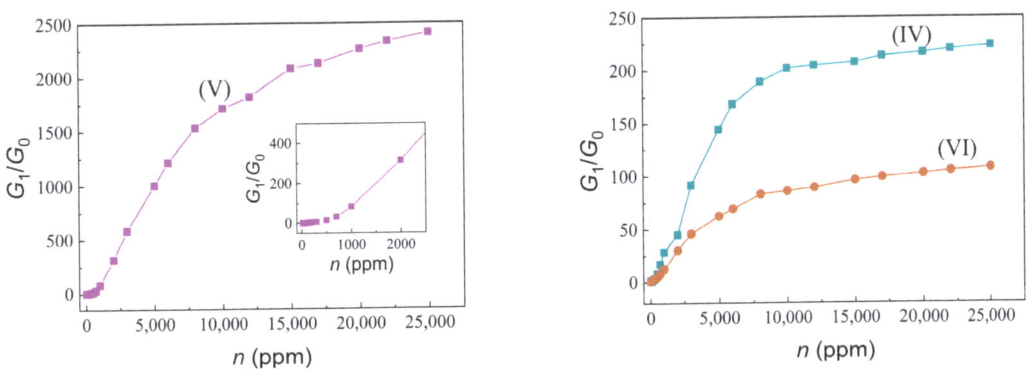

(a) (b)

Figure 4. Concentration dependences of the response to H_2 measured at 670 K for sensors of different compositions: Composition (V) (**a**); Compositions (IV) and (VI) (**b**). The inset shows the region G_1/G_0~exp[$e\varphi_s$].

Figure 4 shows the concentration dependences of the response to hydrogen for the sensors of Compositions (IV)–(VI) measured in the constant heating mode at 670 K. In all cases, exponential segments, G_1/G_0 ~ exp [n], are observed in the concentration range of 20–1500 ppm of H_2 (it is shown in the inset in Figure 4a). In this case, we can assume that the characteristics correspond to a model considering the existence of Schottky barriers at intergrain boundaries [22]. It is possible that H atoms produced due to the dissociative adsorption of H_2 can penetrate into the intergrain boundaries and affect the potential barrier height, $e\varphi_s$. Then $G_1/G_0 \sim \exp[e\varphi_s]$. At n > 1500 ppm, a power dependency is realized. A predominant role is played by the over-barrier component of conductivity.

Of particular interest are the data for the films of Composition (V) Pt/Pd/ In_2O_3 (94.55 at%): Sn (0.5 at%), Dy (4.95 at%) (Figure 4a), whose unique feature is the absence of saturation with an increase in the concentration of H_2 up to high values of 25,000 ppm. In this case, G_1/G_0 reaches 2400–2500. Therefore, for these sensors, there is an unusually high density of hydrogen adsorption centers. For the sensors of Compositions (IV) and (VI) (Figure 4b) at n = 10,000–25,000 ppm, the values of G_1/G_0 increase insignificantly; that is, the density of the adsorption centers is lower.

Since in case of the studied sensors based on modified indium oxide, the over-barrier component of conductivity is predominant under the action of hydrogen, the above method for determining $e\phi_s$ can be used. Based on the results of preliminary studies, the following thermal cycling modes were chosen: in the heating cycle, the temperature T_1 = 673 K and duration t_1 = 8 s; in the cooling cycle, T_2 = 473 K and t_2 = 6 s. The sensor temperature was increased from T_2 up to T_1 in about 0.4 s, while it was decreased from T_1 down to T_2 in about 1 s. In the time dependence of the conductivity, G_0, after a step-wise rise of temperature from T_2 to T_1 at $0.4 < t < 7$ s, a decrease of conductivity (Figure 5a) caused by a decrease in the surface density of adsorbed neutral hydroxyl groups (OH-groups) and subsequent increase of N_i due to the adsorption of O^- ions on the vacated sites are observed. At $t > 6.82$ s, stationary values of $N_i(T_1)$, $e\varphi_s(T_1)$, and $G_0(T_1)$ are established. At the same time, the condition T_2 = 470 K must be satisfied in order for ions O^- to be adsorbed on the SnO_2 surface. According to the method of [25], the value of $e\phi_s$ can be determined by Formula (1):

$$e\phi_s = [kT_1T_2/(T_1\text{-}T_2)] \times \ln[[G_0(T_1)/G_0(T_2)]\,(T_1/T_2)^{0.75}] + kT_1, \quad (1)$$

where k is the Boltzmann's constant. The conductivity–time profiles (CTP) $G_0(t)$ meet the requirements necessary to determine $e\phi_s$ at the grain boundaries of indium oxide. The shape of the CTP practically does not change under the action of hydrogen (Figure 5b), but the response values at the ends of the cooling cycles, $G_1/G_0(T_2)$, are significantly higher than those in the constant heating mode, G_1/G_0, and at the ends of the heating cycles, $G_1/G_0(T_1)$.

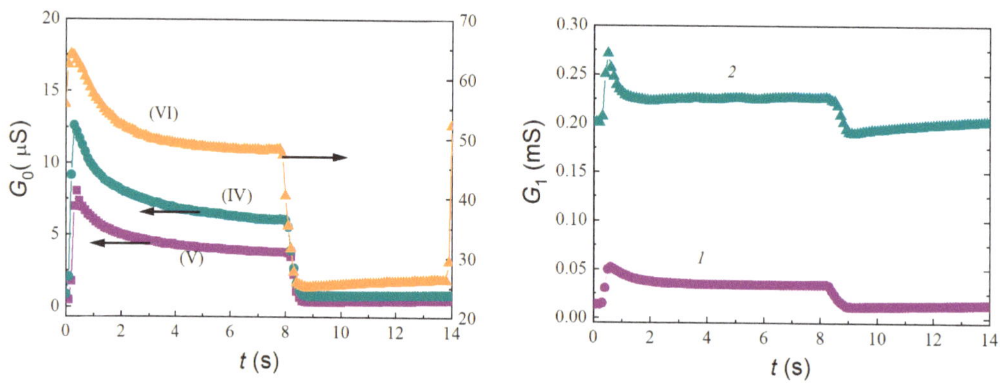

(a)

(b)

Figure 5. Conductivity–time profiles in the thermal cycling mode: (**a**) in pure air for sensors of Compositions (IV)–(VI) and (**b**) for the sensor of Composition (V) exposed to 100 ppm (Curve *1*) and 1000 ppm (Curve 2) of hydrogen.

Table 2 shows the values of $e\varphi_S$ and response measured in the constant and pulsed heating modes for samples from all series at fixed hydrogen concentrations. In the case of Samples (I)–(III), there is a high concentration of donor impurity and a reduced density of chemisorbed oxygen. The condition $2d_0 << d_B$ is satisfied. Therefore, low values of $e\varphi_S$, i.e., $e\varphi_S$ = 0.04–0.06 eV, and response are observed.

Table 2. Band bending values ($e\varphi_S$) and responses (G_1/G_0) in the constant heating mode at 670 K and during thermal cycling at the end of the cooling cycle, $G_1/G_0(T_2)$, for H_2 concentrations of 1000 and 25,000 ppm.

Sample	$e\varphi_S$, eV	G_1/G_0 (1000 ppm)	G_1/G_0 (25,000 ppm)	$G_1/G_0(T_2)$ (1000 ppm)	$G_1/G_0(T_2)$ (25,000 ppm)
(I) 10 Sn	0.04	1.7	-	-	-
(II) 10 Sn	0.04	3.0	-	-	-
(III) 10 Sn + 4 Dy	0.06	3.7	-	-	-
(IV) 1 Sn + 4.95 Dy	0.32	20.7	223	216	28,640
(V) 0.5 Sn + 4.95 Dy	0.37	31.2	2409	785	142,270
(VI) 1 Sn + 2 Dy	0.16	6.9	108	23	240

The decrease in the concentration of tin to 0.5–1 at% and introduction of 4.95 at% dysprosium contribute to the implementation of the condition $2d_0 \leq d_B$, which provides a significant increase in $e\varphi_S$, i.e., $e\varphi_S$ = 0.16–0.37 eV, and G_1/G_0. For these samples, at the end of the cooling cycle, ultrahigh values of the response, $G_1/G_0(T_2)$, are observed at pre-explosive H_2 concentrations of 25,000 ppm.

For all sensors known from the literature based on both thin and thick films of tin dioxide commonly used for detecting hydrogen, the signal saturation is observed at n > 3000–5000 ppm, since there is a limited density of H_2 adsorption centers on the surface. An analysis of semiconductor hydrogen gas analyzers available on the market shows that, in most cases, they are used to detect low H_2 concentrations of 0–2000 ppm. There are no sensors for high hydrogen concentrations. In particular, Figaro (Rolling Meadows, IL, USA), one of the world leaders in this field, produces TGS821 sensors for the range of 30–1000 ppm and TGS600 sensors for the range of 1–30 ppm of H_2 [35].

Previously, novel semiconductor sensors with high selectivity and stability were developed [23] for detecting high pre-explosive concentrations of H_2 (0.1–2.5 vol%). The sensors were based on thin (~100 nm) nanocrystalline SnO_2 films fabricated by the magnetron sputtering of dispersed Pt and Pd layers deposited on the surface, with an addition of 13–14 at% Pt in the bulk. The response values of G_1/G_0 at 25,000 ppm were 220–230. For the samples of Composition (V) studied in this work based on In_2O_3 with optimal concentrations of Sn and Dy, the G_1/G_0 value is an order of magnitude higher and reaches 2400–2500. In the thermal cycling mode, ultrahigh response values are observed at the end of the cooling cycle (Table 2). According to preliminary studies, the sensors are selective for other explosive gases: methane, propane, butane, etc. The response of G_1/G_0 < 15 to 25,000 ppm of methane is observed. It is important that the sensors can be used to detect both low (50–100 ppm) and high pre-explosive hydrogen concentrations.

4. Conclusions

The defectiveness and electrical and gas-sensitive properties of hydrogen sensors based on thin In_2O_3 films obtained via the RF magnetron sputtering of specially designed ceramic targets containing various concentrations of indium, tin, and dysprosium oxides were studied. Targets with a diameter of 75 mm and a thickness of 6 mm were obtained by uniaxial static pressing, followed by sintering of powder compacts prepared by the sol–gel method.

The Raman spectra of films with various concentrations of Sn and Dy contain some peaks characteristic of the In_2O_3 bixbyite syngony. Some of the low-frequency peaks are shifted in frequency, and the maximum of the highest intensity at 130 cm^{-1} is broadened and observed at 119 cm^{-1}. An important feature of the spectra is the presence of additional maxima at 565, 590, and 648 cm^{-1}. The peaks of low intensity at 565 and 590 cm^{-1} may be due to the catalysts deposited on the surface in the form of Pd^{2+} and Pt^{2+} ions. Dysprosium manifests itself as a bright maximum at 648 cm^{-1}, the intensity of which is higher in samples from Series (III) to (V) with the concentration of (4–4.95) at% of Dy.

Regularities have been established for the changes in properties of films measured under constant heating and thermal cycling modes, depending on the ratio of the concentrations of tin and dysprosium oxides in the targets, as well as on the deposition of dispersed Pt/Pd catalysts on the surface of the films. Sensors containing 10 at% of tin, which creates shallow donor centers in indium dioxide, are characterized by low values of the resistance R_0 in pure air, activation energies of temperature dependences of R_0, and response G_1/G_0 to the action of hydrogen. The presence of Pt/Pd on the surface and the introduction of dysprosium into the bulk of the films increase the density of oxygen adsorption centers and the band bending, $e\varphi_s$, at the grain boundaries.

Reducing the concentration of tin to 0.5–1 at% and introducing 4.95 at% of dysprosium contribute to a significant increase in $e\varphi_s$, with $e\varphi_s$ = 0.35–0.37 eV, and G_1/G_0. The presence of the region G_1/G_0~exp[$e\varphi_s$] in the concentration range n = 20–1500 ppm of H_2 indicates the predominant role of the over-barrier component of conductivity. An important feature of sensors based on Pt/Pd/In_2O_3 (94.55 at%): Sn (0.5 at%), Dy (4.95 at%) is the absence of signal saturation when n increases to high values of 25,000 ppm of H_2. In this case, G_1/G_0 reaches 2400–2500.

The mechanism of the dysprosium effect is determined by the higher energy of breaking the bond with oxygen Dy–O compared to In–O and Sn–O bonds. During technological annealing under the film crystallization, impurity atoms segregate on the surface of microcrystals and actively form strong bonds with lattice oxygen, contributing to the formation of superstoichiometric indium and, possibly, tin, which, together with Dy^{3+} ions, sharply increases the density of oxygen and hydrogen adsorption centers. On the basis of original films obtained by the RF magnetron sputtering of oxide targets with an optimal ratio of Sn and Dy impurity concentrations, unique sensors for detecting hydrogen in a wide range of concentrations, including pre-explosive values of 25,000 ppm, can be created.

Based on the results obtained, we plan to further optimize the composition and methods of obtaining films to increase the efficiency of sensors based on In_2O_3, detailed studies of their composition, structure and physicochemical properties, as well as the study of selectivity and testing of the service life.

Author Contributions: Conceptualization, N.M., T.M. and V.S.; methodology, V.Z. and E.C.; validation, N.M., T.M. and V.S.; formal analysis, N.M., T.M. and N.S.; investigation, V.S., N.S., V.Z., E.C. and I.L.; resources, N.M., T.M. and V.S.; writing—original draft preparation, N.M. and T.M.; visualization, V.S., N.S. and I.L.; supervision, N.M., T.M. and V.S.; project administration, N.M. and T.M.; funding acquisition, N.M. All authors have read and agreed to the published version of the manuscript.

Funding: This study was supported by the Tomsk State University Development Programme (Priority-2030).

Institutional Review Board Statement: Not applicable.

Informed Consent Statement: Not applicable.

Data Availability Statement: The data presented in this study are available upon request from the corresponding author.

Acknowledgments: The Raman spectroscopy research was carried out with the equipment of Tomsk Regional Core Shared Research Facilities Center of National Research Tomsk State University. The Center was supported by the Ministry of Science and Higher Education of the Russian Federation, grant no. 075-15-2021-693 (no. 13.RFC.21.0012).

Conflicts of Interest: The authors declare no conflict of interest.

References

1. Sezemsky, P.; Burnat, D.; Kratochvil, J.; Wulff, H.; Kruth, A.; Lechowicz, K.; Janik, M.; Bogdanowicz, R.; Cada, M.; Hubicka, Z.; et al. Tailoring properties of indium tin oxide thin films for their work in both electrochemical and optical label-free sensing systems. *Sens. Actuators B Chem.* **2021**, *343*, 130173. [CrossRef]
2. Frank, G.; Köstlin, H. Electrical properties and defect model of tin-doped indium oxide layer. *Appl. Phys. A* **1982**, *27*, 197–206. [CrossRef]
3. Ray, S.; Banerjee, R.; Basu, N.; Batabyal, A.K.; Barua, A.K. Properties of tin doped indium oxide thin films prepared by magnetron sputtering. *J. Appl. Phys.* **1983**, *54*, 3497–3501. [CrossRef]
4. Stadler, A. Transparent Conducting Oxides—An Up-To-Date Overview. *Materials* **2012**, *5*, 661–683. [CrossRef]
5. Txintxurreta, J.; G-Berasategui, E.; Ortiz, R.; Hernández, O.; Mendizábal, L.; Barriga, J. Indium Tin Oxide Thin Film Deposi-tion by Magnetron Sputtering at Room Temperature for the Manufacturing of Efficient Transparent Heaters. *Coatings* **2021**, *11*, 92. [CrossRef]
6. Zhang, M.; Xu, W.; Shen, H.; Wang, J.; Han, J. New research progress of transparent conductive oxide films. *Yadian Yu Shengguang Piezoelectrics Acoustooptics* **2010**, *32*, 811–815.
7. Hosono, H.; Ueda, K. Transparent Conductive Oxides. In *Book Springer Handbook of Electronic and Photonic Materials*, 2nd ed.; Kasap, S., Capper, P., Eds.; Springer Nature: Cham, Switzerland, 2017; pp. 1391–1404. [CrossRef]
8. Mokrushin, A.S.; Fisenko, N.A.; Gorobtsov, P.Y.; Simonenko, T.L.; Glumov, O.V. Pen plotter printing of ITO thin film as a highly CO sensitive component of a resistive gas sensor. *Talanta* **2021**, *221*, 121455. [CrossRef]
9. Korotcenkov, G.; Cho, B.K. The role of grain size on the thermal instability of nanostructured metal oxides used in gas sensor applications and approaches for grain-size stabilization. *Prog. Cryst. Growth* **2012**, *58*, 167–208. [CrossRef]
10. Zhao, Y.; Wang, S.; Fan, S.; Hua, Z.; Wu, Y.; Tian, X.; Yuan, W. Selective detection of methane by Pd-In_2O_3 sensors with a catalyst filter film. *Sens. Actuators B Chem.* **2021**, *328*, 129030. [CrossRef]
11. Korotcenkov, G.; Boris, I.; Brinzari, V.; Han, S.H.; Cho, B.K.; Lychkovsky, Y.N. In_2O_3: Ga and In_2O_3: P-based one-electrode gas sensors: Comparative study. *Ceram. Int.* **2015**, *41*, 7478–7488. [CrossRef]
12. Ullah, H.; Yamani, Z.H.; Qurashi, A.; Iqbal, J.; Safeen, K. Study of the optical and gas sensing properties of In_2O_3 nanoparticles synthesized by rapid sonochemical method. *J. Mater. Sci. Mater. Electron.* **2020**, *31*, 17474–17481. [CrossRef]
13. Rahamim, G.; Mirilashvili, M.; Nanikashvili, P.; Greenberg, E.; Shpaisman, H.; Grinstein, D.; Welner, S.; Zitoun, D. Hydrogen sensors with high humidity tolerance based on indium-tin oxide colloids. *Sens. Actuators B Chem.* **2020**, *310*, 127845. [CrossRef]
14. Anand, K.; Thangaraj, R.; Kohli, N.; Singh, R.C. Structural, Optical and Ethanol Gas Sensing Properties of In_2O_3 and Dy^{3+}: In_2O_3 Nanoparticles. *AIP Conf. Proc.* **2014**, *1591*, 477–480. [CrossRef]
15. Carreno, N.L.V.; Maciel, A.P.; Leite, E.R.; Lisboa-Filho, P.N.; Longo, E.; Valentino, A.; Probst, L.E.D.; Paiva-Santos, C.O.; Schreiner, W.H. The influence of cation segregation on the methanol decomposition on nanostructured SnO_2. *Sens. Actuators B Chem.* **2002**, *86*, 185–192. [CrossRef]
16. Korotcenkov, G.; Cho, B.K. Metal oxide composites in conductometric gas sensors: Achievements and challenges. *Sens. Actuators B Chem.* **2017**, *244*, 182–210. [CrossRef]
17. Leite, R.; Maciel, A.P.; Weber, I.T.; Lisboa-Filho, P.N.; Longo, E.; Paiva-Santos, C.O.; Andrade, A.V.C.; Pakoscimas, C.A.; Maniette, Y.; Schreiner, W.H. Development of metal oxide nanoparticles with high stability against particle growth using a metas solid solution. *Adv. Mater.* **2002**, *14*, 905–908. [CrossRef]
18. Cheng, L.; Ma, S.Y.; Li, X.B.; Luo, J.; Li, W.Q.; Li, F.M.; Mao, Y.Z.; Wang, T.T.; Li, Y.F. Highly sensitive acetone sensors based on Y-doped SnO_2 prismatic hollow nanofibers synthesized by electrospinning. *Sens. Actuators B Chem.* **2014**, *200*, 181–190. [CrossRef]
19. Cheng, L.; Ma, S.Y.; Wang, T.T.; Luo, J.; Li, X.B.; Li, W.Q.; Mao, Y.Z.; Gz, D.J. Highly sensitive acetic acid gas sensor based on coral-like and Y-doped SnO_2 nanoparticles prepared by electrospinning. *Mater. Lett.* **2014**, *137*, 265–268. [CrossRef]
20. Kazuyoshi, I.; Koki, Y.; Masashi, K. Sputtering Target, Semiconductor Film and Semiconductor. Device. Patent US8333913B2, 18 December 2012.
21. Maksimova, N.K.; Almaev, A.V.; Sevastyanov, E.Y.; Potekaev, A.I.; Chernikov, E.V.; Sergeychenko, N.V.; Korusenko, P.M.; Nesov, S.N. Effect of Additives Ag and Rare-Earth Elements Y and Sc on the Properties of Hydrogen Sensors Based on Thin SnO_2 Films during Long-Term Testing. *Coatings* **2019**, *9*, 423. [CrossRef]
22. Sevastyanov, E.Y.; Maksimova, N.K.; Novikov, V.A.; Rudov, F.V.; Sergeychenko, N.V.; Chernikov, E.V. Effect of Pt, Pd, Au additives on the surface and in the bulk of tin dioxide thin films on the electrical and gas-sensitive properties. *Semiconductors* **2012**, *46*, 801–809. [CrossRef]
23. Maksimova, N.K.; Sevastyanov, E.Y.; Chernikov, E.V.; Korusenko, P.M.; Nesov, S.N.; Kim, S.V.; Biryukov, A.A.; Sergeychenko, N.V.; Davletkildeev, N.A.; Sokolov, D.V. Sensors based on tin dioxide thin films for the detection of pre-explosive hydrogen concentrations. *Sens. Actuators B Chem.* **2021**, *341*, 130020. [CrossRef]
24. Maksimova, N.K.; Khludkova, L.S.; Biryukov, A.A.; Sevast'yanov, E.Y.; Chernikov, E.V.; Kushnarev, B.O. The Stability of Liquid Hydrocarbon Vapor Sensors Based on SnO_2 Thin Films Modified with Various Catalysts. *Tech. Phys.* **2021**, *66*, 999–1008. [CrossRef]
25. Gaman, V.I.; Sevast'yanov, E.Y.; Maksimova, N.K.; Almaev, A.V.; Sergeichenko, N.V. Characteristics of the semiconductor resistive hydrogen sensors in the thermo-cyclic operation mode. *Russ. Phys. J.* **2014**, *56*, 1427–1434. [CrossRef]
26. Malinovskaya, T.; Ghyngazov, S.; Zhek, V. Thermal destruction of coprecipitated hydroxides of indium and dysprosium. *J. Therm. Anal. Calorim.* **2019**, *138*, 1871–1877. [CrossRef]

27. Wang, C.Y.; Dai, Y.; Pezoldt, J.; Lu, B.; Kups, T.; Cimalla, V.; Ambacher, O. Phase stabilization and phonon properties of single crystalline rhombohedral indium oxide. *Cryst. Growth Des.* **2008**, *8*, 1257–1260. [CrossRef]
28. Kranert, C.; Schmidt-Grund, R.; Grundmann, M. Raman active phonon modes of cubic In_2O_3. *Phys. Status Solidi RRL* **2014**, *8*, 554–559. [CrossRef]
29. Gan, J.; Lu, X.; Wu, J.; Xie, S.; Zhai, T.; Yu, M.; Zhang, Z.; Mao, Y.; Wang, S.C.I.; Shen, Y.; et al. Oxygen vacancies promoting photochemical performance of In_2O_3 nanocubes. *Sci. Rep.* **2013**, *3*, 1021. [CrossRef]
30. Berengue, O.M.; Rodrigues, A.D.; Dalmaschio, C.J.; Lanfredi, A.J.C.; Leite, E.R.; Chiquito, A.J. Structural characterization of indium oxide nanostructures: A Raman analysis. *J. Phys. D* **2010**, *43*, 045401. [CrossRef]
31. Lee, H.; Kim, B.; Gao, C.Y.; Choi, H.J.; Ko, J.H.; Seo, C.H.; Park, J. Raman spectroscopy study of solution processed In_2O_3 thin films: Effect of annealing temperature on the characteristics of In_2O_3 semiconductors and thin-film transistors. *Mol. Cryst. Liq. Cryst.* **2019**, *679*, 38–47. [CrossRef]
32. Yadav, K.; Mehta, B.R.; Singh, J.P. Synthesis and Raman spectrum of crystalline indium oxide micro-rods with rectangular cross-section. *Solid State Phys. AIP Conf. Proc.* **2014**, *1591*, 409–410. [CrossRef]
33. Rumyantseva, M.N.; Gaskov, A.M.; Rosman, N.; Pagnier, T.; Morante, J.R. Raman Surface Vibration Modes in Nanocrystalline SnO_2: Correlation with Gas Sensor Performances Interfaces. *Chem. Mater.* **2005**, *17*, 893–901. [CrossRef]
34. Luo, Y.R. *Comprehensive Handbook of Chemical Bond Energies*, 1st ed.; CRC Press: Boca Raton, FL, USA, 2007; Volume 1655. [CrossRef]
35. FIGARO USA, Inc. Gas Sensors & Modules. Available online: https://www.figarosensor.com/product/sensor/ (accessed on 10 November 2022).

chemosensors

MDPI

Article

Effect of Ti_2CT_x MXene Oxidation on Its Gas-Sensitive Properties

Artem S. Mokrushin [1,*], Ilya A. Nagornov [1], Philipp Yu. Gorobtsov [1], Aleksey A. Averin [2], Tatiana L. Simonenko [1], Nikolay P. Simonenko [1], Elizaveta P. Simonenko [1] and Nikolay T. Kuznetsov [1]

1 Kurnakov Institute of General and Inorganic Chemistry, Russian Academy of Sciences, 31 Leninsky pr., 119991 Moscow, Russia

2 Frumkin Institute of Physical Chemistry and Electrochemistry, Russian Academy of Sciences, 31 Leninsky pr., bldg. 4, 199071 Moscow, Russia

* Correspondence: artyom.nano@gmail.com

Abstract: The oxidation process was studied for the synthesized low-layer Ti_2CT_x MXene deposited on a special Al_2O_3/Pt sensor substrate using in situ Raman spectroscopy. It is noted that on the ceramic parts of the substrate (Al_2O_3), the beginning of oxidation (appearance of anatase mod phase) is observed already at 316 °C, in comparison with platinum, for which the appearance of anatase is noted only at 372 °C. At the temperature 447 °C, the initial MXene film is completely oxidized to TiO_2. Using scanning electron microscopy and atomic force microscopy, the microstructure and dispersity of the obtained MXene film were studied. It was found that the obtained films exhibit chemoresistive responses to the detection of a wide group of gases, H_2, CO, NH_3, C_6H_6, C_3H_6O, CH_4, C_2H_5OH and O_2, at room temperature and RH = 50%. The highest sensitivity is observed for NH_3. The partial oxidation of the Ti_2CT_x MXene was shown to favorably affect the gas-sensitive properties.

Keywords: chemoresistive gas sensors; MXenes; T_2C; Titanium dioxide (TiO_2); Raman; 2D-nanomaterials

Citation: Mokrushin, A.S.; Nagornov, I.A.; Gorobtsov, P.Y.; Averin, A.A.; Simonenko, T.L.; Simonenko, N.P.; Simonenko, E.P.; Kuznetsov, N.T. Effect of Ti_2CT_x MXene Oxidation on Its Gas-Sensitive Properties. *Chemosensors* **2023**, *11*, 13. https://doi.org/10.3390/chemosensors11010013

Academic Editors: Kai Xu and Zhong Li

Received: 30 November 2022
Revised: 19 December 2022
Accepted: 20 December 2022
Published: 22 December 2022

1. Introduction

Two-dimensional titanium-containing carbide MXenes with the general formula $Ti_{n+1}C_nT_x$ have attracted increased attention from the scientific community in recent years due to their unique properties—a large surface-to-volume ratio, high electrical conductivity, which can be metallic or semiconducting, and surface functionalization by various groups (primarily −F, −OH, −O), whose composition can be adjusted in some limits by choice of synthesis method. [1–3]. This combination of unique MXene properties opens up the potential for their use in various fields: in lithium/sodium-ion batteries and supercapacitors [4,5], fuel cells [6,7], photocalysis [8,9], as well as chemical gas sensing [10–14] etc.

In contrast to metal oxide semiconductors (MOS), the classical receptor materials for chemoresistive gas sensors, $Ti_{n+1}C_nT_x$ MXenes often have metallic conductivity, their outer surface is completely covered by functional groups, and their morphology refers to 2D-nanomaterials. This combination makes this class of compounds attractive for chemoresistive gas sensors with a high signal-to-noise ratio (SNR) [15], including room temperature operation.

The detection of the MXenes mechanism differs significantly from MOS materials. A universal model describing the interaction of the analyte gas with ion-adsorbed oxygen and the formation of an electron-depleted layer (EDL) on the surface layer of the sensitive material can be used for MOS-gas sensors [16]. At present, the question about the mechanism of gas detection by MXenes is debatable. Nevertheless, theoretical and experimental models allowing the description of the appearance of the chemoresistive signal in the atmosphere of different gases have already been observed [17–19].

MXenes, due to their predisposition to form hydrogen bonds with functional groups, are extremely sensitive to moisture. The harmonic structure characteristic of MXenes multilayer, the metallic type of conductivity, and the fact that water molecules do not form covalent bonds with the surface of MXenes make them extremely sensitive to humidity and allow a reproducible response. In Zhang C. et al. [20], using FT-IR spectroscopy and the study of the wetting angle on the MXenes surface, the authors found that water molecules from the gas phase can be chemically sorbed on the surface and in the interlayer space of $Ti_3C_2T_x$ MXene. The authors proposed a mechanism whereby the hydroxyl groups thus formed on the surface provide an electrostatic field and prevent charge transfer, which is explained by an increase in electrical resistance (positive response: *p*-type) with increasing humidity. The nanoscale interlayer space (~1.01 nm) of $Ti_3C_2T_x$, as well as the surface super hydrophilicity found by the authors, prevents capillary condensation of water and limits the movement of hydroxyl groups, which are characteristic of bulk nanomaterials.

Structural water plays an important role in reducing the overall interaction between the MXenes layers and creating additional space for the adsorption and diffusion of gases with affinity to functional groups. This feature is an advantage over other 2D-layered materials, such as graphene, in which strong Van-der-Waals interactions between two-dimensional sheets and small interlayer distances prevent the intercalation of gas molecules. Koh H-J. et al., in their article [21], in situ using XRD demonstrated that the interlayer space of MXenes contains water fragments (hydroxyl groups or water molecules themselves), which can be removed when the material is exposed to dry nitrogen. The authors showed that ethanol molecules can chemically bind with $Ti_3C_2T_x$ MXene in its interlayer space. As a result of these interactions, the MXene swells, and its interlayer distance increases due to the steric effect. The swollen two-dimensional MXene decreases the number of electrons involved in charge transfer between the layers during ethanol detection, which leads to an increase in electrical resistance and allows the positive (*p*-type) chemoresistance response to be recorded.

The MXenes show the best sensitivity when detecting gases able to form hydrogen bonds. Majhi S.M. et al., in article [18], showed a positive (*p*-type) response when acetone is detected by the $Ti_3C_2T_x$ MXene. Kim S.J. et al. [15] also showed that when used as a receptor material, $Ti_3C_2T_x$ MXene demonstrates the greatest response to gases capable of forming a hydrogen bond (acetone, ethanol, propanal and ammonia) than observed for acid gases (NO_2, SO_2 and CO_2). The authors demonstrated the ability to detect 50–1000 ppb of acetone, ethanol and ammonia with a low SNR value. DFT calculations found that hydroxyl groups (−OH) made the largest contribution to acetone bonding to functional groups on the $Ti_3C_2T_x$ MXene surface compared to oxygen (−O) and fluoride (−F) groups. It was found that for OH-groups, the bonding energy value is significantly more negative compared to −O and −F: −0.774 compared to −0.317 and −0.311 eV, respectively. The authors concluded that the terminal hydroxyl groups on the surface of MXenes play a key role in the detection of various gases to form hydrogen bonds [15].

In most of the papers presented in the literature, a positive chemoresistive response of the *p*-type is recorded when detecting various gases with carbide titanium-containing MXenes. Nevertheless, there are also studies in which the electrical resistance decreases when gases are injected (negative response: *n*-type). For example, Wu M. et al. [22] observed a decrease in electrical resistance when detecting ammonia by the $Ti_3C_2T_x$ MXenes. The authors explain this by the strong functionalization of the surface by chloride groups (−Cl), which were formed on the surface of MXenes due to the peculiarities of the synthesis (etching MAX-phase in the Lewis acid $ZnCl_2$ melt, which is different from the classical liquid-phase methods). This phenomenon is not fully explained in this paper. The authors suggest that chloride groups can affect the MXene band structure, which leads to the reverse chemoresistive response.

Nanocomposites based on them, primarily based on titanium dioxide $Ti_{n+1}C_nT_x/TiO_2$, are no less interesting receptor materials for gas sensors as compared to individual $Ti_{n+1}C_nT_x$ MXenes. Such nanocomposites can be obtained, in particular, by thermal oxidation in an oxygen-containing atmosphere of the initial $Ti_{n+1}C_nT_x$ MXenes. Such $Ti_{n+1}C_nT_x/TiO_2$ nanocomposites can have improved gas-sensitive characteristics. In [23], the authors oxidized a $Ti_3C_2T_x$ film deposited on a special multi-sensor chip in air at 350°C. The resulting nanocomposite demonstrated an increased semiconductor response to ethanol at relatively high temperatures (300–350 °C), and a decrease in the response and recovery time of the sensor was noted. In [24], the authors used oxygen plasma to oxidize MXene. The resulting nanocomposite showed an increased response to ethanol by several orders of magnitude. The formation of $Ti_{n+1}C_nT_x/TiO_2$ nanocomposites is also possible by holding MXene in an aqueous solution, specifically under the influence of light. In [25], the authors obtained $Ti_{n+1}C_nT_x/TiO_2$ by holding the individual components in an ethanol-water mixture. For the synthesized nanocomposite, they were able to increase the sensitivity, as well as lower the threshold of sensitivity to ammonia to 100 ppb. In [26], the authors in situ-oxidized the initial $Ti_3C_2T_x$ MXene using hydrothermal synthesis. The resulting oxidized nanocomposite demonstrated increased sensitivity and selectivity to NO_2 detection. In [27], the authors used a "simple spray method" to spray an aqueous dispersion of TiO_2 onto the surface of $Ti_3C_2T_x$ MXene. The resulting nanocomposite showed improved gas-sensitive characteristics when detecting low NH_3 concentrations at RT, as well as a more stable signal compared to the individual $Ti_3C_2T_x$ MXene.

The present paper is devoted to the study of the oxidation process of the $Ti_3C_2T_x$ MXene film using in situ Raman spectroscopy and to the study of the gas-sensitive chemoresistance properties of the obtained nanomaterials to a wide group of analyte gases.

2. Materials and Methods

2.1. Mxene Synthesis and Film Application

Reagents: powders of metallic titanium (99.9%, 0.5–100 μm, Ruskhim, Moscow, Russia), aluminum (99.2%, 30 μm, Ruskhim), graphite (MPG-8 grade, Ruskhim), potassium bromide KBr (99%, Ruskhim), sodium fluoride NaF (osc. Part 9-2, Reackhim, Moscow, Russia), hydrochloric acid HCl (>99%, Sigma Tech, Moscow, Russia), tetramethylammonium hydroxide solution $(CH_3)_4N(OH)$ (TMAOH, 25%, aqueous solution, Technic, Saint-Denis, France).

The synthesis of Ti_2CT_x MXene was performed by selective etching of aluminum layers contained in the MAX-phase of Ti_2AlC. For this purpose, Ti_2AlC powder was added to a sodium fluoride solution in hydrochloric acid; the Ti_2AlC and Ti_2CT_x synthesis techniques used in this study are described in detail in [28,29].

Briefly, to obtain Ti_2AlC, powders of aluminum, titanium, graphite and potassium bromide were mixed in the ratio n(Ti):n(Al):n(C) = 2:1.2:0.8 and m(Ti+Al+C) = m(KBr), then co-milled, pressed into tablets and heat treated in a muffle furnace at 1000 °C in a protective melt KBr.

To obtain the multilayer Ti_2CT_x MXene, a 1 g sample of the MAX-phase Ti_2AlC mass was slowly added to 20 mL of 6 M HCL solution with 1 g of NaF. The system was then incubated under stirring and at 40 ± 5 °C for 24 h. The resulting powder was separated by centrifugation and repeatedly washed with distilled water until pH ~5–6 was reached. Delamination was performed using tetramethylammonium hydroxide $(CH_3)_4N(OH)$ solution under ultrasonic influence.

The obtained Ti_2CT_x MXene sample was held in an aqueous dispersion at 4–6 °C for 7 days and then used to apply the receptor layer.

The Ti_2CT_x MXene film was deposited by drop casting method on the surface of a specialized Al_2O_3 substrate with platinum interdigital electrodes (from the front side) and a platinum microheater (from the back side). For this purpose, the dispersion system (carbide nanosheets and deionized water in a volume of 50 μL) was applied to the substrate surface in the area of counter-pin electrodes using an automatic dispenser. Then, a step drying in the range of 25–150 °C under reduced pressure was performed to remove the solvent.

2.2. Instrumentation

The microstructure and chemical composition of the film surface were studied by scanning (NVision 40 scanning electron microscope, Carl Zeiss (Oberkochen, Germany, secondary electron detector, accelerating voltage 1–10 kV) and transmission electron microscope (JEOL JEM-1011, Akishima, Japan), as well as X-ray spectral elemental microanalysis (energy-dispersive X-ray (EDX) spectrometer INCA X-MAX 80, Oxford Instruments (Oxford, UK), accelerating voltage 20 kV). The powder phase composition and films were studied by means of X-ray diffraction (XRD) on D8 Advance device (Bruker, Billerica, MA, USA,), CuKα = 1.5418 Å, Ni filter, E = 40 keV, I = 40 mA; the range of 2θ: 5–45°; resolution: 0.02°; signal accumulation time in the point: 0.3 s).

The obtained Ti_2CT_x MXene films were examined by atomic force microscopy (AFM). Surface topography and Kelvin-probe force microscopy (KFM) were studied on a Solver Pro-M scanning probe microscope (NT-MDT production, Zelenograd, Moscow, Russia) using ETALON HA-HR probes with W_2C-based conductive coating (tip curvature <35 nm). The microstructure of the obtained films was studied using SEM and AFM on the Al_2O_3 section between platinum microelectrodes.

Raman spectra were obtained with a Renishaw inVia Reflex Microscope system equipped with a Peltier-cooled CCD. The 532-nm lines of a Nd:YAG laser were used for excitation. The laser light was focused on the sample through a 50× objective to a spot size of ~2 µm. The power of the sample was <0.3 mW. The variable-temperature Raman scattering measurements were made using a THMS600 stage (Linkam Scientific Instruments Ltd., Redhill, UK). The in situ film heating rate was 5 °C/min. After conducting Raman in situ heating experiments on the samples, the temperature was recalibrated on a typical sample using a high-precision Testo 868 thermal imaging camera. Due to the fact that Raman spectroscopy can be destructive for some samples (especially those prone to oxidation, as in this case) due to local heating of the region from which the spectrum is taken after laser exposure, spectra were recorded in different areas of the film, which are nevertheless localized close to each other. This approach avoided the local overheating that could occur with repeated laser exposure, even at low power, which allowed more reliable data to be obtained. All spectra obtained were normalized with respect to the most intense spectrum.

The chemoresistive responses were obtained using a special laboratory setup, a detailed description of which can be found in our earlier papers [30,31]. The gas medium was created in a quartz cell using two Bronkhorst gas flow controllers with a maximum throughput of 100 and 200 mL/min. The quartz cell volume is 7×10^{-5} m^3. The electrical resistance of the obtained oxide films was measured using a Fluke 8846A Digit Precision Multimeter, which has an upper detection limit of 1000 MΩ. The temperature of the sensor was monitored using a platinum microheater pre-calibrated with a high-precision Testo 868 thermal imaging camera. The measurements of gas-sensitive properties, among others, were carried out at room temperature. Before starting the gas-sensitive measurements, the film to be measured was incubated in a baseline gas atmosphere until a stable signal was established.

To measure the signal at different relative humidity (RH), we used a special unit with a bubbler flask; the RH of the gas mixture was controlled by a digital flow-through hygrometer "Excis". The temperature value of the relative humidity was set and then measured at 20 °C.

All gas-sensing measurements were carried out at room temperature (RT) and 50% relative humidity. The response to H_2, CO, NH_3, benzene (C_6H_6), acetone (C_3H_6O), methane (CH_4), ethanol (C_2H_5OH) and oxygen (O_2) was calculated using the following ratio:

$$S_1 = \frac{|R_{BL} - R_g|}{R_{BL}} \times 100\% \quad (1)$$

where R_{BL} is the baseline resistance (nitrogen (99.9999%) and was used as the baseline for oxygen detection and synthetic air for other gases, and R_g is the resistance at a given concentration of analyte gas.

The response to humidity was calculated using the following ratio:

$$S_2 = \frac{|R_{BL} - R_{RH}|}{R_{BL}} \times 100\% \quad (2)$$

where R_{BL} is the resistance at 30% relative humidity, and R_{RH} is a given relative humidity.

3. Results and Discussion

3.1. Investigation of the Phase Composition and Microstructure of the Obtained Ti_2CT_x Mxene

As shown in Figure 1a, the obtained MXene sample contains no significant admixtures of the original Ti_2AlC, and the shift in the reflex position (002) toward lower angles to $2\theta = 4.6°$ indicates the successful delamination of the multilayer aggregates formed by etching the MAX-phase with NaF-HCl. The shift of the reflex associated with the Ti_2CT_x phase for the coating on the specialized substrate toward higher angles to $2\theta = 5.3°$ indicates that the reverse aggregation of the MXene sheets and the decrease in the interlayer distance during vacuum drying occurs during coating formation.

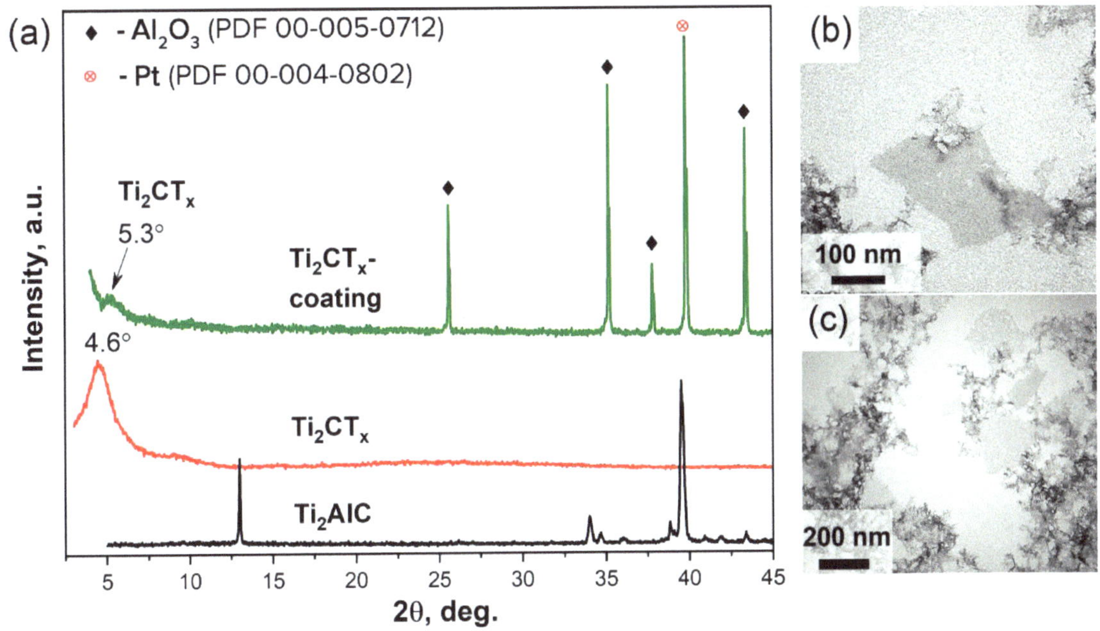

Figure 1. The X-ray patterns of the initial Ti_2AlC, low-layer Ti_2CT_x MXene and its coating on a specialized sensor substrate (**a**), and TEM microphotographs of Ti_2CT_x MXene (**b**,**c**).

The TEM data indicate partial destruction of the MXene sheets (Figure 1b,c) while holding it in aqueous dispersion, as well as the appearance of loose nanoparticles on the sheet boundaries, probably related to oxidation products—titanium dioxide in different crystalline modification.

3.2. Raman Spectroscopy

In situ Raman spectra recording for Ti_2CT_x MXene films on an Al_2O_3 chip with Pt interdigital microelectrodes (Figure 2a) was performed both for platinum microelectrodes and for the Al_2O_3 area between them (Figure 2b). The results of which can be attributed to the inhomogeneity of the prepared film. It should be noted that modes on the Raman spectra characteristic of the MXene phase are more intensive on platinum.

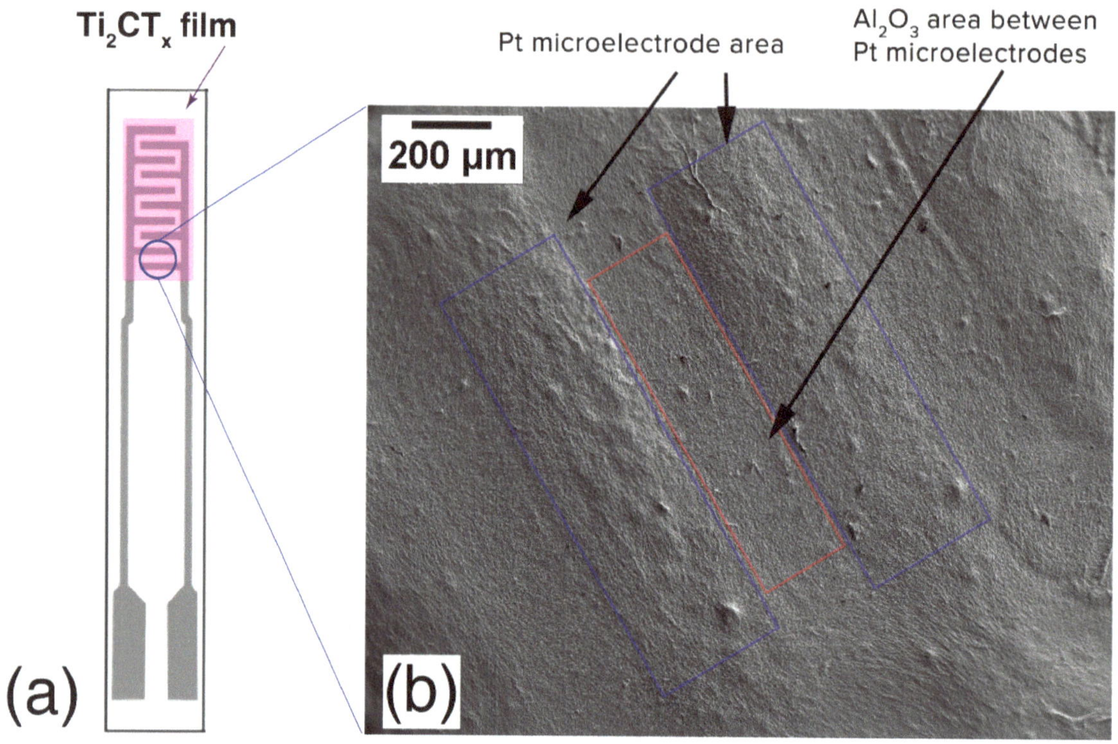

Figure 2. The scheme of an Al_2O_3 substrate with Pt microelectrodes and Ti_2CT_x MXene film (**a**) and a microphotograph of the Ti_2CT_x film on the substrate surface with an indication of the areas where in situ Raman spectra were measured (**b**).

3.2.1. MXene Raman Spectra for the Platinum Electrodes Area

A Ti_2CT_x MXene film is characterized by three normal modes, ω_1, ω_2 and ω_3, in Raman spectra, located at 270, 380 and 697 cm^{-1}, respectively (Figure 3a). Modes ω_1-ω_3 can be attributed both to vibrations of MAX-phase Ti_2AlC, remaining after synthesis, and to Ti_2CT_x MXene itself [32–35]. Analysis of the spectrum obtained at RT also reveals other modes: ω_{R1} and ω_{R2} modes are located at 447 and 821 cm^{-1}, which can be attributed to high convergence to E_g and B_{2g} modes characteristic of rutile TiO_2 [36,37]. For the rutile phase, the largest intensity should be observed for E_g and A_{1g} modes with maxima at ~448 and 613 cm^{-1}, respectively, while B_{2g} should be the least intensive of all first-order modes characteristic of rutile [38]. In this case, however, the ω_{R2} (B_{2g}) mode is the most intensive of all rutile phase modes, which is highly irregular for rutile TiO_2 materials in the literature [38,39]. Such behavior allows us to suppose that the obtained MXene at RT does not include a separate rutile TiO_2 phase, with observed ω_{R1} (E_g) and ω_{R2} (B_{2g}) modes arising from Ti–O bonds, with energies close to analogous bonds in rutile, on the MXene surface. Moreover, the observed rutile species on the surface of Ti_2CT_x MXene likely formed as a result of keeping the synthesized sample in water suspension.

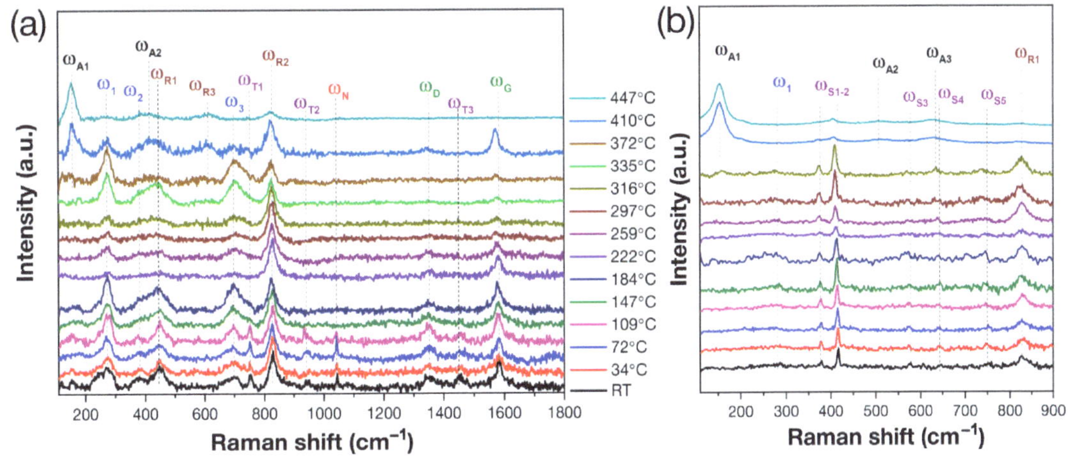

Figure 3. In situ Raman spectra at different temperatures (20–447 °C) of the thin film at different substrate areas: on Pt (**a**) and Al_2O_3 (**b**).

In the more long-wave region of the spectrum, ω_D and ω_G modes are located at 1349 and 1583 cm^{-1}, which are characteristic carbon D- and G-bands [32], which are common for many carbon systems with sp^2 hybridization of the C−C bond. Therefore, they can be attributed to carbon in MXene as well as to graphene layers resulting from the excessive etching of Ti_2AlC by hydrofluoric acid and the partial removal of titanium atoms beside aluminum atoms [40–42].

In addition to the main MXene modes, low-intensity modes ω_{T1}, ω_{T2} and ω_{T3} were found at 754, 935 and 1454 cm^{-1}, respectively. Modes ω_{T1}-ω_{T3} are characteristic of tetramethylammonium hydroxide (TMAOH) [43], fragments of which had likely been captured in MXene interlayer space after delamination and washing. At 1042 cm^{-1}, the ω_N mode is also present, attributed to the nitro group (NO_3^-) [44,45], which could have formed on the MXene surface after the gas-sensing properties studies in an atmosphere containing NO_2. The possibility of gaseous NO_2 transformation into an NO_3- group on the surface of the $Ti_3C_2T_x/TiO_2$ material has been considered in studies [46,47].

The described set of modes is retained by the sample under heating up to 109 °C. At higher temperatures, the spectra undergo significant changes. Starting with 147 °C, no ω_{T1}-ω_{T3} and ω_N modes (characteristic of TMAOH and NO_3^- groups, respectively) appear in the spectra. This can be attributed to their desorption and/or decomposition with the following desorption of gaseous decomposition products from the MXene surface. Heating to $\geq$147 °C results in the overlap of ω_2 mode with the rutile ω_{R1} band, yielding a new widened peak. Further heating leads to the transformation of this peak with maximum shifting to a lower wave number. At 410 °C, the maximum of this phase is located at 420 cm^{-1}, which is close to the B_{1g} mode of the titanium dioxide anatase phase [38]. This might indicate partial restructuring of the crystal lattice. Modes characteristic of Ti_2CT_x (ω_1 and ω_3), as well as of TiO_2 rutile (ω_{R2}), still remain on the film spectra, with their maxima not undergoing any changes.

The most intensive E_g mode at 156 cm^{-1} (ω_{A1}), characteristic of anatase, starts to appear on spectra at 372 °C. The intensity of this mode rises with the temperature, which indicates that MXene is being oxidized to titanium dioxide. The A_{1g} mode at 608 cm^{-1} (ω_{R3}), which is the most intensive mode of the rutile phase [38], also appears on the Raman spectra at temperatures $\geq$ 372 °C as a result of the oxidation process. The Raman spectrum at 447 °C significantly differs from those recorded for the MXene film at lower temperatures, having the following set of bands: ω_{A1} (E_g) and ω_{A2} (B_{1g}) at 156 and 396 cm^{-1} from the anatase phase and ω_{R3} (A_{1g}) and ω_{R2} (B_{2g}) at 615 and 821 cm^{-1} from the rutile phase. The obtained values are in good agreement with the literature [37,38].

D- and G-modes are retained up to the temperature of 447 °C. An increase in temperature gives rise to the systematic shift of the G-mode to lower wave numbers. Thus, at 410 °C, the G-mode maximum is situated at 1570 cm^{-1}. The shift of the maximum position can be related to temperature effects arising from in situ heating of the sample.

Thus, analysis of the recorded Raman spectra for MXene film on platinum allows us to infer that the oxidation of MXene begins at 372 °C. It is at this temperature that the Ti_2CT_x/TiO_2 composite with a large content of rutile and anatase phases of TiO_2 forms. At the temperature of 447 °C, the Ti_2CT_x/TiO_2 film fully oxidizes TiO_2, consisting of rutile and anatase.

3.2.2. Raman Spectra for the Al_2O_3 Area

Considering the features of sensing studies, it is the Raman spectra of MXene on a ceramic Al_2O_3 surface between platinum microelectrodes that are of the most interest since chemoresistive signals of the gas-sensitive film arise from this area of the sample. We have observed that Ti_2CT_x spectra recorded on local areas on the Al_2O_3 surface significantly differ from those recorded for MXene film areas situated on the surface of platinum electrodes. We were able to obtain intensive signals for Ti_2CT_x on the surface of the platinum electrodes related to vibrations of various phases. On the Al_2O_3 surface, however, most modes are of low intensity, and for the initial TI_2CT_x spectrum, large background luminescence was observed, which obstructs registering Raman scattering. For this reason, only the region up to 900 cm^{-1} is shown in the graph with characteristic MXene modes and oxidation products.

On all spectra of the MXene film (Figure 3b), the characteristic set of modes ω_{S1}-ω_{S5}, typical for α-Al_2O_3 substrate, can be seen at 377, 417, 574, 644 and 749 cm^{-1} [48]. At RT, only weak ω_1 mode of MXene at 276 cm^{-1} and intense mode ω_{R2} (B_{2g}) at 828 cm^{-1} from rutile (these modes were the most intense for Raman spectra of MXene on the platinum surface—see Section 3.2.1) are observed. No other modes, characteristic of MXene, were found for the film on Al_2O_3's surface due to their low intensity and background luminescence. The described set of modes is retained up to 316 °C when the E_g (1) mode at 156 cm^{-1} (ω_{A1}) from anatase, as well as E_g (3) at 634 cm^{-1} (ω_{A3}), appear, with the latter mode not observed in spectra for the platinum surface. The obtained data are in good agreement with those in the literature [38]. It should be noted that the start of anatase formation owing to MXene oxidation already starts at 316 °C on the Al_2O_3 substrate, while for the platinum substrate, the same is only observed at 372 °C. Further increases in temperature to 410–447 °C result in full oxidation of MXene to TiO_2. Modes characteristic of anatase can be observed in this case on the Raman spectra: strong E_g (1) at 156 cm^{-1} (ω_{A1}) and weaker $B_{1g}(2)$ + A_{1g} at 510 (ω_{A2}) and $E_g(3)$ at 634 cm^{-1} (ω_{A3}), as well as B_{2g} at 828 cm^{-1} (ω_{R2}) from the rutile phase [38]. It should be noted that on Al_2O_3's surface, the oxidation of Ti_2CT_x MXene can be seen to proceed with the formation of mostly anatase phase. Only the weak B_{2g} mode can be attributed to rutile, which was present from the start. In the case of the final Raman spectra for platinum substrate, ω_{R1} (E_g) and ω_{R3} (A_{1g}) of the rutile phase can be observed. These seeming differences in oxidation behavior of the Ti_2CT_x MXene-receptive layer can be connected to the effect of the substrate material used for MXene film deposition on the recorded Raman spectra.

For further studies of the gas-sensing properties of the prepared Ti_2CT_x/TiO_2 nanocomposite, we have chosen the film oxidized at the minimal temperature of 316 °C.

3.3. Microstructure

SEM micrographs for films on the special Al_2O_3/Pt substrate are given in Figure 4. The initial MXene film (Figure 4a,b) exhibits the folded wavelike microstructure: the surface is reminiscent of sea crests. The length of such crests can reach several microns. It can be seen from the micrograph that the film surface is rather developed, which is important for gas-sensing applications, which require a high specific surface area for better adsorption of gases from the atmosphere. Separate nanoparticles can also be seen on the micrographs, having a mostly spherical shape, with sizes of about 200 nm. Such particles might consist of impurities.

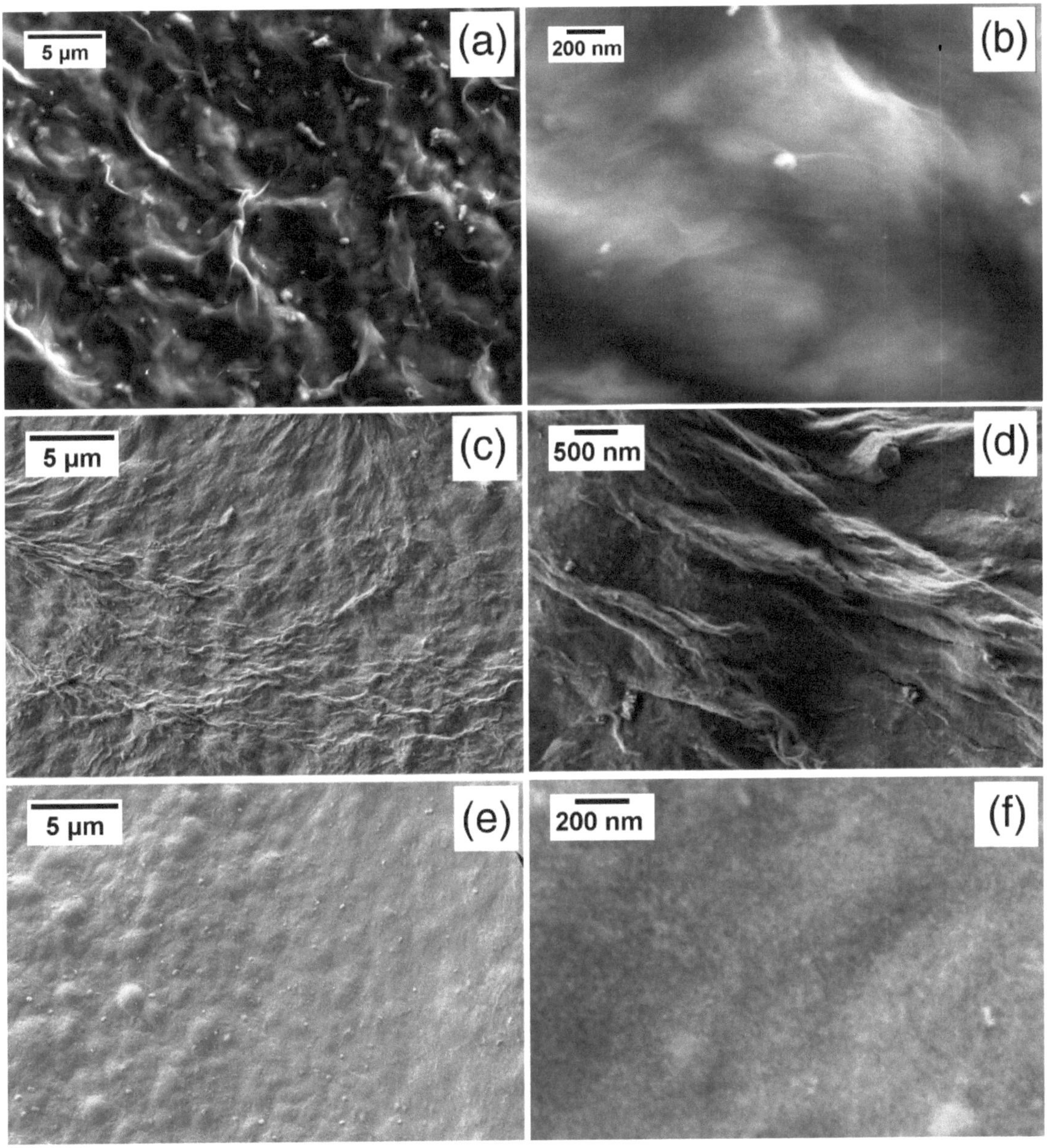

Figure 4. The SEM micrographs of films on a special Al_2O_3/Pt substrate: initial MXene (**a**,**b**), oxidized at 316 °C (**c**,**d**) and 447 °C (**e**,**f**).

Heating up to 316 °C gives rise to significant microstructural changes in the MXene film (Figure 4c,d). It becomes more uniform, although the wavelike microstructure is retained. Spherical particles 1–3 μm in diameter start to appear, most likely being aggregates of

TiO_2 in rutile crystal modification. The enlargement of TiO_2 particles can be attributed to the aggregation of smaller particles, as well as to overall phase transformation due to heat treatment of the films. Judging from Raman spectroscopy data given earlier, a nanocomposite is formed under such temperatures, consisting of the MXene phase and TiO_2.

Further increases in heat treatment temperature to 447 °C result in even more pronounced changes in the microstructure (Figure 4e,f). The wavelike structure fully gives way to a smoother surface, consisting mostly of spherical aggregates 1–3 μm in diameter. Smaller spherical nanoparticles, 20–50 nm in size, can be seen from micrographs with higher magnification (Figure 4f). Raman spectroscopy analysis suggests that under such temperatures, the TiO_2 film forms, consisting of a rutile and anatase phases mixture.

The microstructure of MXene films after their heat treatment at 316 and 447 °C was additionally studied using AFM. Topography scans are given in (Figure 5a,b), as well as the distribution of surface potential for the sample of MXene film oxidized at 316 °C. Topography results from AFM are in good agreement with SEM results (Figure 4c,d). Areas with spherical aggregates 2–4 μm in size, as well as narrow (width of ~100–250 nm) and elongated wavelike structures, roughly 100 nm in size, can be seen. Surface potential values for different film areas differ noticeably, with a work function changing varying from 4.57 to 5.00 eV, which is a consequence of film inhomogeneity. The mean work function value was 4.71 eV, which is rather close to the value for individual Ti_2CT_x MXene with a varyingly functionalized surface: 4.5–4.98 eV [49,50]. Thus, the obtained AFM results correlate well with Raman spectroscopy and SEM data and also confirm the formation of the Ti_2CT_x/TiO_2 nanocomposite.

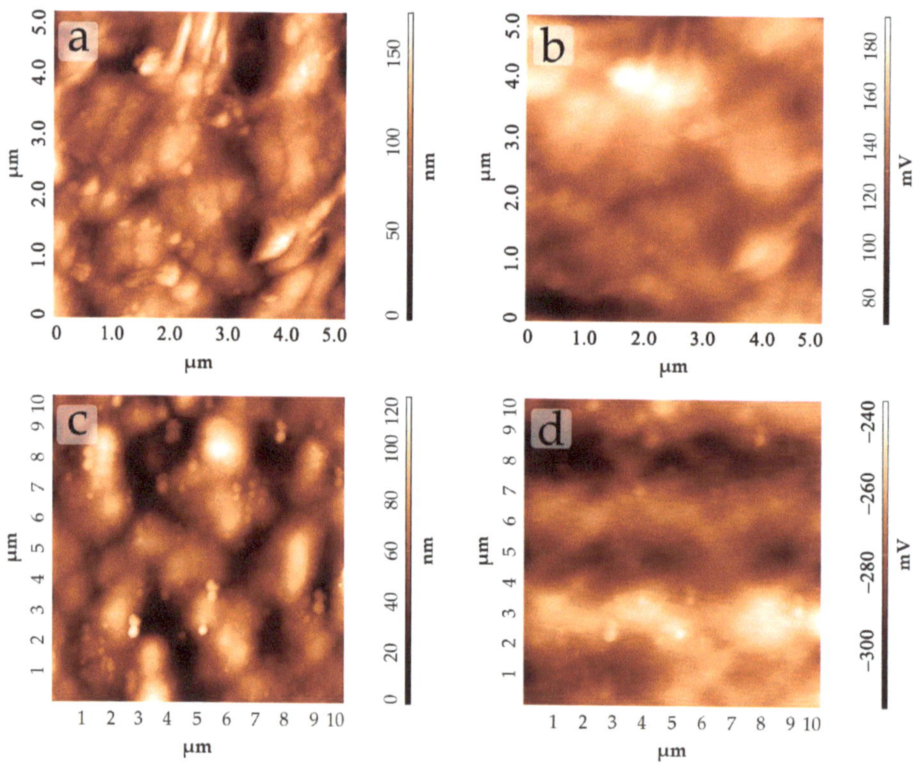

Figure 5. The AFM results: topography (**a**) and surface potential distribution map (**b**) of Ti_2CT_x/TiO_2 MXene film oxidized at 316 °C; topography (**c**) and surface potential distribution map (**d**) of Ti_2CT_x/TiO_2 film oxidized at 447 °C.

The topography scans are given in (Figure 5c,d), as well as the distribution of the surface potential for the sample of the MXene film oxidized at 447 °C. The obtained film surface is relatively smooth, consisting of low (50–80 nm) spherical aggregates 2–4 μm in size, with small spherical agglomerates 200–400 nm in size situated on their edges and in between them. Under higher magnification and in the phase contrast regime, nanoparticles with a shape close to spherical, size of ~45–95 nm, were found, which is in agreement with the SEM results. The mean square roughness on an area of 100 μm^2 is just 20 nm. It can be seen from the map of the potential surface distribution obtained from KPFM (Figure 5d) that charge carriers are spread relatively uniformly on the film surface. The mean work function value for all scanned areas was 5.00 eV. This value is in good agreement with those reported in the literature for anatase [51]. For anatase, the work function lies in the range of 4.94–5.07 eV, while for rutile, it is ~4.80 eV. Thus, AFM also confirms the formation of TiO_2 with mostly anatase crystal structure.

3.4. Gas-Sensing Chemoresistive Properties

The gas-sensitive chemoresistive properties at room temperature were studied for the initial Ti_2CT_x MXene film as well as the Ti_2CT_x/TiO_2 film oxidized in situ in a Raman spectroscopy cell at 316 °C. A further increase in the operating temperature (including in an atmosphere of high humidity) led to a significant increase in electrical resistance (R > 1 GOhm), which made it impossible to measure the gas-sensitive properties.

In the first stage, the obtained Ti_2CT_x and Ti_2CT_x/TiO_2 films had their baseline resistances measured in an atmosphere of dry air and dry nitrogen. In both cases, the resistance of the films exceeded 1 GOm, which did not allow their gas-sensitive properties to be measured in these atmospheres. The Ti_2CT_x and Ti_2CT_x/TiO_2 films showed increased chemoresistance sensitivity to humidity: a significant decrease in the baseline was observed as the RH increased. Figure 6a shows the responses to humidity variations in the air atmosphere. As can be seen, when the RH was increased from 30% to 40%, 50%, 64% and 93%, the response value increased from 0% to 46%, 67%, 83% and 96% and to 50%, 76%, 87% and 91% for Ti_2CT_x and Ti_2CT_x/TiO_2, respectively. The initial Ti_2CT_x film showed a higher sensitivity to humidity at elevated RH values (93%). The high MXenes sensitivity to humidity is typical and is well described in the literature. For further gas-sensitive measurements, a gas atmosphere with constant humidity RH = 50% was chosen.

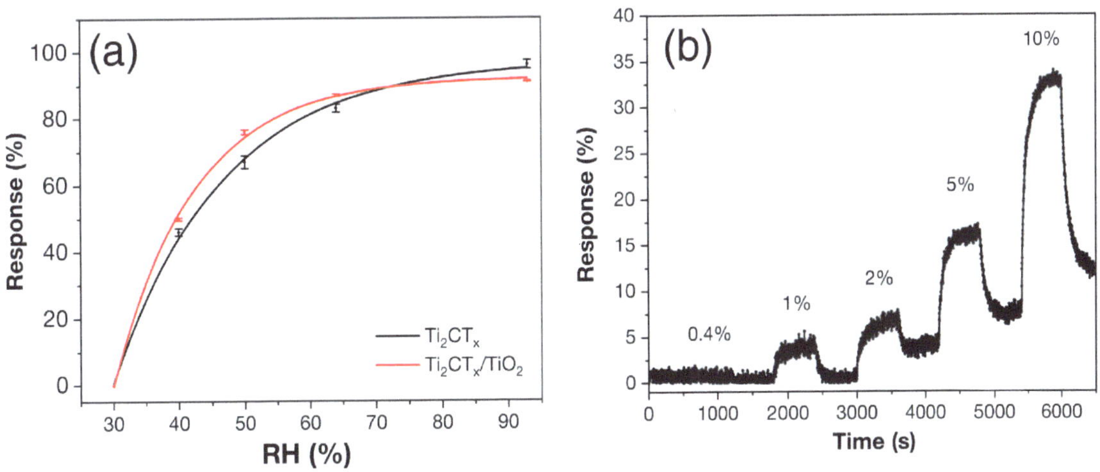

Figure 6. The responses of Ti_2CT_x and Ti_2CT_x/TiO_2 films to humidity (**a**) and Ti_2CT_x films to 0.4–10% O_2 at 50%RH (**b**).

In the following step, the chemoresistive responses were determined for the detection of 100 ppm CO, NH_3, benzene (C_6H_6), acetone (C_3H_6O), ethanol (C_2H_5OH) and 1000 ppm methane (CH_4), H_2. The experimental responses (S1) to the above gases are shown in Figure 7a,b. The selectivity diagrams (Figure 7c,d) are plotted from the obtained response array. The bar chart (Figure 7c) shows the sign of the resulting response: a positive value corresponds to an increase in electrical resistance when the analyte gas is injected (*p*-response), and a negative value corresponds to a decrease (*n*-response). The radar chart (Figure 7d) shows the response values (S1 in %) for all analyzed gases.

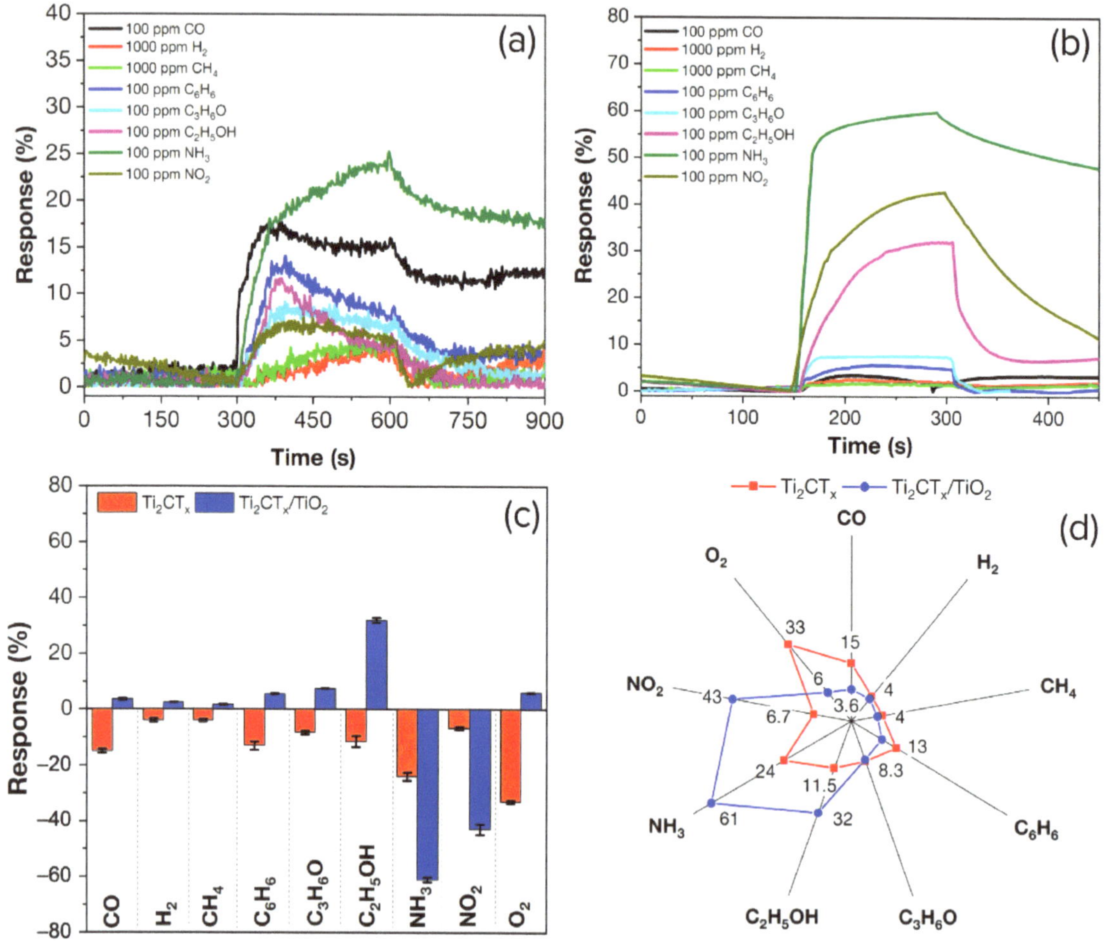

Figure 7. The gas-sensing properties of the samples at RT and 50%RH: responses to 100 and 1000 ppm of different gases of the Ti_2CT_x film (**a**) and Ti_2CT_x/TiO_2 (**b**); selectivity diagrams plotted from the responses to the different gases (**c**,**d**).

For the initial Ti_2CT_x MXene film containing its oxidation products in water dispersion, the highest response (24%) was observed when NH_3 was detected. Notable responses were recorded for CO (15%), C_6H_6 (13%), C_2H_5OH (11.5%) and C_3H_6O (8.3%); the response for H_2 and CH_4 detection was 4%. The sensitivity to 0.4–10% oxygen was also additionally measured (Figure 6b). A gradual increase in the response (S1) from 4% to 33% was observed when the oxygen concentration was increased from 1% to 10%. The response to oxygen by MXenes is practically not described in the literature but was nevertheless recorded by us in our previous study [29]. When all the above gases were injected, regardless of their chemical nature, the resistance decreased, i.e., an *n*-response was observed.

A significant change in gas-sensitive properties was observed for the Ti_2CT_x/TiO_2 composite film obtained by oxidation in air at 316 °C. The highest response (61%) was recorded for NH_3 detection, as for the initial Ti_2CT_x film. Further, notable responses were obtained for the detection of C_2H_5OH (32%) and NO_2 (43%), and the response for all other gases did not exceed 7.6%. As can be clearly seen from the selectivity radar chart (Figure 7d), the oxidized Ti_2CT_x/TiO_2 film showed an increased response to all analyzed gases (except oxygen) compared to the Ti_2CT_x film. Especially noticeable is the sensitivity to NO_2 (the response increased by 10 times), as well as a completely absent response to oxygen. In contrast to the Ti_2CT_x film, for the air-oxidized Ti_2CT_x/TiO_2 film, a different type of response was observed for the detection of almost all gases (except for NH_3 and NO_2, to which the maximum response was observed). When detecting CO, C_6H_6, C_3H_6O, C_2H_5OH, CH_4, H_2 and O_2, a *p*-type response was obtained, and when detecting nitrogen-containing gases NH_3 and NO_2 an *n*-type response was obtained.

Figure 8 shows the response of 4–100 ppm NH_3 and the signal reproducibility when detecting 10 ppm NH_3 by the Ti_2CT_x/TiO_2 composite film. As can be seen (Figure 8a), with an increase in NH_3 concentration from 4 to 100 ppm, there is a gradual increase in the response (S1) from 16% to 61% with a noticeable drift of the baseline (which can be explained by the high interaction energy of the ammonia molecule and receptor material), and the response itself is well reproduced (Figure 8b)

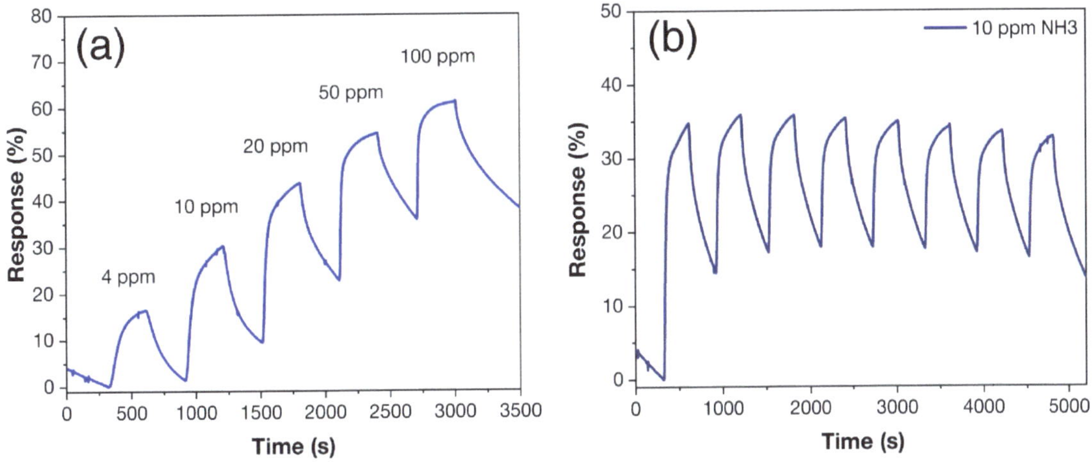

Figure 8. The gas-sensitive properties of the samples at RT and 50%RH. Responses of Ti_2CT_x/TiO_2 films to 4–100 ppm NH_3 (**a**) and 10 ppm NH_3 (**b**).

During the consideration of the mechanisms of the detection of various gases by the obtained materials, it is necessary to take into account the fact that all measurements were carried out in an atmosphere of 50% RH. Therefore, it can be assumed that the MXene surface, in this case, contains a large amount of sorbed water, as well as hydroxyl groups (OH^-), i.e., hydroxyl groups can be directly involved in reactions with various gases that precede the chemoresistive response.

In the present work, only the *n*-type response was observed for the initial Ti_2CT_x sample. When detecting the reducing gases (all gases in this work except NO_2 and O_2), this behavior is typical for *n*-type metal oxide semiconductors [52] but is not typical for $Ti_{n+1}C_nT_x$ MXenes, for which a *p*-type response is most often observed [15,18,53]. Raman spectroscopy and TEM data revealed that the surface of the initial MXene sheets contains oxygen groups, and an impurity of the rutile TiO_2 phase was noted. Thus, in addition to the hydroxyl functional groups of MXene, as well as sorbed water molecules, TiO_2 particles may also be involved in the detection mechanism. The presence of the *n*-type response detected in our study is not fully understood at this time and requires further investigation.

After oxidation of the initial MXene film in air to Ti_2CT_x/TiO_2 composite, a change in the response character from *n*- to *p*-type is observed when detecting CO, H_2 and VOC's. According to Raman spectroscopy data, it is found that a single *n*-type TiO_2 semiconductor phase is formed for this sample. The surface microstructure of the resulting composite also changes. As a rule, large-band-gap semiconductors, such as TiO_2, have poor electrical conductivity at room temperature. At room temperature, it is the MXene phase, which is characterized by *p*-type conductivity, that makes the main contribution to the gas sensitivity. The increase in sensing signals can be related to the fact that at the Ti_2CT_x/TiO_2 interface, a Schottky barrier appears, the height of which will change during the adsorption of different gases [23].

Table 1 provides information on the gas-sensitive properties of composites of MXenes $Ti_3C_2T_x/TiO_2$ and Ti_2CT_x/TiO_2 operating at room temperature for chemoresistive gas sensors reported in the literature. As can be seen, in the field of gas-sensing composites based on $Ti_3C_2T_x$, MXenes are mainly used, and the sensory properties of Ti_2CT_x MXenes and nanocomposite Ti_2CT_x/TiO_2 are practically not studied. For many of the composites presented in Table 1, it is ammonia that shows the greatest response. This corresponds to the data of theoretical calculations [54] for the maximum oxidized compound, Ti_2CO_2, which is associated with high adsorption energy and the calculated value of charge transfer. A comparison of the experimental data on the gas-sensitive properties shows that the materials obtained in this work have a sufficiently high response (to a wide range of NH_3 concentrations), significantly exceeding most of the literature analogs. The increased response can be attributed to the more flexible heat treatment conditions used in this work under the control of in situ Raman spectroscopy for the controlled formation of heterojunctions between Ti_2CT_x and TiO_2. In addition, the responses obtained in this work were recorded at relative humidity RH = 50% (closest to the real conditions), which has not been noted in other works for TiO_2-modified MXenes.

Table 1. Comparison of gas-sensitive characteristics of chemoresistive gas sensors based on $Ti_3C_2T_x/TiO_2$ and Ti_2CT_x/TiO_2 MXenes operating at room temperature that is presented in the literature.

No.	Year	Composition	Target Gas	Conc., ppm	Response	RH, %	Ref.
1	2019	$Ti_3C_2T_x/TiO_2$	NH_3	10 ppm	3%	60	[27]
2	2020	Ti_2CT_x/TiO_2	NH_3	10 ppm	1.9%	0	[25]
3	2020	$Ti_3C_2T_x/TiO_2$	NO_2	5 ppm	16.05%	0	[46]
4	2021	$Ti_3C_2T_x/TiO_2$	C_2H_5OH	100 ppm	22.47%	0	[55]
5	2021	$Ti_3C_2T_x/TiO_2$	hexanal	100 ppm	8.8%	0	[56]
6	2021	$Ti_3C_2T_x/TiO_2$	NO_2	100 ppm	4%	0	[26]
7	2021	$Ti_3C_2T_x$ (N-doped)/TiO_2	NH_3	200 ppb	7.3%	0	[57]
8	2022	$Ti_3C_2T_x/TiO_2$	C_2H_5OH	90 ppm	91 a.u.	0	[24]
9	2022	Ti_2CT_x/TiO_2	NH_3	4–100 ppm	16–61%	50	This work

4. Conclusions

The in situ oxidation of the film of layered Ti_2CT_x MXene synthesized as a result of the exposure of the MAX-phase of Ti_2AlC to sodium fluoride solution in hydrochloric acid, delaminated by tetramethylammonium hydroxide and held in water dispersion, has been studied, in detail, using Raman spectroscopy. The phenomenon of substrate (Al_2O_3 ceramics or platinum) influence on the intensity of Ti_2CT_x MXene phase modes, as well as the modes formed during oxidation in the air atmosphere of rutile and anatase phases, has been noted. It is shown that the formation of a new oxidation product phase (TiO_2 with anatase structure) on the ceramic Al_2O_3 substrate can be fixed at a lower temperature (316 °C) than it is typical for the Ti_2CT_x layer on the platinum electrodes of the specialized substrate (372 °C). It was found that at 447 °C the full oxidation of the initial Ti_2CT_x MXene under the conditions of heating in an air atmosphere occurs.

The coating microstructure was studied using a complex of methods (SEM, TEM and AFM), and the work function was determined for both the Ti_2CT_x/TiO_2 composite obtained at 316 °C and the fully oxidized composition using Kelvin-probe force microscopy.

It was found that the obtained Ti_2CT_x and Ti_2CT_x/TiO_2 films exhibit chemoresistive responses when detecting a wide group of gases, H_2, CO, NH_3, C_6H_6, C_3H_6O, CH_4, C_2H_5OH and O_2, at room temperature and 50% relative humidity. It was shown that partial oxidation of MXene favors the gas-sensitive properties, and the most sensitivity was observed for NH_3.

Author Contributions: Conceptualization, A.S.M. and E.P.S.; Methodology, I.A.N. and E.P.S.; Validation, E.P.S.; Formal analysis, P.Y.G., A.A.A., T.L.S. and N.P.S.; Investigation, A.S.M.; Writing—original draft, A.S.M.; Writing—review & editing, E.P.S. and N.T.K.; Visualization, A.S.M.; Supervision, N.T.K.; Funding acquisition, N.P.S. All authors have read and agreed to the published version of the manuscript.

Funding: This work was supported by the Russian Science Foundation, project No. 21-73-10251, https://rscf.ru/en/project/21-73-10251/ (accessed on 29 November 2022).

Institutional Review Board Statement: Not applicable.

Informed Consent Statement: Not applicable.

Data Availability Statement: Not applicable.

Conflicts of Interest: The authors declare that they have no known competing financial interest or personal relationships that could have appeared to influence the work reported in this paper.

References

1. Naguib, M.; Mashtalir, O.; Carle, J.; Presser, V.; Lu, J.; Hultman, L.; Gogotsi, Y.; Barsoum, M.W. Two-Dimensional Transition Metal Carbides. *ACS Nano* **2012**, *6*, 1322–1331. [CrossRef] [PubMed]
2. Naguib, M.; Mochalin, V.N.; Barsoum, M.W.; Gogotsi, Y. 25th Anniversary Article: MXenes: A New Family of Two-Dimensional Materials. *Adv. Mater.* **2014**, *26*, 992–1005. [CrossRef] [PubMed]
3. Naguib, M.; Kurtoglu, M.; Presser, V.; Lu, J.; Niu, J.; Heon, M.; Hultman, L.; Gogotsi, Y.; Barsoum, M.W. Two-Dimensional Nanocrystals Produced by Exfoliation of Ti 3AlC 2. *Adv. Mater.* **2011**, *23*, 4248–4253. [CrossRef] [PubMed]
4. Kumar, K.S.; Choudhary, N.; Jung, Y.; Thomas, J. Recent Advances in Two-Dimensional Nanomaterials for Supercapacitor Electrode Applications. *ACS Energy Lett.* **2018**, *3*, 482–495. [CrossRef]
5. Aslam, M.K.; Xu, M. A Mini-Review: MXene Composites for Sodium/Potassium-Ion Batteries. *Nanoscale* **2020**, *12*, 15993–16007. [CrossRef]
6. Zhong, Y.; Xia, X.H.; Shi, F.; Zhan, J.Y.; Tu, J.P.; Fan, H.J. Transition Metal Carbides and Nitrides in Energy Storage and Conversion. *Adv. Sci.* **2016**, *3*, 1500286. [CrossRef]
7. Khan, K.; Tareen, A.K.; Aslam, M.; Zhang, Y.; Wang, R.; Ouyang, Z.; Gou, Z.; Zhang, H. Recent Advances in Two-Dimensional Materials and Their Nanocomposites in Sustainable Energy Conversion Applications. *Nanoscale* **2019**, *11*, 21622–21678. [CrossRef]
8. Li, K.; Zhang, S.; Li, Y.; Fan, J.; Lv, K. MXenes as Noble-Metal-Alternative Co-Catalysts in Photocatalysis. *Chin. J. Catal.* **2020**, *42*, 3–14. [CrossRef]
9. Xie, X.; Zhang, N. Positioning MXenes in the Photocatalysis Landscape: Competitiveness, Challenges, and Future Perspectives. *Adv. Funct. Mater.* **2020**, *30*, 2002528. [CrossRef]
10. Zhao, Q.; Zhou, W.; Zhang, M.; Wang, Y.; Duan, Z.; Tan, C.; Liu, B.; Ouyang, F.; Yuan, Z.; Tai, H.; et al. Edge-Enriched $Mo_2TiC_2T_x/MoS_2$ Heterostructure with Coupling Interface for Selective NO_2 Monitoring. *Adv. Funct. Mater.* **2022**, *32*, 2203528. [CrossRef]
11. Wang, H.; Shi, X.; Liu, F.; Duan, T.; Sun, B. Non-Invasive Rapid Detection of Lung Cancer Biomarker Toluene with a Cataluminescence Sensor Based on the Two-Dimensional Nanocomposite Pt/$Ti_3C_2T_x$-CNT. *Chemosensors* **2022**, *10*, 333. [CrossRef]
12. Choi, S.J.; Kim, I.D. Recent Developments in 2D Nanomaterials for Chemiresistive-Type Gas Sensors. *Electron. Mater. Lett.* **2018**, *14*, 221–260. [CrossRef]
13. Lee, E.; Kim, D.-J. Review—Recent Exploration of Two-Dimensional MXenes for Gas Sensing: From a Theoretical to an Experimental View. *J. Electrochem. Soc.* **2020**, *167*, 037515. [CrossRef]
14. Devaraj, M.; Rajendran, S.; Hoang, T.K.A.; Soto-Moscoso, M. A Review on MXene and Its Nanocomposites for the Detection of Toxic Inorganic Gases. *Chemosphere* **2022**, *302*, 134933. [CrossRef]

15. Kim, S.J.; Koh, H.J.; Ren, C.E.; Kwon, O.; Maleski, K.; Cho, S.Y.; Anasori, B.; Kim, C.K.; Choi, Y.K.; Kim, J.; et al. Metallic $Ti_3C_2T_x$ MXene Gas Sensors with Ultrahigh Signal-to-Noise Ratio. *ACS Nano* **2018**, *12*, 986–993. [CrossRef]
16. Ji, H.; Zeng, W.; Li, Y. Gas Sensing Mechanisms of Metal Oxide Semiconductors: A Focus Review. *Nanoscale* **2019**, *11*, 22664–22684. [CrossRef]
17. Khakbaz, P.; Moshayedi, M.; Hajian, S.; Soleimani, M.; Narakathu, B.B.; Bazuin, B.J.; Pourfath, M.; Atashbar, M.Z. Titanium Carbide MXene as NH_3 Sensor: Realistic First-Principles Study. *J. Phys. Chem. C* **2019**, *123*, 29794–29803. [CrossRef]
18. Majhi, S.M.; Ali, A.; Greish, Y.E.; El-Maghraby, H.F.; Qamhieh, N.N.; Hajamohideen, A.R.; Mahmoud, S.T. Accordion-like-Ti_3C_2 MXene-Based Gas Sensors with Sub-Ppm Level Detection of Acetone at Room Temperature. *ACS Appl. Electron. Mater.* **2022**, *4*, 4094–4103. [CrossRef]
19. Wang, J.; Xu, R.; Xia, Y.; Komarneni, S. Ti_2CT_x MXene: A Novel p-Type Sensing Material for Visible Light-Enhanced Room Temperature Methane Detection. *Ceram. Int.* **2021**, *47*, 34437–34442. [CrossRef]
20. Zhang, C.; Zhang, Y.; Cao, K.; Guo, Z.; Han, Y.; Hu, W.; Wu, Y.; She, Y.; He, Y. Ultrasensitive and Reversible Room-Temperature Resistive Humidity Sensor Based on Layered Two-Dimensional Titanium Carbide. *Ceram. Int.* **2021**, *47*, 6463–6469. [CrossRef]
21. Koh, H.J.; Kim, S.J.; Maleski, K.; Cho, S.Y.; Kim, Y.J.; Ahn, C.W.; Gogotsi, Y.; Jung, H.T. Enhanced Selectivity of MXene Gas Sensors through Metal Ion Intercalation: In Situ X-Ray Diffraction Study. *ACS Sens.* **2019**, *4*, 1365–1372. [CrossRef] [PubMed]
22. Wu, M.; An, Y.; Yang, R.; Tao, Z.; Xia, Q.; Hu, Q.; Li, M.; Chen, K.; Zhang, Z.; Huang, Q.; et al. V_2CT_x and $Ti_3C_2T_x$ MXenes Nanosheets for Gas Sensing. *ACS Appl. Nano Mater.* **2021**, *4*, 6257–6268. [CrossRef]
23. Pazniak, H.; Plugin, I.A.; Loes, M.J.; Inerbaev, T.M.; Burmistrov, I.N.; Gorshenkov, M.; Polcak, J.; Varezhnikov, A.S.; Sommer, M.; Kuznetsov, D.V.; et al. Partially Oxidized $Ti_3C_2T_x$ MXenes for Fast and Selective Detection of Organic Vapors at Part-per-Million Concentrations. *ACS Appl. Nano Mater.* **2020**, *3*, 3195–3204. [CrossRef]
24. Wang, Z.; Wang, F.; Hermawan, A.; Zhu, J.; Yin, S. Surface Engineering of $Ti_3C_2T_x$ MXene by Oxygen Plasma Irradiation as Room Temperature Ethanol Sensor. *Funct. Mater. Lett.* **2022**, *15*, 2251007. [CrossRef]
25. Sun, Q.; Wang, J.; Wang, X.; Dai, J.; Wang, X.; Fan, H.; Wang, Z.; Li, H.; Huang, X.; Huang, W. Treatment-Dependent Surface Chemistry and Gas Sensing Behavior of the Thinnest Member of Titanium Carbide MXenes. *Nanoscale* **2020**, *12*, 16987–16994. [CrossRef]
26. Liu, S.; Wang, M.; Liu, G.; Wan, N.; Ge, C.; Hussain, S.; Meng, H.; Wang, M.; Qiao, G. Enhanced NO_2 Gas-Sensing Performance of 2D Ti_3C_2/TiO_2 Nanocomposites by In-Situ Formation of Schottky Barrier. *Appl. Surf. Sci.* **2021**, *567*, 150747. [CrossRef]
27. Tai, H.; Duan, Z.; He, Z.; Li, X.; Xu, J.; Liu, B.; Jiang, Y. Enhanced Ammonia Response of $Ti_3C_2T_x$ Nanosheets Supported by TiO_2 Nanoparticles at Room Temperature. *Sens. Actuators B Chem.* **2019**, *298*, 126874. [CrossRef]
28. Simonenko, E.P.; Simonenko, N.P.; Nagornov, I.A.; Simonenko, T.L.; Mokrushin, A.S.; Sevastyanov, V.G.; Kuznetsov, N.T. Synthesis of MAX Phases in the Ti_2AlC-V_2AlC System as Precursors of Heterometallic MXenes $Ti_{2-x}V_xC$. *Russ. J. Inorg. Chem.* **2022**, *67*, 705–714. [CrossRef]
29. Simonenko, E.P.; Simonenko, N.P.; Nagornov, I.A.; Simonenko, T.L.; Gorobtsov, P.Y.; Mokrushin, A.S.; Kuznetsov, N.T. Synthesis and Chemoresistive Properties of Single-Layer MXene Ti_2CT_x. *Russ. J. Inorg. Chem.* **2022**, *67*, 1838–1847. [CrossRef]
30. Mokrushin, A.S.; Simonenko, T.L.; Simonenko, N.P.; Yu, P.; Bocharova, V.A.; Kozodaev, M.G.; Markeev, A.M.; Lizunova, A.A.; Volkov, I.A.; Simonenko, E.P.; et al. Microextrusion Printing of Gas-Sensitive Planar Anisotropic NiO Nanostructures and Their Surface Modification in an H_2S Atmosphere. *Appl. Surf. Sci.* **2022**, *578*, 151984. [CrossRef]
31. Mokrushin, A.S.; Nagornov, I.A.; Simonenko, T.L.; Simonenko, N.P.; Yu, P.; Khamova, T.V.; Kopitsa, G.P.; Evzrezov, A.N.; Simonenko, E.P.; Sevastyanov, V.G.; et al. Chemoresistive Gas-Sensitive ZnO/Pt Nanocomposites Films Applied by Microplotter Printing with Increased Sensitivity to Benzene and Hydrogen. *Mater. Sci. Eng. B* **2021**, *271*, 115233. [CrossRef]
32. Melchior, S.A.; Raju, K.; Ike, I.S.; Erasmus, R.M.; Kabongo, G.; Sigalas, I.; Iyuke, S.E.; Ozoemena, K.I. High-Voltage Symmetric Supercapacitor Based on 2D Titanium Carbide (MXene, Ti_2CT_x)/Carbon Nanosphere Composites in a Neutral Aqueous Electrolyte. *J. Electrochem. Soc.* **2018**, *165*, A501–A511. [CrossRef]
33. Habib, I.; Ferrer, P.; Ray, S.C.; Ozoemena, K.I. Interrogating the Impact of Onion-like Carbons on the Supercapacitive Properties of MXene (Ti_2CT_X). *J. Appl. Phys.* **2019**, *126*, 134301. [CrossRef]
34. Lioi, D.B.; Neher, G.; Heckler, J.E.; Back, T.; Mehmood, F.; Nepal, D.; Pachter, R.; Vaia, R.; Kennedy, W.J. Electron-Withdrawing Effect of Native Terminal Groups on the Lattice Structure of $Ti_3C_2T_x$ MXenes Studied by Resonance Raman Scattering: Implications for Embedding MXenes in Electronic Composites. *ACS Appl. Nano Mater.* **2019**, *2*, 6087–6091. [CrossRef]
35. Spanier, J.E.; Gupta, S.; Amer, M.; Barsoum, M.W. Vibrational Behavior of the $M_{n+1}AX_n$ Phases from First-Order Raman Scattering (M = Ti, V, Cr, A = Si, X = C, N). *Phys. Rev. B* **2005**, *71*, 012103. [CrossRef]
36. Ma, R.; Fukuda, K.; Sasaki, T.; Osada, M.; Bando, Y. Structural Features of Titanate Nanotubes/Nanobelts Revealed by Raman, X-Ray Absorption Fine Structure and Electron Diffraction Characterizations. *J. Phys. Chem. B* **2005**, *109*, 6210–6214. [CrossRef]
37. Ma, H.L.; Yang, J.Y.; Dai, Y.; Zhang, Y.B.; Lu, B.; Ma, G.H. Raman Study of Phase Transformation of TiO_2 Rutile Single Crystal Irradiated by Infrared Femtosecond Laser. *Appl. Surf. Sci.* **2007**, *253*, 7497–7500. [CrossRef]
38. Frank, O.; Zukalova, M.; Laskova, B.; Kürti, J.; Koltai, J.; Kavan, L. Raman Spectra of Titanium Dioxide (Anatase, Rutile) with Identified Oxygen Isotopes (16, 17, 18). *Phys. Chem. Chem. Phys.* **2012**, *14*, 14567–14572. [CrossRef]
39. Zhang, J.Z.; Shen, Y.D.; Li, Y.W.; Hu, Z.G.; Chu, J.H. Composition Dependence of Microstructure, Phonon Modes, and Optical Properties in Rutile TiO_2:Fe Nanocrystalline Films Prepared by a Nonhydrolytic Sol-Gel Route. *J. Phys. Chem. C* **2010**, *114*, 15157–15164. [CrossRef]

40. Gan, Y.; Zhu, F.; Shi, Y.; Wen, Q. Single Frequency Fiber Laser Base on MXene with KHz Linewidth. *J. Mater. Chem. C* **2021**, *9*, 2276–2281. [CrossRef]
41. Murugan, N.; Jerome, R.; Preethika, M.; Sundaramurthy, A.; Sundramoorthy, A.K. 2D-Titanium Carbide (MXene) Based Selective Electrochemical Sensor for Simultaneous Detection of Ascorbic Acid, Dopamine and Uric Acid. *J. Mater. Sci. Technol.* **2021**, *72*, 122–131. [CrossRef]
42. Liu, X.; Ji, L.; Zhu, F.; Gan, Y.; Wen, Q. Linear-Cavity-Based Single Frequency Fiber Laser with a Loop Mirror and Ti_2CT_x Quantum Dots. *Opt. Mater.* **2021**, *122*, 111686. [CrossRef]
43. Tan, Z.; Sato, K.; Ohara, S. Synthesis of Layered Nanostructured TiO_2 by Hydrothermal Method. *Adv. Powder Technol.* **2015**, *26*, 296–302. [CrossRef]
44. Waterland, M.R.; Stockwell, D.; Kelley, A.M. Symmetry Breaking Effects in NO^{3-}: Raman Spectra of Nitrate Salts and Ab Initio Resonance Raman Spectra of Nitrate-Water Complexes. *J. Chem. Phys.* **2001**, *114*, 6249–6258. [CrossRef]
45. Waterland, M.R.; Kelley, A.M. Far-Ultraviolet Resonance Raman Spectroscopy of Nitrate Ion in Solution. *J. Chem. Phys.* **2000**, *113*, 6760–6773. [CrossRef]
46. Choi, J.; Kim, Y.J.; Cho, S.Y.; Park, K.; Kang, H.; Kim, S.J.; Jung, H.T. In Situ Formation of Multiple Schottky Barriers in a Ti_3C_2 MXene Film and Its Application in Highly Sensitive Gas Sensors. *Adv. Funct. Mater.* **2020**, *30*, 2003998. [CrossRef]
47. Wang, Z.; Wang, F.; Hermawan, A.; Asakura, Y.; Hasegawa, T.; Kumagai, H.; Kato, H.; Kakihana, M.; Zhu, J.; Yin, S. SnO-SnO_2 Modified Two-Dimensional MXene $Ti_3C_2T_x$ for Acetone Gas Sensor Working at Room Temperature. *J. Mater. Sci. Technol.* **2021**, *73*, 128–138. [CrossRef]
48. Roy, A.; Sood, A.K. Phonons and Fractons in Sol-Gel Alumina: Raman Study. *Pramana* **1995**, *44*, 201–209. [CrossRef]
49. Xu, J.; Shim, J.; Park, J.H.; Lee, S. MXene Electrode for the Integration of WSe_2 and MoS_2 Field Effect Transistors. *Adv. Funct. Mater.* **2016**, *26*, 5328–5334. [CrossRef]
50. Khazaei, M.; Arai, M.; Sasaki, T.; Ranjbar, A.; Liang, Y.; Yunoki, S. OH-Terminated Two-Dimensional Transition Metal Carbides and Nitrides as Ultralow Work Function Materials. *Phys. Rev. B* **2015**, *92*, 075411. [CrossRef]
51. Mansfeldova, V.; Zlamalova, M.; Tarabkova, H.; Janda, P.; Vorokhta, M.; Piliai, L.; Kavan, L. Work Function of TiO_2 (Anatase, Rutile, and Brookite) Single Crystals: Effects of the Environment. *J. Phys. Chem. C* **2021**, *125*, 1902–1912. [CrossRef]
52. Deng, Y. Sensing mechanism and evaluation criteria of semiconducting metal oxides gas sensors. In *Semiconducting Metal Oxides for Gas Sensing*; Springer: Singapore, 2019; pp. 23–51. ISBN 9789811358524.
53. Lee, E.; Vahidmohammadi, A.; Prorok, B.C.; Yoon, Y.S.; Beidaghi, M.; Kim, D.J. Room Temperature Gas Sensing of Two-Dimensional Titanium Carbide (MXene). *ACS Appl. Mater. Interfaces* **2017**, *9*, 37184–37190. [CrossRef]
54. Yu, X.F.; Li, Y.C.; Cheng, J.B.; Liu, Z.B.; Li, Q.Z.; Li, W.Z.; Yang, X.; Xiao, B. Monolayer Ti_2CO_2: A Promising Candidate for NH_3 Sensor or Capturer with High Sensitivity and Selectivity. *ACS Appl. Mater. Interfaces* **2015**, *7*, 13707–13713. [CrossRef]
55. Hou, M.; Guo, S.; Yang, L.; Gao, J.; Hu, T.; Wang, X.; Li, Y. Improvement of Gas Sensing Property for Two-Dimensional $Ti_3C_2T_x$ Treated with Oxygen Plasma by Microwave Energy Excitation. *Ceram. Int.* **2021**, *47*, 7728–7737. [CrossRef]
56. Kuang, D.; Wang, L.; Guo, X.; She, Y.; Du, B.; Liang, C.; Qu, W.; Sun, X.; Wu, Z.; Hu, W.; et al. Facile Hydrothermal Synthesis of $Ti_3C_2T_x$-TiO_2 Nanocomposites for Gaseous Volatile Organic Compounds Detection at Room Temperature. *J. Hazard. Mater.* **2021**, *416*, 126171. [CrossRef]
57. Zhou, Y.; Wang, Y.; Wang, Y.; Yu, H.; Zhang, R.; Li, J.; Zang, Z.; Li, X. MXene $Ti_3C_2T_x$-Derived Nitrogen-Functionalized Heterophase TiO_2 Homojunctions for Room-Temperature Trace Ammonia Gas Sensing. *ACS Appl. Mater. Interfaces* **2021**, *13*, 56485–56497. [CrossRef]

chemosensors

MDPI

Article

Tetrafluorosubstituted Metal Phthalocyanines: Study of the Effect of the Position of Fluorine Substituents on the Chemiresistive Sensor Response to Ammonia

Darya Klyamer [1,*], Dmitry Bonegardt [1], Pavel Krasnov [2], Alexander Sukhikh [1], Pavel Popovetskiy [1] and Tamara Basova [1]

1 Nikolaev Institute of Inorganic Chemistry SB RAS, 3 Lavrentiev Pr., 630090 Novosibirsk, Russia
2 International Research Center of Spectroscopy and Quantum Chemistry, Siberian Federal University, 26 Kirensky St., 660074 Krasnoyarsk, Russia
* Correspondence: klyamer@niic.nsc.ru

Abstract: A comparative analysis of the chemiresistive sensor response of thin films of a series of tetrasubstituted phthalocyanines of various metals with F-substituent in peripheral ($MPcF_4$-p, M = Cu, Co, Zn, Pb, VO) and non-peripheral ($MPcF_4$-np) positions in macroring to low concentrations of ammonia (1–50 ppm) was carried out. It was found that $MPcF_4$-p films exhibit a higher sensor response than $MPcF_4$-np ones. A $CoPcF_4$-p film demonstrated a calculated LOD of 0.01 ppm with a recovery time of 215 s, while a $VOPcF_4$-p film had LOD of 0.04 ppm and the recovery time of 270 s. The selectivity test showed that CO_2, ethanol, acetone, benzene, and formaldehyde did not interfere with the determination of ammonia, while H_2S at a concentration of more than 10 ppm could act as an interfering gas. It was shown that, as a result of quantum-chemical calculations, the observed regularities are best described by the interaction of NH_3 with phthalocyanines through the formation of hydrogen bonds between NH_3 and side atoms of the macroring. In the case of $MPcF_4$-p, the NH_3 molecule approaches the macrocycle more closely and binds more strongly than in the case of $MPcF_4$-np. The stronger binding leads to a stronger effect of the ammonia molecule on the electronic structure of phthalocyanine and, as a consequence, on the chemiresistive sensor response of the films to ammonia.

Keywords: metal phthalocyanines; fluorine substituents; ammonia; chemiresistive sensors; DFT calculations

Citation: Klyamer, D.; Bonegardt, D.; Krasnov, P.; Sukhikh, A.; Popovetskiy, P.; Basova, T. Tetrafluorosubstituted Metal Phthalocyanines: Study of the Effect of the Position of Fluorine Substituents on the Chemiresistive Sensor Response to Ammonia. *Chemosensors* **2022**, *10*, 515. https://doi.org/10.3390/chemosensors10120515

Academic Editor: Elisabetta Comini

Received: 6 November 2022
Accepted: 2 December 2022
Published: 4 December 2022

1. Introduction

Ammonia (NH_3) is one of the most toxic gases in nature. It has a toxic effect on the human body, an irritating effect on the mucous membranes and skin, and causes pulmonary edema and severe damage to the nervous system [1,2]. Ammonia is one of the main products in the chemical industry; for example, it is widely used in the production of fertilizers, explosives, medicine and agriculture, as well as the production of polymers, nitric acid, and also as a refrigerant [3,4]. Normally, the gas content in the ambient air should not exceed 2 mg/m^3. The human sense of smell is able to detect the characteristic sharp ammonia smell only at a gas concentration of 37 mg/m^3, when it poses a direct danger to health and life [5,6]. Therefore, it is important to use gas sensors to monitor the ammonia content in the atmosphere [6,7]. The release of large amounts of NH_3 can lead to serious poisoning. Thus, constant monitoring of the amount of ammonia in enterprises or in the surrounding atmosphere is necessary to prevent environmental disasters. This fact prompted the search for new materials for the rapid, selective, and sensitive detection. At the same time, the sensors must be able to operate at variable humidity and temperature.

In addition to the determination of ammonia in the surrounding atmosphere, one of the important directions where analysis of NH_3 concentrations is required is the non-invasive diagnosis of various diseases by the composition of exhaled air. Conclusions about changes in metabolism or the presence of a disease can be made by changing the ratio of substances released during exhalation. For example, an ammonia concentration of more than 1 ppm indicates renal insufficiency in nephritis, atherosclerosis of the renal arteries, or toxic kidney lesions [8,9]. The test for the content of ammonia in exhaled air can also be used to monitor hemodialysis in acute or chronic renal failure [10]. Therefore, the search for new materials that are selective with respect to ammonia and sensitive to such low concentrations is also an urgent task.

Several types of gas analyzers are used to control the concentration of ammonia [11]. Among them, electrochemical ammonia sensors have become the most popular due to the best combination of price and reliability of results [12]. The disadvantage of such ammonia sensors is the dependence of their readings on temperature and humidity. Semiconductor ammonia sensors have a low cost, but the main disadvantages of this type of ammonia sensors include selectivity. Therefore, the search for new materials that can act as active layers of sensor devices is constantly underway.

Thin films of metal oxides (SnO_2, In_2O_3, Ga_2O_3, WO_3, V_2O_5, etc.) [13,14], carbon nanomaterials [15,16], conducting organic polymers [17,18], and metal phthalocyanines [19,20] are most often used as materials of sensitive layers of chemiresistive sensors to detect ammonia. Among these materials, phthalocyanines are distinguished by high sensitivity, exceptional chemical and thermal stability, and the ability to form ordered thin films during deposition by both physical vapor deposition (PVD) [21] and solution methods (e.g., spin coating, Langmuir Blodgett, and drop casting) [22]. The properties of phthalocyanines can be regulated by introducing various central metals and substituents into their aromatic macroring.

In a number of early studies, it was shown that the introduction of electron-acceptor substituents, such as fluorine, led to an increase in the sensitivity of the sensor to electron-donating gases, which include ammonia [21,23,24]. The introduction of fluorine substituents leads to a decrease in the electron density of the aromatic macrocycle and an increase in the oxidative potential of the phthalocyanine molecules [25,26]. Apart from this, the introduction of F-substituents leads to a lower level of LUMO energy, which promotes the injection of electrons and, consequently, affects their semiconductor properties [27,28].

In our previous studies, the chemiresistive sensor response of unsubstituted (MPc) phthalocyanines, tetra- ($MPcF_4$) and hexadecafluorosubstituted ($MPcF_{16}$) phthalocyanines of Cu, Zn, and Co to ammonia was studied and it was shown that the sensor response increased in the following order: MPc < $MPcF_{16}$ < $MPcF_4$. $MPcF_4$ (M = Zn, Co) layers were shown to demonstrate the best sensor performance, with the calculated detection limit of NH_3 reaching to 0.01 ppm and an average response time of 15 sec [21]. It has been mentioned that thin films of tetrafluorinated phthalocyanine of other metals (viz. with M = Pb, VO, Fe) also exhibited sensor response to ammonia, but the detailed characteristics of the sensors were not investigated. In addition, it is known that in the case of tetrafluorosubstituted MPc substituents can be introduced into both peripheral ($MPcF_4$-p) and non-peripheral ($MPcF_4$-np) positions of the aromatic ring of phthalocyanine (Figure 1). At the same time, it should be noted that there are only sporadic works on the study of the sensor properties of $MPcF_4$-np films in the literature [29].

This work is aimed at a comparative analysis of the sensor response of thin films of a series of tetrasubstituted phthalocyanine of various metals with F-substituent in peripheral ($MPcF_4$-p, M = Cu, Co, Zn, Pb, VO) and non-peripheral ($MPcF_4$-np) positions in macroring to low concentrations of ammonia (1–50 ppm). The quantum-chemical calculations are performed to study the interaction nature between the phthalocyanine and analyte molecules and to explain the observed regularities. This contributes to a more complete understanding of the relationship between structure and properties, and allows the selection of phthalocyanine films with the most attractive properties (sensitivity, detection limits, response and recovery times) for the development of active layers of chemiresistive gas sensors for the determination of ammonia.

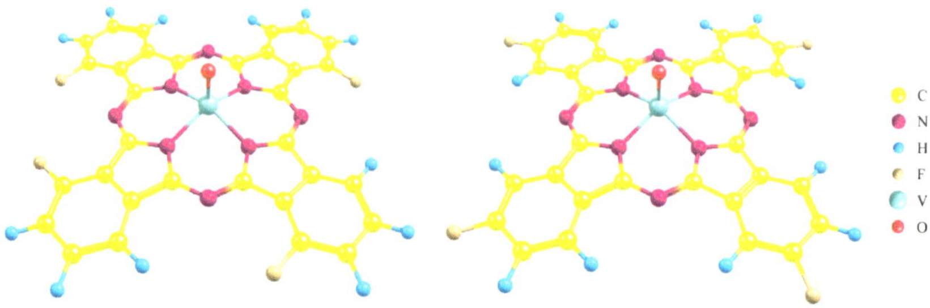

Figure 1. Structure of tetrafluorosubstituted metal phthalocyanines with F-substituents in non-peripheral ($MPcF_4$-np, **left**) and peripheral ($MPcF_4$-p, **right**) positions.

2. Materials and Methods

2.1. Synthesis of $MPcF_4$ and Deposition of Their Films

$MPcF_4$-p and $MPcF_4$-np derivatives were synthesized by the method of template synthesis by fusing a mixture (1:4) of a salt (chloride or acetate) of the corresponding metal with 4-fluorophthalonitrile (Sigma-Aldrich, CAS 65610-14-2) or 3-fluorophthalonitrile (Fluorochem, CAS 65610-13-1), respectively, as was described in our previous works in more detail [30,31]. The prepared $MPcF_4$-p and $MPcF_4$-np powders were purified by vacuum (10^{-5} Torr) gradient sublimation; the sublimation temperature was 420–450 °C.

Thin films of the phthalocyanines with a nominal thickness of about 100 nm were deposited by a PVD method in vacuum of 10^{-5} Torr. Glass slides with pre-deposited interdigitated Pt electrodes (IDE, G-IDEPT10, Metrohm, DropSens, Spain) were used as substrates; the substrate temperature was 60 °C. The IDE has the cell constant of 0.0188 cm^{-1}, the number of digits of 125 × 2 and the digit length of 6760 μm.

The XRD study of thin films was carried out using a Bruker D8 advance powder diffractometer (vertical θ-θ goniometer in Bragg–Brentano geometry, Cu-anode sealed tube 40 mA at 40 kV, LYNXEYE XE-T compound silicon strip detector, motorized divergence slit in fixed sample illumination area mode) with 2θ scan step of 0.01023°, and the acquisition time of 2 s/step.

The morphology of the $MPcF_4$-p and $MPcF_4$-np films was studied by atomic force microscopy (AFM) in the semicontact mode using the Ntegra Prima II nanolaboratory (NT-MDT, Moscow, Russia). The parameters of the HA_NC probe were as follows: thickness—3 μm, probe length—123 μm, width—34 mm, force constant—17 N/m, resonant frequency—230 kHz. A Nova SPM software (standards ISO 4287-1, ISO 4287, and ASME B46) was used to calculate roughness parameters.

2.2. Investigation of the Chemiresistive Sensor Response

Sensor properties of the investigated films were studied by measuring the change in the resistance when exposed to ammonia in a wide range of concentrations (1–100 ppm). The change in the resistance was measured using a Keithley 236 universal electrometer by applying a constant DC voltage (10 V). The normalized sensor response was calculated as $R_n = (R - R_o)/R_o$; where R is the steady resistance of the $MPcF_4$ film at a certain NH_3 concentration and R_o is the baseline resistance of the film in air. Three different samples were used to calculate the standard deviation of the values of sensor response. The scheme of the installation for studying sensor properties and a photograph of a flowing gas cell were presented in our previous publication [29].

Pure commercial NH_3 gas (Company "Chistye Gasy", Novosibirsk, Russia) was used as an analyte source; air was used as a carrier and diluent gas. The gas flow was regulated using mass flow regulators and passed through the cell at the constant flow rate of 300 mL·min^{-1}.

2.3. Quantum-Chemical Calculations

Quantum-chemical calculations of the geometric structure and total energies of $MPcF_4$-p and $MPcF_4$-np with ammonia molecules were carried out by the DFT method using the BP86 GGA functional [32,33], def2-SVP basis set of atomic orbitals [34], DFT-D3 dispersion interaction corrections [35,36], RI approximation [37–42], and the corresponding Def2/J auxiliary basis set [43]. The ORCA software package (version 5.0.2) was used for the calculations [44,45]. After optimizing the geometry of the compounds under consideration, their vibrational spectra were calculated to make sure that there were no negative frequencies and, as a result, that the calculated geometries corresponded to the ground states. The choice of the BP86 functional and the def2-SVP basis set was due to the fact that this combination allowed us to accurately describe the geometric structure of 3D metal complexes [46]. Visualization of the structure of the compounds under consideration was carried out using the ChemCraft program and the AIMStudio module of the AIMAll software package [47].

The location of four fluorine atoms in the peripheral and non-peripheral positions of cobalt phthalocyanine corresponded to the C_s symmetry point group of both $CoPcF_4$-p and $CoPcF_4$-np molecules, since it was previously shown that the formation of these isomers was preferable to the molecules with D_{2h}, C_{2v}, or C_{4h} symmetries [48]. However, symmetry constraints were not used during the calculations. The location of four fluorine atoms in the case of $VOPcF_4$-np and $VOPcF_4$-p was analogous (Figure 1). The spin multiplicity of all compounds, including those containing an ammonia molecule, was equal to two, so the calculations were performed in the framework of the spin-unrestricted Kohn–Sham theory (UKS).

After optimizing the geometry of the considered aggregates, the binding energy E_b of the NH_3 molecule with phthalocyanine was calculated from the difference in the total energies of both molecules separately, the geometry of which was also optimized, as well as their aggregate, as

$$E_b = E_{NH_3} + E_{MPcF_4-(p)np} - E_{MPcF_4-(p)np/NH_3} - \Delta E_{BSSE} \quad (1)$$

where ΔE_{BSSE} is the correction to the binding energy, taking into account the basis set superposition error [49,50], which in turn was estimated as follows:

$$\Delta E_{BSSE} = \left(E^*_{MPcF_4-(p)np} + E^*_{NH_3}\right) - \left(E^*_{MPcF_4-(p)np^*} + E^*_{NH_3^*}\right) \quad (2)$$

Here, an asterisk in the superscript indicates that $MPcF_4$-(p)np and NH_3 geometries, respectively, were taken from an optimized aggregate of phthalocyanine and ammonia molecules, while their geometry optimization was not carried out, but only the total energies calculation was performed. An asterisk in the subscript means that the indicated fragment of the whole aggregate was considered, while the points described by the corresponding basis sets of atomic orbitals were considered instead of the second fragment atoms.

Next, the electron wave functions of the optimized $MPcF_4$-(p)np/NH_3 aggregates were calculated using the cc-pVTZ basis set of atomic orbitals [51]. Using these wave functions and the AIMAll software package [47], a topological analysis of the electron density distribution $\rho(\mathbf{r})$ in the considered structures was performed in the framework of the QTAIM theory [52–54]. In particular, the values of $\rho(\mathbf{r})$ and its Laplacian $\nabla^2\rho(\mathbf{r})$ were obtained at bond critical points (BCPs), characterizing the interaction of ammonia molecules with phthalocyanine atoms.

3. Results

3.1. Characterization of $MPcF_4$ Films

XRD patterns of $MPcF_4$-p and $MPcF_4$-np thin films are shown in Figure 2. Diffraction patterns of all thin films except $VOPcF_4$-np contain single strong diffraction peak in 6÷7 °2θ area, which may indicate a preferred orientation of crystallites within thin film. For the films of Cu, Co and Zn phthalocyanines (both –p and –np), the positions of diffraction peaks match well with the first peak on their powder diffraction patterns calculated from respective crystal structure data [31,55,56]. An additional weak (0–12) diffraction peak is also visible on the diffraction pattern of a $CuPcF_4$-p film, indicating that its preferred orientation is weaker compared to others. The diffraction pattern of the $VOPcF_4$-p film also contains a second peak, but its corresponding interplanar distance is half of the position of the first peak, which means that both of these peaks belong to the same group of lattice planes, which is often observed for films with a strong preferred orientation. The positions of these peaks do not coincide with the calculated diffraction pattern, indicating that the $VOPcF_4$-p film consists of another crystalline phase with an unknown crystal structure [56]. On the other hand, the position of the diffraction peak observed on the $VOPcF_4$-np diffraction pattern coincides with the calculated diffraction patterns of both known $VOPcF_4$-np polymorphs (peak (−222) or (004)). However, both of these $VOPcF_4$-np polymorphs have a similar structure, with molecules packed in 2D layers, and these diffraction peaks correspond to the distance between these molecular layers.

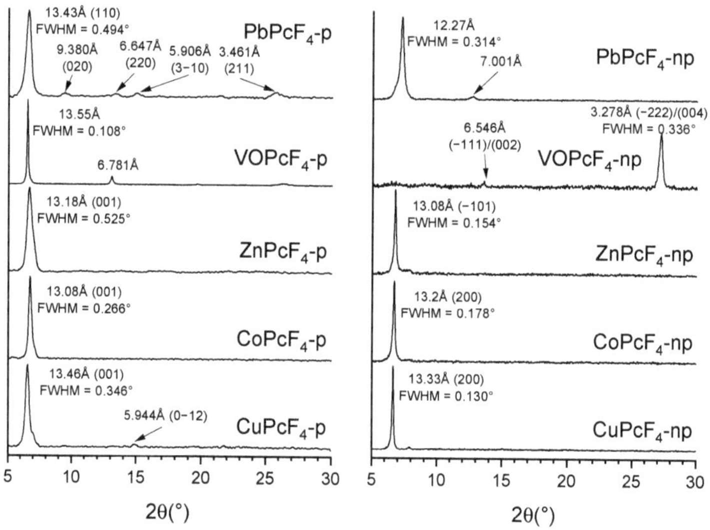

Figure 2. XRD patterns of $MPcF_4$-p and $MPcF_4$-np thin films in the 2θ range of 5–30°.

The X-ray pattern of the $PbPcF_4$-p film has several weak diffraction peaks in addition to the main peak, which coincide with the data of the single crystal [20]. This indicates a weak preferred orientation in the thin film. Finally, the diffractogram of the $PbPcF_4$-np film has one strong diffraction peak and one weak diffraction peak, neither of which correspond to the data of the single crystal. This means that the $PbPcF_4$-np film consists of an unknown crystalline phase but has a more preferred orientation compared to $PbPcF_4$-p.

Using the FWHM values of the observed diffraction peak, it is possible to estimate the size of the coherent scattering region for all investigated films using the Scherrer equation. Taking into account the instrumental broadening of the peak, which is 0.05° for the Bruker D8 advance powder diffractometer (measured using LaB_6 SRM-660a powder as a reference), the coherent scattering regions are 30 nm for $CuPcF_4$-p, 41 nm for $CoPcF_4$-p, 19 nm for $ZnPcF_4$-p, 150 nm for $VOPcF_4$-p, and 20 nm for $PbPcF_4$-p. For $CuPcF_4$-np, $CoPcF_4$-np, $ZnPcF_4$-np, $VOPcF_4$-np, and $PbPcF_4$-np films, these values are 110 nm, 69 nm, 85 nm, 32 nm, and 34 nm, respectively.

Finally, knowing the plane of preferred orientation and the crystal structure of the film, it is possible to calculate the angle between the substrate surface and the phthalocyanine molecules inside the crystallites. The obtained inclination angles are 82.72° for $CuPcF_4$-p, 81.07° for $CoPcF_4$-p, 80.28° for $ZnPcF_4$-p, 90° for $PbPcF_4$-p (due to the tetragonal space group), 83.36° and 89.08° for $CuPcF_4$-np, 78.15° and 80.99° for CoPcF4-np, 78.51° for $ZnPcF_4$-np and 0.39° for $VOPcF_4$-np. Two values are given for $CuPcF_4$-np and $CoPcF_4$-np, since their crystal structures contain two symmetrically unequal molecules. A visual representation of the orientation of the molecules relative to the substrate surface is shown in Figure 3.

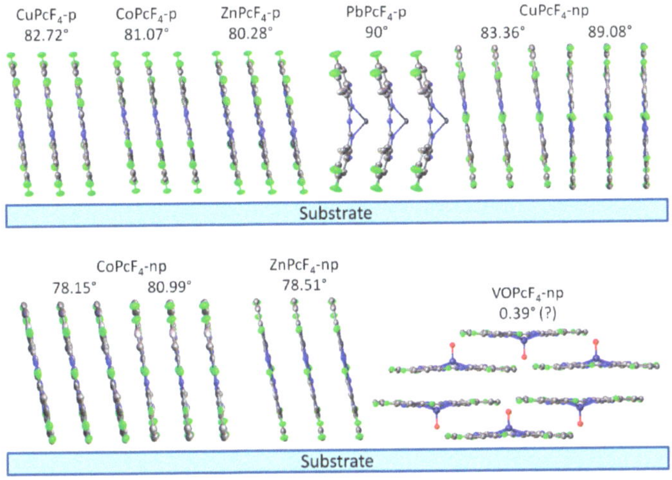

Figure 3. Orientation of $MPcF_4$-p and $MPcF_4$-np molecules relative to the substrate surface.

The morphology of the films was studied by atomic force microscopy. The AFM images of films of tetrafluorosubstituted phthalocyanines of cobalt, vanadyl and zinc are shown in Figure 4 as examples. The morphology of $CoPcF_4$-np and $CoPcF_4$-p films is very similar. The surface of $CoPcF_4$ films is formed by elongated crystallites with a width of 50 to 100 nm and a length up to 200 nm. The root mean square (RMS) roughness is 4.04 nm and 4.18 nm for the $CoPcF_4$-p and $CoPcF_4$-np films, respectively. The $VOPcF_4$-p film is also formed by elongated crystallites that are oriented parallel to the substrate surface with their larger face, while in the $VOPcF_4$-np film, the crystallites are oriented almost perpendicular to the substrate surface. The RMS roughness of $VOPcF_4$-p and $VOPcF_4$-np films are 7.6 and 5.9 nm, respectively. The surface of both films of $ZnPcF_4$ consists of slightly rounded crystallites with the size reaching 70 nm. The RMS roughness of $ZnPcF_4$-np and $ZnPcF_4$-np films are 9.54 and 4.44 nm, respectively.

3.2. Sensor Properties of MPcF4 Films

3.2.1. Comparative Study of the Sensor Response of MPcF4-p and MPcF4-np Films to Ammonia

A typical chemiresistive sensor response, which is an increase in the resistance of the film when ammonia is injected into the cell and the resistance returns to its original value, after purging the cell with air, is shown in Figure 5 for the film of $CoPcF_4$-p as an example.

The films of all investigated $MPcF_4$-p and $MPcF_4$-np films demonstrate an increase in the resistance of the film when ammonia is injected into the cell, which is typical for metal phthalocyanines, exhibiting p-type semiconductor behavior [57]. When NH_3 adsorption occurs on the surface of a phthalocyanine film, the electron density transfers from the NH_3 molecule to the film. This reduces the concentration of the main charge carriers and increases the resistance. After purging the gas cell with air, NH_3 removes from the film surface, which leads to a return of resistance to its initial value. The sensor response to ammonia in the investigated concentration range from 1 to 50 ppm is completely reversible at room temperature.

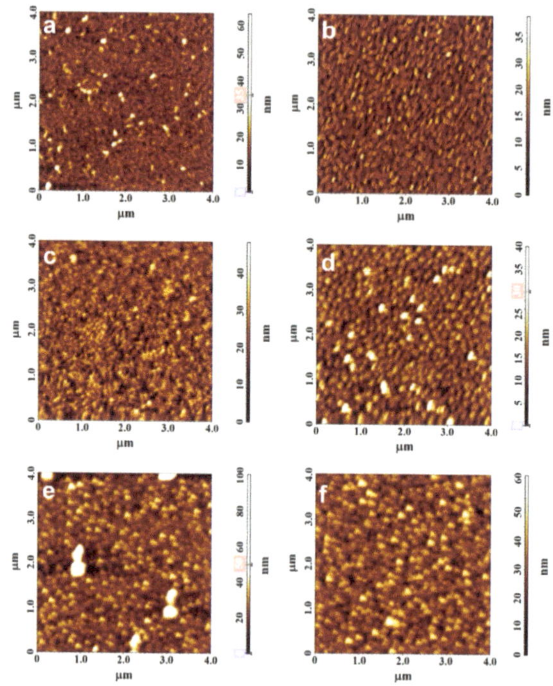

Figure 4. AFM images of CoPcF$_4$-p (**a**), CoPcF$_4$-np (**b**), VOPcF$_4$-p (**c**), VOPcF$_4$-np (**d**), ZnPcF$_4$-p (**e**), and ZnPcF$_4$-np (**f**) films.

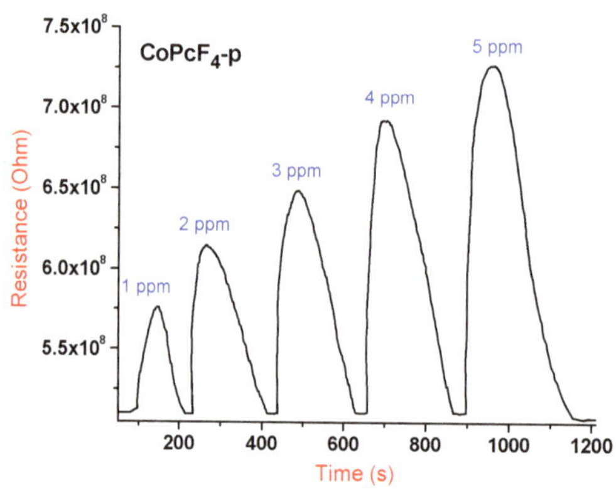

Figure 5. Typical chemiresistive sensor response of a CoPcF$_4$-p film to ammonia (1–5 ppm), measured at room temperature.

To compare the sensor properties of MPcF$_4$ with different central metals and positions of F-substituents, the sensor response of all films was measured at the same experimental conditions. The resulted dependencies of the value of the sensor response of all investigated films on ammonia concentration in the range from 1 to 50 ppm are shown in Figure 6.

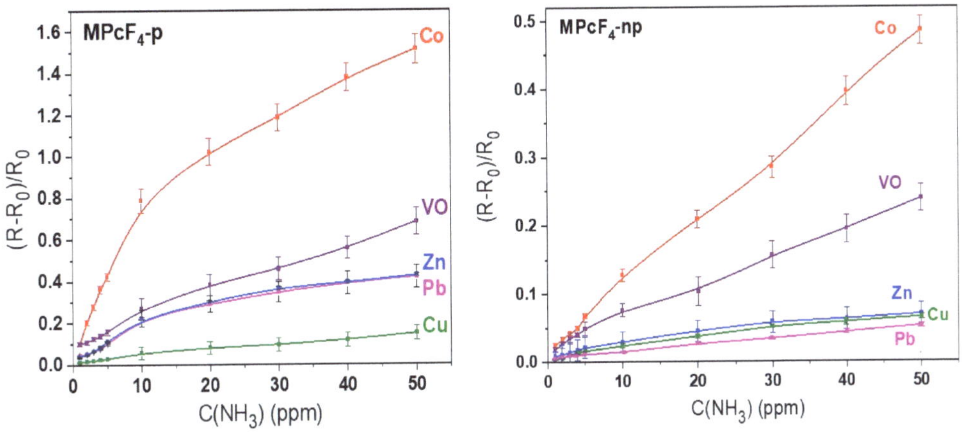

Figure 6. Dependence of the sensor response of $MPcF_4$-p and $MPcF_4$-np films on ammonia concentration (1–50 ppm), measured at room temperature.

Figure 6 shows that, in the case of $MPcF_4$-p, the sensor response increases in the order $CuPcF_4$-p < $PbPcF_4$-p ~ $ZnPcF_4$-p < $VOPcF_4$-p < $CoPcF_4$-p, while in the case of $MPcF_4$-np the order is slightly different: $PbPcF_4$-np < $CuPcF_4$-np ~ $ZnPcF_4$-np < $VOPcF_4$-np < $CoPcF_4$-np. For both types of substitution, cobalt phthalocyanine derivatives demonstrate the maximal value of the response to ammonia, followed by vanadyl phthalocyanines. The sensor response of cobalt phthalocyanine derivatives to 10 ppm NH_3 is more than 3–14 times higher than in the case of other $MPcF_4$-p, and 1.5–10 times higher than in the case of other $MPcF_4$-np, depending on the type of central metal ion. When comparing phthalocyanines with different positions of F-substituents in the macroring, $MPcF_4$-p films exhibit a higher sensor response than $MPcF_4$-np ones. For example, the sensor response of $CoPcF_4$-p to 10 ppm NH_3 is about six times higher than that of $CoPcF_4$-np.

All investigated phthalocyanine films demonstrate linear dependencies of the sensor response on ammonia concentration in the range from 1 to 10 ppm. The limits of ammonia detection (LOD), calculated as $3\sigma/m$ (σ is the standard deviation of the sensor response to 1 ppm NH_3 and m is the slope of the calibration plot in the linear region of 1–10 ppm) are summarized in Table 1. The values of the response and recovery time determined upon exposure of the films with 5 ppm NH_3 (Figure 7) are also given in Table 1.

Table 1. Limits of ammonia detection (LOD), response and recovery times of $MPcF_4$-p and $MPcF_4$-np films.

Sensing Layer	LOD, ppm	Response Time, s (at 5 ppm)	Recovery Time, s (at 5 ppm)
$CoPcF_4$-p	0.01	55	215
$VOPcF_4$-p	0.04	48	270
$ZnPcF_4$-p	0.01	45	210
$PbPcF_4$-p	0.65	40	250
$CuPcF_4$-p	0.75	40	240
$CoPcF_4$-np	0.11	60	215
$VOPcF_4$-np	0.87	62	350
$ZnPcF_4$-np	0.08	35	155
$PbPcF_4$-np	1.82	40	110
$CuPcF_4$-np	1.49	55	90

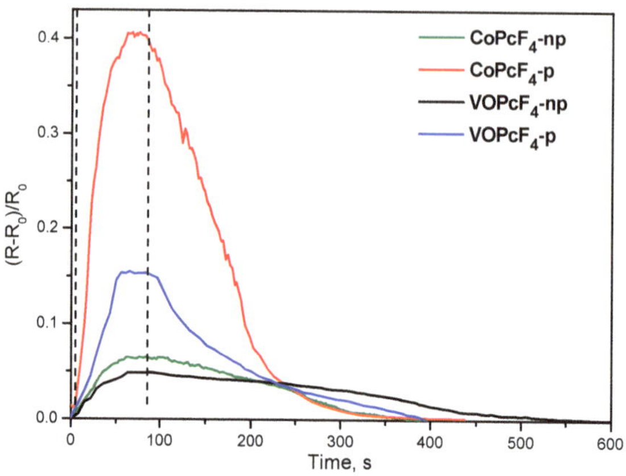

Figure 7. Sensor response of $MPcF_4$-p and $MPcF_4$-np films (M = Co, VO) to ammonia (5 ppm), measured at room temperature.

3.2.2. Quantum-Chemical Modeling of the Interaction of the NH_3 Molecule with Phthalocyanines

To explain the effect of the stronger sensor response of $MPcF_4$-p phthalocyanines compared to $MPcF_4$-np toward ammonia, quantum-chemical modeling of the interaction of the NH_3 molecule with them was carried out. In this case, the complexes of cobalt and vanadyl, which are examples of phthalocyanines with a planar and non-planar structure, were considered.

Despite the fact that the bond of ammonia with the central metal in the phthalocyanine molecule is stronger, we considered the interaction of NH_3 with the side atoms of the macrocycles (Figures 8 and 9). This choice of sites for the adsorption of NH_3 molecules was due to two reasons. First, the strong binding of ammonia to metal atoms with the formation of a coordination bond complicates its desorption under normal conditions, and, as a consequence, can lead to an irreversible sensor response, which contradicts experimental data [29]. Second, the typical distance between phthalocyanine molecules in a stack is about 3.4 Å [55]. This is slightly larger than the van der Waals diameter of the nitrogen atom, and approximately equal to the diameter of the carbon atom, which is due to the π-π-interaction between macrocycles. Therefore, there is not enough space for the ammonia molecule between two phthalocyanine macrocycles. That is precisely why the probability of the formation of a strong bond between a metal atom and the NH_3 molecule is low. Moreover, Chia et al. [58] demonstrated that ammonia did not interact with the metal center of the unsubstituted copper phthalocyanine, using the methods of in situ X-ray absorption spectroscopy (XAS) and EXAFS. The EXAFS study showed that the interaction occurred at the benzene ring or bridging nitrogen atom of the macrocycle. We have previously proposed and considered the mechanism of the sensor response of hybrids of carbon nanotubes with phthalocyanines, which is associated with a decrease in their electrical conductivity during the formation of hydrogen bonds between the ammonia molecule and side atoms of phthalocyanines [59].

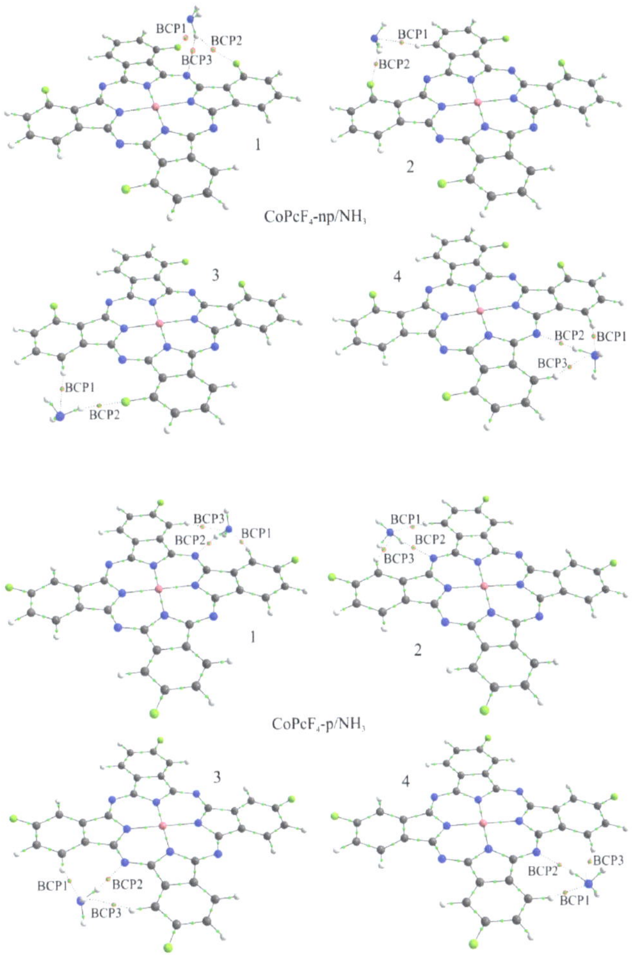

Figure 8. Structure of $CoPcF_4$-np/NH_3 and $CoPcF_4$-p/NH_3 aggregates and bond critical points (BCP) in them. Red circles indicate BCPs characterizing the interaction of the ammonia molecule with phthalocyanine atoms.

In the process of geometry optimization, it was found that in all cases, except for $VOPcF_4$-np/NH_3-1 and $CoPcF_4$-np/NH_3-1, the NH_3 molecule occupies an equilibrium position in such a way that its nitrogen atom is approximately in the same plane as the phthalocyanine molecule. In this case, the distance of ammonia from the macrocycle can be characterized using the distance *d* between the nitrogen atom of the NH_3 molecule and the nearest bridging nitrogen atom of phthalocyanine (Tables 2 and 3). In the case of $VOPcF_4$-np/NH_3-1 and $CoPcF_4$-np/NH_3-1, the ammonia molecule occupies an equilibrium position above the macrocycle plane, which contradicts the above idea that this molecule cannot fit into the space between two adjacent phthalocyanines in the stack. Therefore, these two aggregates were excluded from further consideration.

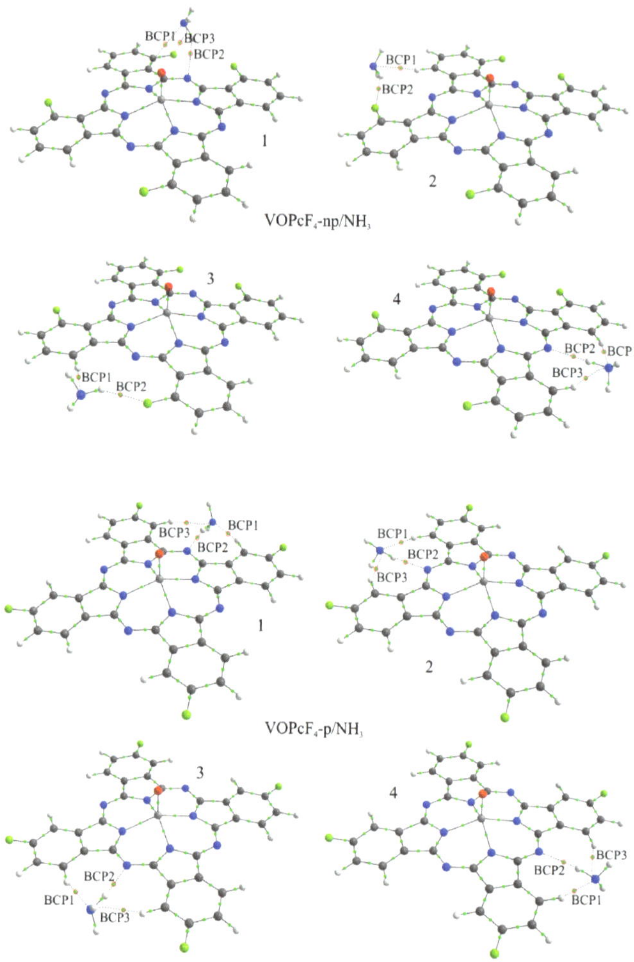

Figure 9. Structure of VOPcF$_4$-np/NH$_3$ and VOPcF$_4$-p/NH$_3$ aggregates and bond critical points in them. Red circles indicate BCPs characterizing the interaction of the ammonia molecule with phthalocyanine atoms.

Table 2. The interaction parameters of the NH$_3$ molecule with cobalt phthalocyanines.

Compound	E_b, eV	d, Å	BCP	Atoms *	ρ(r), e/Å^3	$\nabla^2\rho$(r), e/Å^5
CoPcF$_4$-np/NH$_3$-2	0.036	4.251	1 2	H-N F-H	0.083 0.066	0.880 1.058
CoPcF$_4$-np/NH$_3$-3	0.035	4.270	1 2	H-N F-H	0.083 0.066	0.885 1.060
CoPcF$_4$-np/NH$_3$-4	0.130	3.236	1 2 3	H-N N-H H-N	0.137 0.123 0.099	1.526 1.238 1.070
CoPcF$_4$-p/NH$_3$-1	0.160	3.192	1 2 3	H-N N-H H-N	0.146 0.135 0.094	1.567 1.334 1.009

Table 2. *Cont.*

Compound	E_b, eV	*d*, Å	BCP	Atoms *	ρ(r), e/Å^3	∇^2ρ(r), e/Å^5
$CoPcF_4$-p/NH_3-2	0.161	3.209	1 2 3	H-N N-H H-N	0.137 0.131 0.095	1.502 1.300 1.023
$CoPcF_4$-p/NH_3-3	0.162	3.207	1 2 3	H-N N-H H-N	0.137 0.131 0.095	1.504 1.304 1.023
$CoPcF_4$-p/NH_3-4	0.158	3.203	1 2 3	H-N N-H H-N	0.138 0.133 0.093	1.516 1.312 1.005

* The first atom belongs to the phthalocyanine molecule and the second one belongs to the ammonia molecule.

Table 3. The interaction parameters of the NH_3 molecule with vanadyl phthalocyanines.

Compound	E_b, eV	*d*, Å	BCP	Atoms *	ρ(r), e/Å^3	∇^2ρ(r), e/Å^5
$VOPcF_4$-np/NH_3-2	0.035	4.213	1 2	H-N F-H	0.085 0.065	0.904 1.050
$VOPcF_4$-np/NH_3-3	0.034	4.220	1 2	H-N F-H	0.085 0.065	0.904 1.053
$VOPcF_4$-np/NH_3-4	0.144	3.227	1 2 3	H-N N-H H-N	0.137 0.125 0.096	1.519 1.256 1.038
$VOPcF_4$-p/NH_3-1	0.174	3.179	1 2 3	H-N N-H H-N	0.148 0.139 0.091	1.570 1.360 0.979
$VOPcF_4$-p/NH_3-2	0.173	3.196	1 2 3	H-N N-H H-N	0.138 0.135 0.093	1.502 1.327 0.997
$VOPcF_4$-p/NH_3-3	0.173	3.195	1 2 3	H-N N-H H-N	0.138 0.135 0.092	1.504 1.328 0.996
$VOPcF_4$-p/NH_3-4	0.170	3.190	1 2 3	H-N N-H H-N	0.140 0.137 0.091	1.518 1.338 0.980

* The first atom belongs to the phthalocyanine molecule and the second one belongs to the ammonia molecule.

The binding energy in $CoPcF_4$-p/NH_3-x (x = 1–4, Figures 8 and 9) aggregates varies from 0.158 eV to 0.161 eV, which exceeds the values of 0.035–0.130 eV obtained for $CoPcF_4$-np/NH_3-x (x = 2–4) (Tables 2 and 3). Stronger binding of the NH_3 molecule has a stronger effect on the electronic structure of phthalocyanine and, as a result, on the conductivity of the macrocycle stack. At the same time, in the case of the peripheral position of F-substituents, the ammonia molecule is located deeper in the cavity between adjacent benzene rings and closer to the bridging nitrogen atom; that is, d varies in the range of 3.192–3.209 Å. In the case of a non-peripheral position of the F-substituents, this distance is larger and equal to 3.236–4.270 Å.

The $CoPcF_4$-np/NH_3-4 aggregate, in which there are no fluorine atoms in the cavity where the ammonia molecule is embedded, should be separately noted. This arrangement allows the ammonia to approach the phthalocyanine nitrogen bridge atom in such a way that the corresponding distance *d* is 3.236 Å, and the binding energy is 0.130 eV. These values are close to those obtained for all four $CoPcF_4$-p/NH_3-x aggregates. This means that fluorine atoms in non-peripheral positions, due to their size and the larger carbon-fluorine distance compared to the carbon–hydrogen distance, prevent the NH_3 molecule from approaching the macrocycle closely, which causes a weaker binding of ammonia and its lesser effect on the electronic phthalocyanine structure.

The topological analysis of the electron density distribution in aggregates with $CoPcF_4$ made it possible to find the bond critical points between the atoms of ammonia and phthalocyanine (Figure 8). In particular, it was shown that in all cases of $CoPcF_4$-p/NH_3-x there are three such points, two of which (BCP1 and BCP3) are located between the hydrogen atoms of phthalocyanine and the nitrogen atom of the NH_3 molecule, and one (BCP2), on the contrary, between one of the ammonia hydrogen atoms and the bridging nitrogen atom of phthalocyanine. A similar situation is observed in the case of $CoPcF_4$-np/NH_3-4. The values of $\rho(\mathbf{r})$ and $\nabla^2\rho(\mathbf{r})$ at these points are in the ranges of 0.093–0.146 $e/Å^3$ and 1.005–1.567 $e/Å^5$ (Table 1), respectively. This indicates the formation of hydrogen bonds between the ammonia and phthalocyanine molecules since it was previously shown that the criterion for this is the values of the electron density and its Laplacian in the ranges of 0.013–0.236 $e/Å^3$ and 0.578–3.350 $e/Å^5$ [60].

Since the fluorine atoms in the considered phthalocyanine molecules are located asymmetrically relative to the rotation axis passing perpendicular to the macrocycle plane through the metal atom, there are four nonequivalent sites for attaching the NH_3 molecule in each phthalocyanine. These sites are located between adjacent benzene rings opposite the bridging nitrogen atom. Consequently, in the case of each phthalocyanine, all four positions of the attachment of ammonia molecule were considered, and the corresponding aggregates were further designated as $MPcF_4$-p/NH_3-x or $MPcF_4$-np/NH_3-x, where x = 1, 2, 3, or 4 (Figures 8 and 9).

In the case of the $CoPcF_4$-np/NH_3-2 and $CoPcF_4$-np/NH_3-3 aggregates, the interaction of the NH_3 molecule with phthalocyanines can be characterized by the presence of two BCPs: the first one (BCP1) is between the phthalocyanine hydrogen atom and the ammonia nitrogen atom, and the second one (BCP2) is between the phthalocyanine fluorine atom and one of the ammonia hydrogen atoms (Figure 8). The values of $\rho(\mathbf{r})$ and $\nabla^2\rho(\mathbf{r})$ at these points are significantly lower than in the case of $CoPcF_4$-p/NH_3-x, which is consistent with the lower E_b values, but they also indicate the formation of hydrogen bonds.

The results of quantum-chemical calculations of the $VOPcF_4$-np/NH_3 and $VOPcF_4$-p/NH_3 aggregates are almost completely equivalent to those obtained for the cobalt complexes, namely the binding energy of NH_3 with $VOPcF_4$-p is higher than in the case of $VOPcF_4$-np derivative. The topological analysis of the electron density distribution in the $VOPcF_4$-np/NH_3 and $VOPcF_4$-p/NH_3 aggregates showed that the values of $\rho(\mathbf{r})$ and $\nabla^2\rho(\mathbf{r})$ at the bond critical points between the ammonia and phthalocyanine atoms are close to those obtained in the case of cobalt complexes. Consequently, the formation of hydrogen bonds is also observed during the interaction of the NH_3 molecule with the side atoms of vanadyl phthalocyanines. However, it is necessary to mention that the binding energy values in the $VOPcF_4$-p/NH_3-x and $VOPcF_4$-np/NH_3-4 aggregates (Table 3) are slightly higher than in the case of the corresponding cobalt phthalocyanine derivatives.

If we compare the calculation data with the data of the experimental study of the sensor response, it can be noted that the sensor response of $VOPcF_4$ derivatives, on the contrary, is lower than $CoPcF_4$, whereas their relaxation time is higher in the case of $CoPcF_4$. It should be noted here that the bond strength is not the only factor affecting the sensor response. The proposed model is applicable only to explain the effect of the position of substituents on the value of the sensor response; however, it does not allow for taking into account the influence of the central metal. Perhaps this was due to the fact that an isolated phthalocyanine molecule was considered. In crystals and films, phthalocyanine molecules are packed in stacks and the conductivity of the films and charge transfer are determined by intermolecular interactions. Taking into account the different packing of cobalt and vanadyl phthalocyanines, ways of overlapping π-orbitals of neighboring macrocycles, and the nature and location of the central metals, it can be assumed that the charge transfer processes in these structures as a result of ammonia adsorption will differ. This issue requires further in-depth consideration, and, apparently, quantum chemical calculations of individual molecules will not be enough. Therefore, in the future, in order to consider the influence of various central metals on the sensor response, it will be necessary to model stacks of phthalocyanine molecules in the form of large clusters or periodic structures.

These statements are also confirmed by additional quantum-chemical calculations of the binding energy of the ammonia molecule with other phthalocyanines $MPcF_4$-p and $MPcF_4$-np, where M = Cu, Zn, and Pb, performed using the same approach. The results are similar to those observed in the case of cobalt and vanadyl phthalocyanines. Firstly, the peripheral positions of fluorine atoms are a more favorable factor for the stronger binding of the NH_3 molecule than the non-peripheral positions (Table 4). Secondly, the binding energies of the ammonia molecule with $MPcF_4$-p and $MPcF_4$-np where M = Cu, Zn, and Pb are close to those obtained in the case of cobalt and vanadyl phthalocyanines. Thirdly, when passing from one metal phthalocyanine to another, the tendency of changing E_b values does not agree with the experimentally established order, changing the value of the sensor response.

Table 4. The binding energies (in eV) of the NH_3 molecule with $MPcF_4$ (M = Cu, Zn, Pb) phthalocyanines.

Compound	NH_3 Position	M = Cu	M = Zn	M = Pb
$MPcF_4$-np/NH_3	2	0.037	0.037	-
	3	0.036	0.036	-
	4	0.145	0.151	0.144
$MPcF_4$-p/NH_3	1	0.172	0.179	0.173
	2	0.172	0.179	0.172
	3	0.173	0.179	0.172
	4	0.169	0.176	0.169

It is necessary to mention that, in the case of lead phthalocyanines, the planar structure of the macrocycle is strongly distorted. For this reason, the size of the cavities in which the location of ammonia is considered decreases. As a result, the NH_3 molecule cannot be localized in positions two and three, where fluorine atoms are present, and in the process of geometry optimization, a structure with NH_3 above the macrocycle is formed as shown in Figure 10. For this reason, the binding energies of NH_3 in $PbPcF_4$-np/NH_3-2 and $PbPcF_4$-np/NH_3-3 aggregates were not calculated (Table 4).

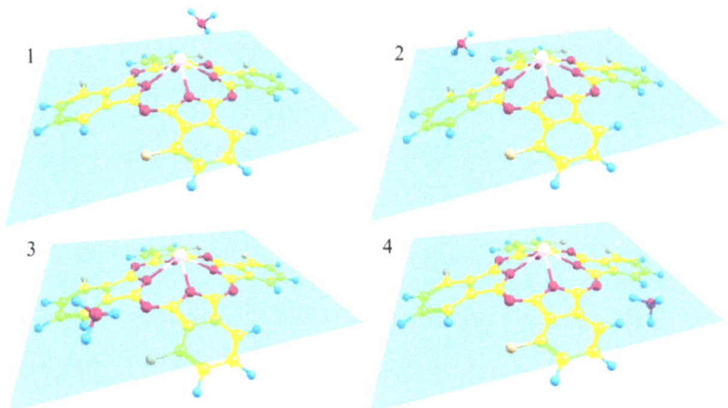

Figure 10. Structure of $PbPcF_4$-np/NH_3 aggregates. The indicated plane passes through bridge nitrogen atoms to show the position of the NH_3 molecule relative phthalocyanine surface, numbers 1–4 indicate type of aggregate.

Thus, there are significantly fewer places for ammonia adsorption in $PbPcF_4$-np than in all other $MPcF_4$-np, and attachment can occur only at position four (Figure 10), at which there are no fluorine atoms in the considered cavity. This result is consistent with the fact that $PbPcF_4$-np films have the lowest sensor response in the experiment.

3.2.3. Detailed Study of the Sensor Characteristics of MPcF$_4$-p (M = Co, VO) Films to Ammonia

For the films of CoPcF$_4$-p and VOPcF$_4$-p, which demonstrate the highest sensor response among those investigated, the more detailed investigation of the sensor properties was carried out. Figure 11 shows the sensor response of CoPcF$_4$-p and VOPcF$_4$-p films to ammonia (3 ppm) and the change in their resistance at a different relative humidity (RH). An increase in RH causes a growth of the films' resistance, since, like ammonia, water is an electron donating molecule. At the same time, an increase in RH up to 40% does not affect the base resistance and the response, while an increase in humidity up to 80% results in a noticeable (1.6 time for a CoPcF$_4$-p film and 4.7 time for a VOPcF$_4$-p film) increase in the sensor response.

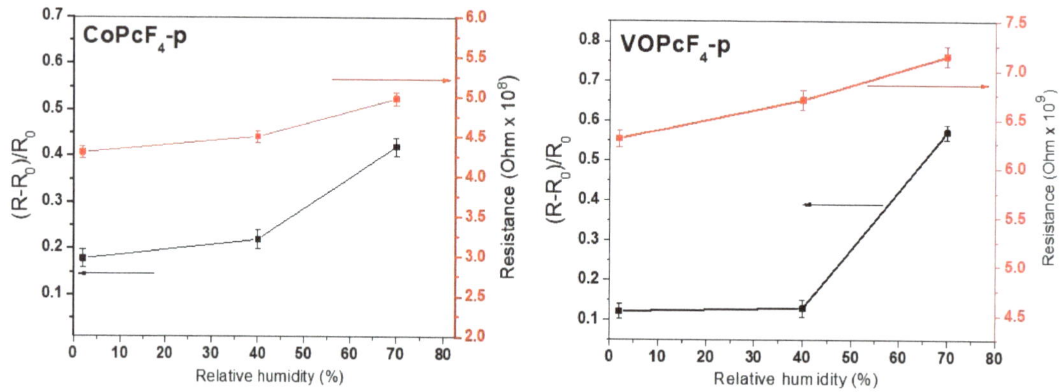

Figure 11. The sensor response of CoPcF$_4$-p and VOPcF$_4$-p films to ammonia (3 ppm) and the change in their resistance at different humidity.

The operating temperature is a unique characteristic of each sensor. The sensor response of cobalt and vanadyl phthalocyanine layers to ammonia was tested at 25, 40, and 80 °C. An increase in temperature leads to an increase in response (Figure 12). This behavior appears to be associated with an increase in the concentration of charge carriers in p-type semiconductors with an increase in temperature [61]. The increase in the concentration of charge carriers is also confirmed by a slight decrease in base films resistance of the films with temperature. An increase in the operating temperature activates the desorption process, which leads to a decrease in response time. For example, in the case of CoPcF$_4$-p, when the operating temperature increases from 25 to 80 °C, the recovery time decreases from 215 to 145 s.

The stability and repeatability of the sensor based on cobalt and vanadyl phthalocyanines were also tested. The sensor response of a fresh CoPcF$_4$-p film to 3 ppm of ammonia and the same film after 7, 14, 28, and 120 days is shown in Figure 13a as an example. The change in both the sensor response and the base resistance does not exceed the measurement error, which indicates good repeatability and long-term stability of the investigated sensing layer.

To investigate the selectivity with respect to NH_3, the response of MPcF$_4$ layers toward other gaseous analytes was studied (Figure 13b). Pure dry NH_3, CO_2, and H_2S were used as analytes. Ethanol, acetone, benzene, and formaldehyde vapors were generated by evaporation of their liquids using an evaporator located inside the test chamber.

It was found that, among reducing gases and volatile organic vapors, the layers exhibited the maximal response to ammonia, but at a concentration of more than 10 ppm, hydrogen sulfide can act as an interfering gas.

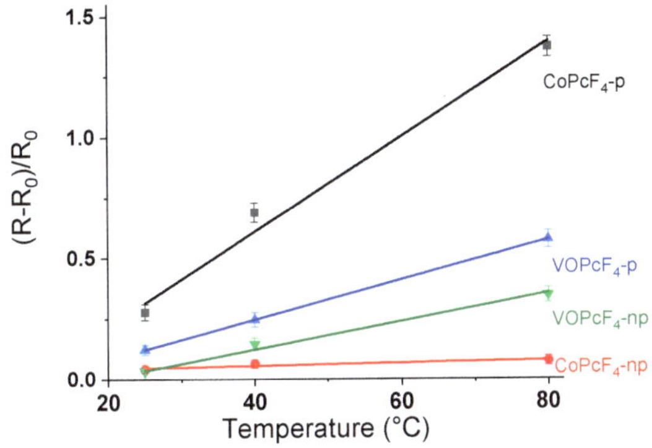

Figure 12. Sensor response of an $MPcF_4$-p and $MPcF_4$-np (M = Co, VO) layer to ammonia (3 ppm) measured at 25, 40 and 80 °C.

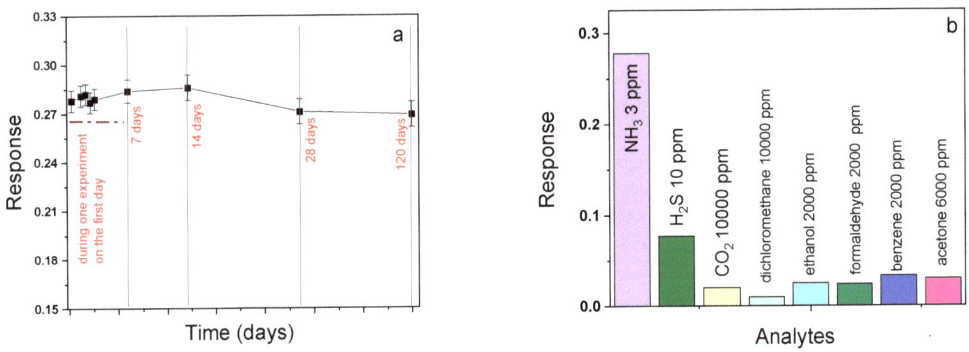

Figure 13. (**a**) Sensor response of a $CoPcF_4$-p film to ammonia (3 ppm) after 7, 14, 28, and 120 days. (**b**) Diagram of sensitivity of a $CoPcF_4$-p film to various gaseous analytes and volatile organic vapors.

Table 5 shows the sensor properties of $CoPcF_4$ films in comparison with the characteristics of some chemiresistive sensors to ammonia described in the literature [62–65]. The presented sensor characteristics of a $CoPcF_4$-p film are comparable to or even exceed the characteristics of sensors based on semiconductor oxides and nanomaterials listed in Table 5. Thus, active layers based on $CoPcF_4$-p and $VOPcF_4$-p are promising materials for creating sensors for the detection of low concentrations of ammonia in the air.

Table 5. Sensor characteristics of various chemiresistive sensors to ammonia.

Material	Concentration Range, ppm	LOD, ppm	Response, %	Response/Recovery Time, s	Ref.
polyimide-SnO_2/rGO	50–800	15	5.16 (100 ppm)	94/57	[62]
MXene/Graphene composite	0.5–100	0.056	25 (100 ppm)	26/148	[63]
ZnO and $WO_3 \cdot H_2O$ composite	2–100	0.76	~15 (10 ppm)	2.4/1.2	[64]
SWCNT-TiOPc	5–50	n/a	1.76 (50 ppm)	120 (fixed)/~40	[65]
$CoPcF_4$-p	1–50	0.01	42 (5 ppm)	55/215	Our work
$VOPcF_4$-p	1–50	0.04	15 (5 ppm)	48/270	Our work

4. Conclusions

A comparative analysis of the chemiresistive sensor response of thin films of a series of tetrasubstituted phthalocyanine of various metals with F-substituent in peripheral ($MPcF_4$-p, M = Cu, Co, Zn, Pb, VO) and non-peripheral ($MPcF_4$-np) positions in macroring to low concentrations of ammonia (1–50 ppm) was carried out. It was found that, in the case of $MPcF_4$-p, the sensor response increases in the order $CuPcF_4$-p < $PbPcF_4$-p ~ $ZnPcF_4$-p < $VOPcF_4$-p < $CoPcF_4$-p, while in the case of $MPcF_4$-np the order is slightly different: $PbPcF_4$-np < $CuPcF_4$-np ~ $ZnPcF_4$-np < $VOPcF_4$-np < $CoPcF_4$-np. When comparing phthalocyanines with different positions of F-substituents in the macroring, $MPcF_4$-p films exhibit a higher sensor response than $MPcF_4$-np ones. The sensor response to ammonia was completely reversible at room temperature. For the films of $CoPcF_4$-p and $VOPcF_4$-p, which demonstrate the highest sensor response among those investigated, the more detailed investigation of the sensor properties was carried out. It was shown that a $CoPcF_4$-p film demonstrated the calculated LOD of 0.01 ppm with the recovery time of 215 s, while a $VOPcF_4$-p film had a LOD of 0.04 ppm and the recovery time of 270 s. Both sensors had good repeatability and long-term stability. The selectivity test showed that CO_2, ethanol, acetone, benzene, and formaldehyde did not interfere with the determination of ammonia, while H_2S at the concentration more than 10 ppm could act as an interfering gas.

It was shown as a result of quantum-chemical modeling that the observed regularities are best described by the interaction of an ammonia molecule with cobalt and vanadyl phthalocyanines through the formation of hydrogen bonds between NH_3 and side atoms of the phthalocyanine. In the case of phthalocyanine with a peripheral position of F-substituents, the NH_3 molecule approaches the macrocycle more closely and binds more strongly than in the case of phthalocyanines with non-peripheral positions of F-substituents. The stronger binding leads to a stronger effect of the ammonia molecule on the electronic structure of phthalocyanine and, as a consequence, on the electrical conductivity of the film. This, in turn, causes a stronger sensor response of $MPcF_4$-p films to ammonia than in the case of $MPcF_4$-np ones. However, the proposed model is applicable only for explaining the effect of the position of substituents on the value of the sensor response; it does not allow for taking into account the influence of the central metal. Perhaps this was due to the fact that an isolated phthalocyanine molecule was considered. Therefore, in the future, in order to consider the effect of central metals on the sensor response, it will be necessary to model stacks of phthalocyanine molecules in the form of large clusters or periodic structures.

Author Contributions: Conceptualization, P.K. and T.B.; methodology, P.K. and T.B.; software, P.K.; validation, P.K. and D.K.; formal analysis, P.K., D.K. and. A.S.; investigation, P.K., A.S., D.K., P.P. and D.B.; resources, D.K.; writing—original draft preparation, D.K., P.P. and T.B.; writing—review and editing, D.K., P.P. and T.B.; visualization, D.B. and P.K.; supervision, T.B.; project administration, D.K.; funding acquisition, D.K. All authors have read and agreed to the published version of the manuscript.

Funding: The study of the films of lead and vanadyl phthalocyanines derivatives was funded by the Russian Science Foundation, grant number 22-73-00145. The study of the cobalt and copper phthalocyanine films was funded by the Ministry of Science and Higher Education of the Russian Federation, N 121031700314-5 and N 121031700313-8.

Institutional Review Board Statement: Not applicable.

Informed Consent Statement: Not applicable.

Data Availability Statement: Not applicable.

Conflicts of Interest: The authors declare no conflict of interest.

References

1. Wang, H.; Zhang, Y.; Han, Q.; Xu, Y.; Hu, G.; Xing, H. The inflammatory injury of heart caused by ammonia is realized by oxidative stress and abnormal energy metabolism activating inflammatory pathway. *Sci. Total Environ.* **2020**, *742*, 140532. [CrossRef] [PubMed]
2. He, K.; Luo, X.; Wen, M.; Wang, C.; Qin, C.; Shao, J.; Gan, L.; Dong, R.; Jiang, H. Effect of acute ammonia toxicity on inflammation, oxidative stress and apoptosis in head kidney macrophage of *Pelteobagrus fulvidraco* and the alleviation of curcumin. *Comp. Biochem. Physiol. Part C Toxicol. Pharmacol.* **2021**, *248*, 109098. [CrossRef] [PubMed]
3. Vuppaladadiyam, A.K.; Antunes, E.; Vuppaladadiyam, S.S.V.; Baig, Z.T.; Subiantoro, A.; Lei, G.; Leu, S.Y.; Sarmah, A.K.; Duan, H. Progress in the development and use of refrigerants and unintended environmental consequences. *Sci. Total Environ.* **2022**, *823*, 153670. [CrossRef] [PubMed]
4. Insausti, M.; Timmis, R.; Kinnersley, R.; Rufino, M.C. Advances in sensing ammonia from agricultural sources. *Sci. Total Environ.* **2020**, *706*, 135124. [CrossRef]
5. Kwak, D.; Lei, Y.; Maric, R. Ammonia gas sensors: A comprehensive review. *Talanta* **2019**, *204*, 713–730. [CrossRef]
6. Wyer, K.E.; Kelleghan, D.B.; Blanes-Vidal, V.; Schauberger, G.; Curran, T.P. Ammonia emissions from agriculture and their contribution to fine particulate matter: A review of implications for human health. *J. Environ. Manage.* **2022**, *323*, 116285. [CrossRef]
7. Kumar, V.; Mirzaei, A.; Bonyani, M.; Kim, K.H.; Kim, H.W.; Kim, S.S. Advances in electrospun nanofiber fabrication for polyaniline (PANI)-based chemoresistive sensors for gaseous ammonia. *TrAC Trends Anal. Chem.* **2020**, *129*, 115938. [CrossRef]
8. Brannelly, N.T.; Hamilton-Shield, J.P.; Killard, A.J. The Measurement of Ammonia in Human Breath and its Potential in Clinical Diagnostics. *Crit. Rev. Anal. Chem.* **2016**, *46*, 490–501. [CrossRef]
9. Song, G.; Jiang, D.; Wu, J.; Sun, X.; Deng, M.; Wang, L.; Hao, C.; Shi, J.; Liu, H.; Tian, Y.; et al. An ultrasensitive fluorescent breath ammonia sensor for noninvasive diagnosis of chronic kidney disease and helicobacter pylori infection. *Chem. Eng. J.* **2022**, *440*, 135979. [CrossRef]
10. Shetty, S.S.; Jayarama, A.; Bhat, S.; Karunasagar, I.; Pinto, R. A review on metal-oxide based trace ammonia sensor for detection of renal disease by exhaled breath analysis. *Mater. Today Proc.* **2022**, *55*, 113–117. [CrossRef]
11. Kim, K.-H.; Jahan, S.A.; Kabir, E. A review of breath analysis for diagnosis of human health. *TrAC Trends Anal. Chem.* **2012**, *33*, 1–8. [CrossRef]
12. Wang, G.; Gao, J.; Sun, B.; He, D.; Zhao, C.; Suo, H. Enhanced ammonia sensitivity electrochemical sensors based on PtCu alloy nanoparticles in-situ synthesized on carbon cloth electrode. *J. Electroanal. Chem.* **2022**, *922*, 116721. [CrossRef]
13. Tsai, J.-H.; Niu, J.-S.; Shao, W.-C.; Liu, W.-C. Characteristics of chemiresistive-type ammonia sensor based on Ga2O3 thin film functionalized with platinum nanoparticles. *Sens. Actuators B Chem.* **2022**, *371*, 132589. [CrossRef]
14. Krishna, K.G.; Parne, S.; Pothukanuri, N.; Kathirvelu, V.; Gandi, S.; Joshi, D. Nanostructured metal oxide semiconductor-based gas sensors: A comprehensive review. *Sens. Actuators A Phys.* **2022**, *341*, 113578. [CrossRef]
15. Ma, J.; Fan, H.; Li, Z.; Jia, Y.; Yadav, A.K.; Dong, G.; Wang, W.; Dong, W.; Wang, S. Multi-walled carbon nanotubes/polyaniline on the ethylenediamine modified polyethylene terephthalate fibers for a flexible room temperature ammonia gas sensor with high responses. *Sens. Actuators B Chem.* **2021**, *334*, 129677. [CrossRef]
16. Ganesan, S.; Kalimuthu, R.; Kanagaraj, T.; Kulandaivelu, R.; Nagappan, R.; Pragasan, L.A.; Ponnusamy, V.K. Microwave-assisted green synthesis of multi-functional carbon quantum dots as efficient fluorescence sensor for ultra-trace level monitoring of ammonia in environmental water. *Environ. Res.* **2022**, *206*, 112589. [CrossRef]
17. Kanaparthi, S.; Singh, S.G. Solvent-free fabrication of a room temperature ammonia gas sensor by frictional deposition of a conducting polymer on paper. *Org. Electron.* **2019**, *68*, 108–112. [CrossRef]
18. Gao, R.; Ma, X.; Liu, L.; Gao, S.; Zhang, X.; Xu, Y.; Cheng, X.; Zhao, H.; Huo, L. In-situ deposition of POMA/ZnO nanorods array film by vapor phase polymerization for detection of trace ammonia in human exhaled breath at room temperature. *Anal. Chim. Acta* **2022**, *1199*, 339563. [CrossRef]
19. Gai, S.; Wang, B.; Wang, X.; Zhang, R.; Miao, S.; Wu, Y. Ultrafast NH_3 gas sensor based on phthalocyanine-optimized non-covalent hybrid of carbon nanotubes with pyrrole. *Sens. Actuators B Chem.* **2022**, *357*, 131352. [CrossRef]
20. Kuprikova, N.M.; Klyamer, D.D.; Sukhikh, A.S.; Krasnov, P.O.; Mrsic, I.; Basova, T.V. Fluorosubstituted lead phthalocyanines: Crystal structure, spectral and sensing properties. *Dye. Pigment.* **2020**, *173*, 107939. [CrossRef]
21. Klyamer, D.; Sukhikh, A.; Gromilov, S.; Krasnov, P.; Basova, T. Fluorinated metal phthalocyanines: Interplay between fluorination degree, films orientation, and ammonia sensing properties. *Sensors* **2018**, *18*, 2141. [CrossRef] [PubMed]
22. Valli, L. Phthalocyanine-based Langmuir-Blodgett films as chemical sensors. *Adv. Colloid Interface Sci.* **2005**, *116*, 13–44. [CrossRef] [PubMed]
23. Schöllhorn, B.; Germain, J.P.; Pauly, A.; Maleysson, C.; Blanc, J.P. Influence of peripheral electron-withdrawing substituents on the conductivity of zinc phthalocyanine in the presence of gases. Part 1: Reducing gases. *Thin Solid Film.* **1998**, *326*, 245–250. [CrossRef]
24. Brinkmann, H.; Kelting, C.; Makarov, S.; Tsaryova, O.; Schnurpfeil, G.; Wöhrle, D.; Schlettwein, D. Fluorinated phthalocyanines as molecular semiconductor thin films. *Phys. Status Solidi Appl. Mater. Sci.* **2008**, *205*, 409–420. [CrossRef]

25. Germain, J.P.; Pauly, A.; Maleysson, C.; Blanc, J.P.; Schöllhorn, B. Influence of peripheral electron-withdrawing substituents on the conductivity of zinc phthalocyanine in the presence of gases. Part 2: Oxidizing gases. *Thin Solid Film.* **1998**, *333*, 235–239. [CrossRef]
26. Hesse, K.; Schlettwein, D. Spectroelectrochemical investigations on the reduction of thin films of hexadecafluorophthalocyaninatozinc (F16PcZn). *J. Electroanal. Chem.* **1999**, *476*, 148–158. [CrossRef]
27. Shao, X.; Wang, S.; Li, X.; Su, Z.; Chen, Y.; Xiao, Y. Single component p-, ambipolar and n-type OTFTs based on fluorinated copper phthalocyanines. *Dye. Pigment.* **2016**, *132*, 378–386. [CrossRef]
28. Kuzumoto, Y.; Matsuyama, H.; Kitamura, M. Partially fluorinated copper phthalocyanine toward band engineering for high-efficiency organic photovoltaics. *Jpn. J. Appl. Phys.* **2014**, *53*, 01AB03. [CrossRef]
29. Bonegardt, D.; Klyamer, D.; Sukhikh, A.; Krasnov, P.; Popovetskiy, P.; Basova, T. Fluorination vs. Chlorination: Effect on the Sensor Response of Tetrasubstituted Zinc Phthalocyanine Films to Ammonia. *Chemosensors* **2021**, *9*, 137. [CrossRef]
30. Klyamer, D.; Bonegardt, D.; Krasnov, P.; Sukhikh, A.; Popovetskiy, P.; Khezami, K.; Durmuş, M.; Basova, T. Halogen-substituted zinc(II) phthalocyanines: Spectral properties and structure of thin films. *Thin Solid Film.* **2022**, *754*, 139301. [CrossRef]
31. Klyamer, D.D.; Sukhikh, A.S.; Gromilov, S.A.; Kruchinin, V.N.; Spesivtsev, E.V.; Hassan, A.K.; Basova, T.V. Influence of fluorosubstitution on the structure of zinc phthalocyanine thin films. *Macroheterocycles* **2018**, *11*, 304–311. [CrossRef]
32. Becke, A.D. Density-functional exchange-energy approximation with correct asymptotic behavior. *Phys. Rev. A* **1988**, *38*, 3098–3100. [CrossRef] [PubMed]
33. Perdew, J.P. Density-functional approximation for the correlation energy of the inhomogeneous electron gas. *Phys. Rev. B* **1986**, *33*, 8822–8824. [CrossRef] [PubMed]
34. Weigend, F.; Ahlrichs, R. Balanced basis sets of split valence, triple zeta valence and quadruple zeta valence quality for H to Rn: Design and assessment of accuracy. *Phys. Chem. Chem. Phys.* **2005**, *7*, 3297–3305. [CrossRef]
35. Grimme, S.; Ehrlich, S.; Goerigk, L. Effect of the damping function in dispersion corrected density functional theory. *J. Comput. Chem.* **2011**, *32*, 1456–1465. [CrossRef] [PubMed]
36. Grimme, S.; Antony, J.; Ehrlich, S.; Krieg, H. A consistent and accurate ab initio parametrization of density functional dispersion correction (DFT-D) for the 94 elements H-Pu. *J. Chem. Phys.* **2010**, *132*, 154104. [CrossRef]
37. Baerends, E.J.; Ellis, D.E.; Ros, P. Self-consistent molecular Hartree-Fock-Slater calculations I. The computational procedure. *Chem. Phys.* **1973**, *2*, 41–51. [CrossRef]
38. Dunlap, B.I.; Connolly, J.W.D.; Sabin, J.R. On some approximations in applications of Xα theory. *J. Chem. Phys.* **1979**, *71*, 3396–3402. [CrossRef]
39. Van Alsenoy, C. Ab initio calculations on large molecules: The multiplicative integral approximation. *J. Comput. Chem.* **1988**, *9*, 620–626. [CrossRef]
40. Kendall, R.A.; Früchtl, H.A. The impact of the resolution of the identity approximate integral method on modern ab initio algorithm development. *Theor. Chem. Acc.* **1997**, *97*, 158–163. [CrossRef]
41. Eichkorn, K.; Treutler, O.; Öhm, H.; Häser, M.; Ahlrichs, R. Auxiliary basis sets to approximate Coulomb potentials. *Chem. Phys. Lett.* **1995**, *242*, 652–660, Erratum in *Chem. Phys. Lett.* **1995**, *242*, 652–660. [CrossRef]
42. Eichkorn, K.; Weigend, F.; Treutler, O.; Ahlrichs, R. Auxiliary basis sets for main row atoms and transition metals and their use to approximate Coulomb potentials. *Theor. Chem. Acc.* **1997**, *97*, 119–124. [CrossRef]
43. Weigend, F. Accurate Coulomb-fitting basis sets for H to Rn. *Phys. Chem. Chem. Phys.* **2006**, *8*, 1057–1065. [CrossRef] [PubMed]
44. Neese, F. The ORCA program system. *Wiley Interdiscip. Rev. Comput. Mol. Sci.* **2012**, *2*, 73–78. [CrossRef]
45. Neese, F. Software update: The ORCA program system, version 4.0. *Wiley Interdiscip. Rev. Comput. Mol. Sci.* **2017**, *8*, 73–78. [CrossRef]
46. Bühl, M.; Kabrede, H. Geometries of transition-metal complexes from Density-Functional Theory. *J. Chem. Theory Comput.* **2006**, *2*, 1282–1290. [CrossRef]
47. Keith, T.A. *AIMAll*, version 19.10.12; TK Gristmill Software: Overland Park, KS, USA, 2019. Available online: aim.tkgristmill.com (accessed on 5 September 2022).
48. Bonegardt, D.; Klyamer, D.; Krasnov, P.; Sukhikh, A.; Basova, T. Effect of the position of fluorine substituents in tetrasubstituted metal phthalocyanines on their vibrational spectra. *J. Fluor. Chem.* **2021**, *246*, 109780. [CrossRef]
49. Boys, S.F.; Bernardi, F. The calculation of small molecular interactions by the differences of separate total energies. Some procedures with reduced errors. *Mol. Phys.* **1970**, *19*, 553–566. [CrossRef]
50. Simon, S.; Duran, M. How does basis set superposition error change the potential surfaces for hydrogen-bonded dimers? *J. Chem. Phys.* **1996**, *105*, 11024–11031. [CrossRef]
51. Dunning, T.H. Gaussian basis sets for use in correlated molecular calculations. I. The atoms boron through neon and hydrogen. *J. Chem. Phys.* **1989**, *90*, 1007–1023. [CrossRef]
52. Bader, R.F.W.; Essén, H. The characterization of atomic interactions. *J. Chem. Phys.* **1984**, *80*, 1943–1960. [CrossRef]
53. Bader, R.F.W. A quantum theory of molecular structure and its applications. *Chem. Rev.* **1991**, *91*, 893–928. [CrossRef]
54. Bader, R.F.W. *Atom in Molecules: A Quantum Theory*; Oxford University Press: Oxford, UK, 1994.
55. Sukhikh, A.; Bonegardt, D.; Klyamer, D.; Basova, T. Effect of non-peripheral fluorosubstitution on the structure of metal phthalocyanines and their films. *Dye. Pigment.* **2021**, *192*, 109442. [CrossRef]

56. Klyamer, D.D.; Sukhikh, A.S.; Trubin, S.V.; Gromilov, S.A.; Morozova, N.B.; Basova, T.V.; Hassan, A.K. Tetrafluorosubstituted Metal Phthalocyanines: Interplay between Saturated Vapor Pressure and Crystal Structure. *Cryst. Growth Des.* **2020**, *20*, 1016–1024. [CrossRef]
57. Wisitsoraat, A.; Tuantranont, A.; Comini, E.; Sberveglieri, G.; Wlodarski, W. Characterization of n-type and p-type semiconductor gas sensors based on NiOx doped TiO2 thin films. *Thin Solid Film.* **2009**, *517*, 2775–2780. [CrossRef]
58. Chia, L.S.; Hua Du, Y.; Palale, S.; See Lee, P. Interaction of Copper Phthalocyanine with Nitrogen Dioxide and Ammonia Investigation Using X-ray Absorption Spectroscopy and Chemiresistive Gas Measurements. *ACS Omega* **2019**, *4*, 10388–10395. [CrossRef]
59. Ivanova, V.; Klyamer, D.; Krasnov, P.; Kaya, E.N.; Kulu, I.; Tuncel Kostakoğlu, S.; Durmuş, M.; Basova, T. Hybrid materials based on pyrene-substituted metallo phthalocyanines as sensing layers for ammonia detection: Effect of the number of pyrene substituents. *Sens. Actuators B Chem.* **2023**, *375*, 132843. [CrossRef]
60. Bushmarinov, I.S.; Lyssenko, K.A.; Antipin, M.Y. Atomic energy in the "Atoms in Molecules" theory and its use for solving chemical problems. *Russ. Chem. Rev.* **2009**, *78*, 283–302. [CrossRef]
61. Bejaoui, A.; Guerin, J.; Aguir, K. Modeling of a p-type resistive gas sensor in the presence of a reducing gas. *Sens. Actuators B Chem.* **2013**, *181*, 340–347. [CrossRef]
62. Chen, Y.; Yuan, M.; Zhang, Y.; Wang, X.; Ke, F.; Wang, H. One-pot synthesis of tin oxide/reduced graphene oxide composite coated fabric for wearable ammonia sensor with fast response/recovery rate. *J. Alloy. Compds.* **2023**, *931*, 167585. [CrossRef]
63. Li, Q.; Xu, M.; Jiang, C.; Song, S.; Li, T.; Sun, M.; Chen, W.; Peng, H. Highly sensitive graphene-based ammonia sensor enhanced by electrophoretic deposition of MXene. *Carbon* **2023**, *202*, 561–570. [CrossRef]
64. Zhang, L.; Zhang, H.; Chen, C.; Hu, Z.; Wang, J. Preparation and mechanism of high-performance ammonia sensor based on tungsten oxide and zinc oxide composite at room temperature. *Curr. Appl. Phys.* **2023**, *45*, 30–36. [CrossRef]
65. Ridhi, R.; Gautam, S.; Saini, G.S.S.; Tripathi, S.K.; Rawat, J.S.; Jha, P. Study of the effect of orbital on interaction behaviour of SWCNT-metal phthalocyanines composites with ammonia gas. *Sens. Actuators B Chem.* **2021**, *337*, 129767. [CrossRef]

chemosensors

MDPI

Article

Prediction of Diamene-Based Chemosensors

Danil W. Boukhvalov [1,2,*] and Vladimir Yu. Osipov [3]

[1] College of Science, Institute of Materials Physics and Chemistry, Nanjing Forestry University, Nanjing 210037, China
[2] Institute of Physics and Technology, Ural Federal University, Mira 19 Str., 620002 Yekaterinburg, Russia
[3] Ioffe Institute, Polytechnicheskaya 26, 194021 St. Petersburg, Russia
* Correspondence: danil@njfu.edu.cn

Abstract: This paper presents the results of systematic studies of the atomic structure of the layered bulk, bilayer, and monolayer of diamene (a two-dimensional diamond monolayer recently synthesized by various methods) functionalized with fluorine and hydroxyl groups with the chemical formulas C_2F and C_2OH. The results of our calculations show that both types of diamene under discussion have a wide optical gap corresponding to the absorption of light in the UV spectral range. The formation of a boundary between these two types of diamene layers leads to a significant decrease in the band gap. Therefore, this layered material, with an interface between fluorinated and hydroxylated diamenes (C_2F/C_2OH structures), can be considered a suitable material for converting UV radiation into visible light in the orange-yellow part of the spectrum. The adsorption of acetone or water on the C_2F/C_2OH structures results in visible changes in the band gap. The effect on photoemission is different for different detected analytes. The presence of formaldehyde in water ensures the appearance of distinct peaks in the absorption spectra of structures based on C_2F/C_2OH. Our simulation results suggest that the simulated C_2F/C_2OH structures can be used as chemically stable, lightweight materials composed of common elements for a highly selective chemical sensor in liquid and air.

Keywords: 2D materials; volatile organic compounds; formaldehyde; acetone; humidity; downconversion

Citation: Boukhvalov, D.W.; Osipov, V.Y. Prediction of Diamene-Based Chemosensors. *Chemosensors* **2022**, *10*, 480. https://doi.org/10.3390/chemosensors10110480

Academic Editor: Eleonora Alfinito

Received: 12 October 2022
Accepted: 8 November 2022
Published: 15 November 2022

1. Introduction

In the last decade, luminescent sensor materials have been widely discussed both experimentally [1–3] and theoretically [4–6]. The advantages of this class of materials are the stability of the sensing characteristics in external electric and magnetic fields and the possibility of separating the sensitive and detecting parts of the device [7,8]. The latter property makes it possible to use luminescent sensor materials in remote areas and/or in hazardous environments. Complex compounds with lanthanides are usually considered promising for use as luminescent sensor materials [7,9]. In addition to the rarity and cost of the lanthanides, another disadvantage of these compounds is the deterioration of perceptual characteristics in a humid environment. Thus, it is necessary to develop an alternative effective luminescent sensing material that is reliable in various harsh and humid conditions and is built from common elements.

When stability is essential, diamonds are an obvious choice for transparent materials. Bulk diamonds are less commonly used as sensors. However, thin diamond layers and submicron diamond particles can be used to design sensors. Thus, diamond particles 50–100 nm in size with nitrogen-vacancy fluorescent centers have recently been used for local determination of the pH of the medium inside biological cells in real time using the method of optically detected magnetic resonance [10]. Thin layers based on diamond-like carbon or nanocrystalline diamond can be used as carriers of grafted molecular structures—sensitive molecules that can be used as sensors. For example, a diamond surface terminated by atomic hydrogen can be used to design gas sensors [11]. In the latter case, the change in the electrical properties of the surface during the sorption of external agents from the atmosphere is monitored. The surface of a diamond can also be a carrier of grafted hydrophobic polymer structures that facilitate the separation of the studied analyte in the sensor from the aqueous medium [12]. In this case, infrared absorption spectroscopy in the total internal reflection mode is used to study the analyte absorbed inside the polymer. In turn, diamond doped with boron can act as an electrode for a number of inert electrochemical sensors with nanomolar threshold sensitivity [13,14]. The advantage of the diamond lattice is its inertness to external agents and environments.

Two-dimensional diamonds (see Figure 1a,b) or diamenes were initially predicted theoretically [15] and, decades later, synthesized in several experiments by different research groups [16,17] (for review, see Ref. [18]). Functional groups on the surface play essential roles in the stabilization of 2D-diamond morphology. Experimental works demonstrate that hydrogenation of the surface allows for the formation of structures that are stable only under high pressure [19]. On the contrary, diamene, with its surface functionalized with fluorine and hydroxyl groups, is rather stable material [18]. Theoretical simulations [15] and experimental measurements [17] demonstrate that diamenes have a wide energy gap. The combination of the energy gap corresponding with the UV part of the spectra with stable and uniform atomic structure makes diamene a prospective wide-gap material for the optical [20] and chemical [21] sensors.

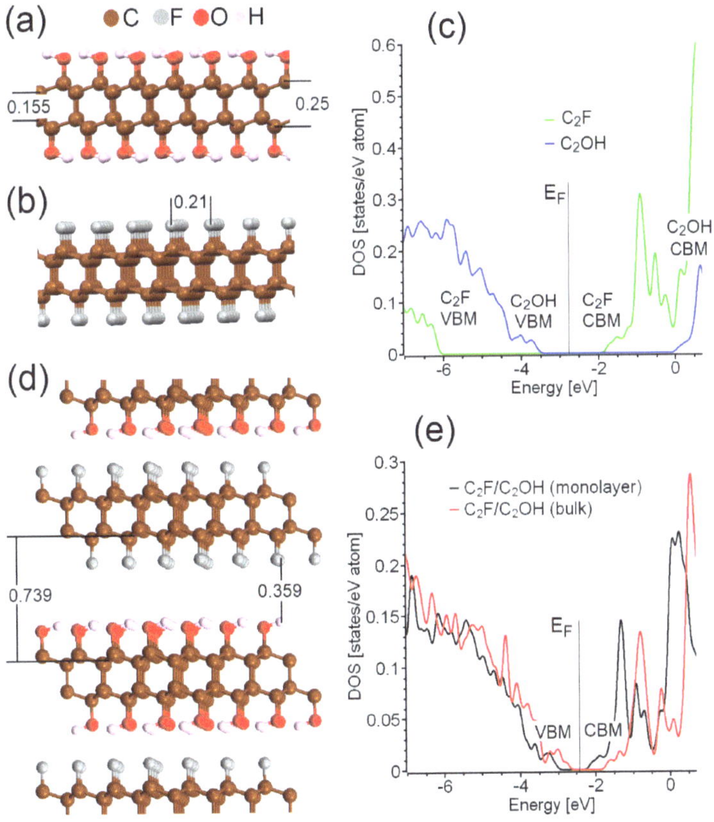

Figure 1. Side view of optimized atomic structure of diamenes C_2OH (**a**), C_2F (**b**), and part of bulk diamene with alternating C_2F/C_2OH structure (**d**). On the right are corresponding total densities of states of C_2OH and C_2F monolayers (**c**) and C_2F/C_2OH-structure in bulk (**e**) and single C_2F/C_2OH-layer form. All distances on panels (**a**,**b**,**d**) are reported in nanometers. E_F is a Fermi energy.

In this paper, we report the results of the simulations of the atomic and electronic structure of the interface between the layers of diamene functionalized by fluorine and the layers of diamene functionalized by hydroxyl groups. We studied the suitability of these compounds for probing in air and liquid. Because the presence of acetone in exhaled air is important for diagnosing early stage diabetes [22], we modeled the adsorption of acetone from the air on diamene-based compounds. To test the effect of moisture and liquid media, the interaction of these compounds with water was simulated. To evaluate the effectiveness of diamene-based compounds as sensors in liquid media, we simulated the interaction of formaldehyde [23] in aqueous media with diamene-based compounds.

2. Computational Methods

Theoretical modeling was carried out using SIESTA pseudopotential code [24] employing the generalized gradient approximation (GGA-PBE) [25] for the exchange-correlation potential in a spin-polarized mode. Van der Waals correction was also taken into account using the approach proposed by Dion et. al. [26] implemented in SIESTA code. A full optimization of the atomic positions was carried out, during which the electronic ground state was consistently found using norm-conserving pseudopotentials [27] for the cores with a double-ξ-plus polarization basis for the localized orbitals of nonhydrogen atoms and a double ξ for hydrogen atoms. Note that the used basis corresponded with a minimal basis set superposition error (BSSE) [28]. Note that the typical van der Waals bonds contribution from BSSE is about 5 K or 0.42 meV [29], which is three orders smaller than the values of energy calculated for these systems. The force and total energy were optimized with an accuracy of 0.04 eV Å^{-1} and 1.0 meV/cell (or less than 0.02 meV/atom), respectively. The values of transferred charge were calculated by Bader analysis.

In this paper, we discuss the relationships between atomic structures of various simulated systems and the value of the energy gap between the valence band maximum (VBM) and conductive band minimum (CBM) in the analyzed periodic systems. The value of this band gap determines the wavelength of the characteristic fluorescence/luminescence—more precisely, its lower limit (in the approximation of the absence of the energy states associated with intrinsic defects in the band gap). In practice, luminescence is excited by optical short-wavelength laser radiation. This kind of luminescence is called photoluminescence (PL). The standard DFT realized in used computational code provides an underestimation of the values of the bandgap. [30] In the mentioned work, the simple linear relationship between the measured (E_{exp}) and calculated within standard DFT (E_{DFT}) values of the bandgap (in the range from 0 to 7 eV) were reported. Relationships between empirical values of bandgaps and those calculated within DFT frameworks can be approximated by a semiempirical formula:

$$E_{expt} = 0.8\ \text{eV} + 1.1E_{DFT} \quad (1)$$

We used this formula to estimate the correct values of the bandgaps based on the results of our DFT calculations.

The values of interlayer binding energies were calculated by the following standard formula:

$$E_{bind} = (E_{m1} + E_{m1} - E_{inter})/n \quad (2)$$

where E_{m1} and E_{m2} are the calculated total energies of the free-standing isolated monolayers forming the interface when in contact, E_{inter} is the calculated total energy of the resulting interface, and n is the number of interlayer interactions for each unit of monolayer. The value of n is equal to one for a bilayer and two for a monolayer. Note that positive binding energy corresponds to stable interfaces. The enthalpy of adsorption was calculated by a similar standard formula:

$$E_{ads} = E_{host+guest} - (E_{host} + E_{guest}) \quad (3)$$

where E_{host}, E_{guest}, and $E_{host+guest}$ are the total energies of the substrate, analyte, and substrate with adsorbed analyte, respectively.

3. Results and Discussion

3.1. Atomic Structure and Physical Properties of C_2F/C_2OH Interface in Air

The first step in our modeling was the construction of models of diamond-like C_2F and C_2OH monolayers. For this, we used a unit cell of two-layer graphene with AB stacking under periodic boundary conditions. Two carbon atoms located one above the other were initially slightly (by 0.1 Å) shifted toward each other to initiate the formation of an interlayer covalent bond. The carbon atoms belonging to another sublattice were displaced in opposite directions with the addition of fluorine atoms or hydroxyl groups to these carbon atoms. After that, complete optimization of the lattice parameters and the positions of atoms was carried out. The results of the DFT calculations are shown in Figure 1a,b. All carbon atoms in these structures are located in a tetrahedral environment, which corresponds to sp^3 hybridization. Half of the carbon atoms in each sublayer of the diamene are bonded to only carbon atoms. The other half of the carbons consist of three carbons from the same sublayer and one surface functional group. The distance between the nearest average parallel lines passing through the centers of functional groups is about 0.21 nm, and the interlayer distance is about 0.74 nm with an interlayer distance of 0.36 nm (see Figure 1a,b,d). Note that the first value (0.21 nm) matches well with the period of straight bands observed in some carbon nanostructures [31]. Thus, we also assumed the possibility of the formation of some diamene-like structures during the production of carbon dots. Thus, the atomic structure of some carbon dots known from the literature should be considered, taking into account the possibility of the formation of diamene-like structures. Note that the methods used to produce the mentioned carbon dots are considerably simpler than the modern methods for the fabrication of diamenes (see Introduction).

Because almost all carbon atoms in these systems have sp^3 hybridization, the electronic structure of both systems has a diamond-like character (see Figure 2c). The key difference between the two discussed systems (C_2F and C_2OH) lies in the absolute positions of the valence band maximum (VBM) and conduction band minimum (CBM) on the energy scale. The formation of the interface between these two compounds allows for the transfer of photogenerated electrons from CBM C_2OH (about 0 eV) through the interface into the C_2F conductive band with a CBM of about -1.8 eV. A photogenerated hole from the top of the C_2F valence band (about -6 eV) moves through the interface to the C_2OH valence bands with a maximum of about -3.5 eV (see Figure 1c), and further radiative recombination occurs. To estimate the energy characteristics of the emitted photons, we simulated the boundary between C_2F- and C_2OH-terminated atomic layers (see Figure 1d) and estimated the energy gap of the material under discussion. The calculated density of states of the C_2F/C_2OH interface (see Figure 1e) confirmed this initial assumption. The calculated value of the band gap is about 1.2 eV. Taking into account the underestimation of the band gap in the standard DFT framework, estimated by Formula (1), the energy of the emitted photon is about 2.1 eV, which corresponds to a wavelength of about 590 nm and lies in the orange-yellow part of the visible spectrum. Therefore, C_2F/C_2OH interfaces can be considered suitable materials for converting absorbed UV radiation into visible light in the orange-yellow part of the spectrum. Thus, diamene, like other wide-gap [15] materials, can be used to convert UV radiation into visible radiation.

The implementation of the scenario considered above is possible only in the presence of charge transfer between parts of the interface (for a detailed discussion and experimental evidence, see Ref. [32]). Similar to the work cited above, we calculated the change in the charge density after the formation of the interface between the C_2F and C_2OH monolayers. The calculation results (Figure 2a) demonstrate a visible change in the charge density map across the interface, which is direct evidence of charge transfer between the C_2F and C_2OH monolayers. The value of the transferred charge is about -0.03 e^- per unit cell. Note that the hopping of the electrons across the interface requires overlap between CBM-related bands of neighboring layers in the coordinate space. The visualization of these bands (see Figure 2b) demonstrates the presence of this overlap. An almost identical picture (omitted due to similarity) could be observed for the overlap between the bands corresponding with VBM of both constituents of the interface. Therefore, the conversion of UV/blue light to the longer wavelength of visible light discussed above is possible.

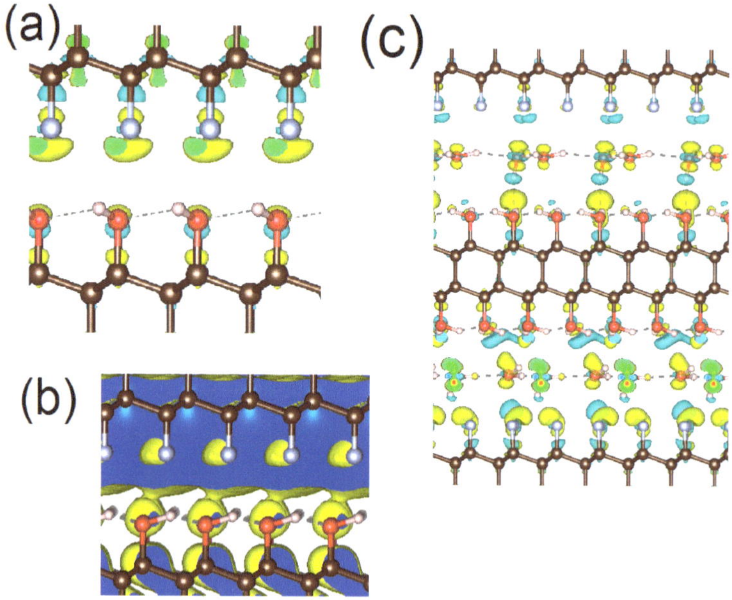

Figure 2. Changes in charge density after formation of C_2F/C_2OH interface (**a**), the sequence charge density of conductive band minimum orbitals of C_2F and C_2OH in C_2F/C_2OH interface (**b**), and changes in charge density after formation of multilayered water–C_2F–water–C_2OH–water structure (**c**). Cyan and yellow clouds corresponding with decreasing and increasing charge density. The color scale of the atoms is the same as in Figure 1.

The next step in our simulation was to test the effect of the formation of a bulk structure from alternating C_2F and C_2OH monolayers (see Figure 1d). The calculated atomic structure of the C_2F/C_2OH bulk structure is similar to that obtained from the calculations for a single layer. The formation of an infinite alternation of C_2F and C_2OH layers leads to insignificant changes in bandwidth values, so the color of the emitted light does not depend on the number of layers in the system. Because the fabrication of the bulk structures based on C_2F/C_2OH by precipitation is much easier than that of single-layer C_2F/C_2OH, the production of bulk devices based on diamene can be considered a realistic proposal.

The next step in our modeling of C_2F/C_2OH interfaces was the assessment of the chemical and structural stability of the system. The calculated energies of removal of one fluorine atom or a single hydroxyl group from the surface of the C_2F/C_2OH are approximately 2.5 eV and 1.8 eV, respectively. Therefore, thermal degradation of C_2F/C_2OH-based materials requires a temperature of hundreds of degrees Celsius (see reference discussion in [33]). The binding energies calculated by Formula (2) between the C_2F and C_2OH monolayers are 0.22 and 0.46 meV/unit for the C_2F/C_2OH bilayer and bulk, respectively. Note that the energies obtained are typical for hydrogen bonds (about 0.15 eV/bond) [34] and higher than those calculated for interlayer bonds in graphite (0.04 eV/bond) [35]. Thus, spontaneous delamination of layers in structures based on C_2F/C_2OH at room temperature can be considered unlikely at temperatures below several hundred degrees Celsius.

3.2. Atomic Structure and Physical Properties of C_2F/C_2OH-Interface in Liquid Media

The C_2F/C_2OH interface described in the previous section can be considered an interface between hydrophobic (C_2F) and hydrophilic (C_2OH) surfaces. Known experimental results demonstrate the possibility of water intercalation between such interfaces, with the formation of hexagonal ice in the interlayer space [36,37]. To test this scenario, we increased the supercell along the layers and then quasirandomly placed several water molecules between and above the layers. Similar to previous works [36,38], the formation of a monolayer of hexagonal ice was observed (see Figure 3a,b). Note that the ordering of water molecules in hexagonal ice was obtained by optimizing quasirandomly distributed water molecules without any initial guidance. The formation of these "bonding clusters" (also called "iceberg layers") on surfaces is a fairly common phenomenon and does not affect the accessibility of the surface to other molecules, which is important for sensors [39]. The incorporation of an ordered layer of water into the interlayer slit affects the charge transfer between the C_2F and C_2OH surfaces (see Figure 2b) and affects the band gap for C_2F/C_2OH based structures (two-layer and bulk; see Figure 3c). The wavelength of the characteristic PL of C_2F/C_2OH-based structures in humid environments, estimated by Formula (1), is about 400 nm, which corresponds to the violet part of the visible light spectrum. The blueshift of the PL of structures based on C_2F/C_2OH in water and a humid environment makes it possible to use these compounds in humidity sensors, as well as to detect various chemical compounds in water. Note that the charge transfer (about -0.01 e^- per unit cell) shown in Figure 2c corresponds to the formation of C_2F–H_2O–C_2OH electrostatic bonds; therefore, C_2F/C_2OH interfaces can be considered stable in wet and liquid media. The smaller value of the charge transfer between the layers is caused by an increase in the distance between the layers and by significant changes in the charge density on water molecules (see Figure 2c). Note that the effect of water on the diamene-based sensors described in this section demonstrates the advantage of these compounds over lanthanide-based luminescent sensor materials, where coordination of active sites by water could quench the luminescence [7].

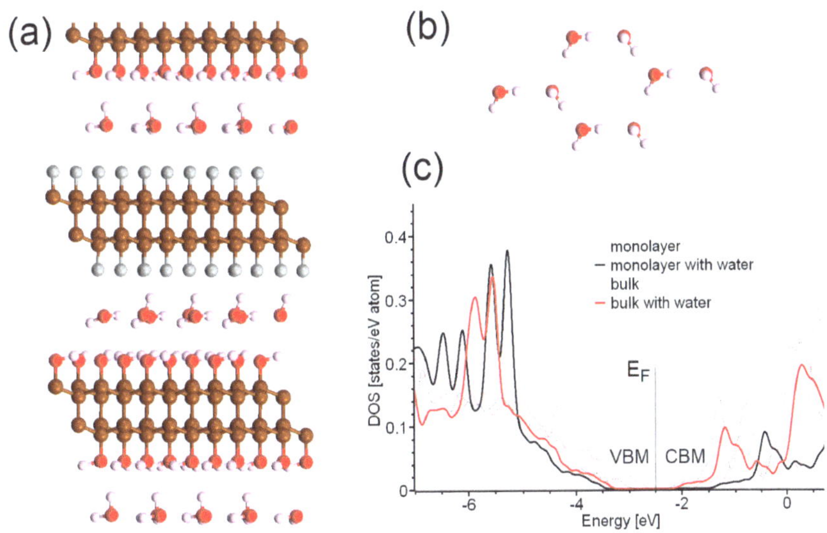

Figure 3. (**a**) Side view of optimized atomic structure of bulk C_2F/C_2OH system intercalated by water, and (**b**) top view of hexagonal patterns formed by water in the interlayer space of the structure presented in panel (**a**). (**c**) Total densities of states of C_2F/C_2OH single layer and bulk intercalated by water (solid lines), and total densities of states of corresponding structures without water. The color scale of the atoms in panel (**a**) is the same as in Figure 1.

3.3. Sensing Properties of C_2F/C_2OH Structures

The sensitivity of the optical properties of structures based on C_2F/C_2OH to a humid environment, considered in the previous section, demonstrates the possibility of using these materials as sensors. To explore this possibility, we modeled the detection of acetone (AC) in air and formaldehyde (FA) in water (see Figures 4a and 5a). Note that the termination of structures based on C_2F/C_2OH can be equally probable with a monolayer of C_2F or C_2OH. Thus, to complete the picture, we considered the adsorption of AC both on the C_2F-terminated side and on the C_2OH-terminated sides of the C_2F/C_2OH bilayers. The results of our calculations show that AC adsorption on both the fluorine and hydroxyl ends of the C_2F/C_2OH bilayer leads to a decrease in the calculated band gap from 1.2 to 0.7 eV. The last value, converted by Equation (1), corresponds to radiation in the red region of the spectrum. The enthalpy of adsorption of AC on the discussed substrate, calculated by Equation (3), is about −34 kJ/mol. Note that the magnitude of this value is close to the value of acetone vaporization enthalpy (about 32 kJ/mol at 300 K) [40]. Based on the formulas derived from our previous works (see, for example, Refs. [18,41]), the estimated temperature of the desorption of AC from C_2F/C_2OH is about 45 °C. Therefore, C_2F/C_2OH can be cleaned from adsorbed AC by simple mild heating. Thus, the redshift in C_2F/C_2OH fluorescent radiation may indicate the presence of AC in the air, and this therefore suggests that C_2F/C_2OH can be used in a device to detect the presence of AC in exhaled air. Note that a fairly small amount of material is required for the efficient operation of sensors of this type. As a rule, a thin layer of the material can be deposited on an area of the same order as the size of a focused excitation laser beam (3–5 µm), which is not overly intense, so that there is no heating of intercalating liquids or burning of a carbon substrate, and luminescence registration can be carried out for several minutes to accumulate signals in photon counting mode.

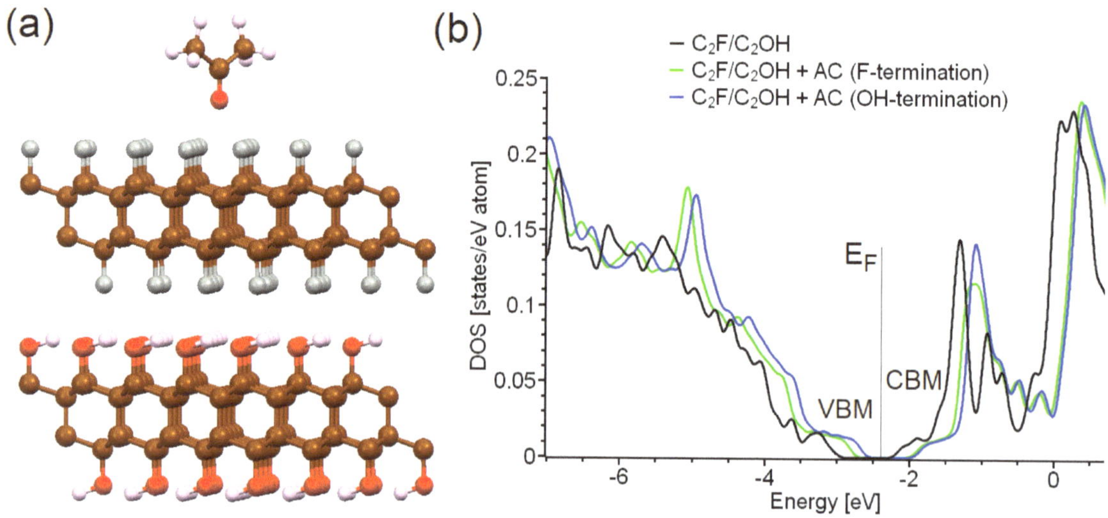

Figure 4. Side view of optimized atomic structure of C_2F/C_2OH single layer with acetone molecule physically adsorbed on F-terminated side (**a**), and total densities of states of the C_2F/C_2OH single layers before and after adsorption of acetone on different sides (terminations) (**b**). The color scale of the atoms in panel (**a**) is the same as in Figure 1.

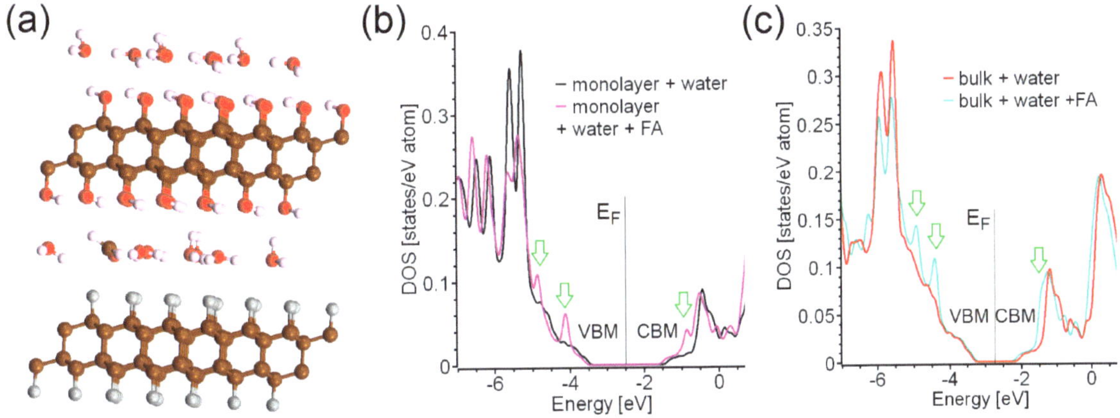

Figure 5. Side view of optimized atomic structure of bulk C_2F/C_2OH structure intercalated by water with molecule of formaldehyde (**a**), and total densities of states of C_2F/C_2OH single layer intercalated by water (**b**) and bulk intercalated by water (**c**) with and without molecules of formaldehyde (see panel (**a**)). The arrows show features in electronic structure corresponding with additional adsorption in near-UV part of spectra. The color scale of the atoms in panel (**a**) is the same as in Figure 1.

To evaluate the sensor properties of the systems based on C_2F/C_2OH in liquid media, we simulated the interaction of formaldehyde (FA) molecules with bulk and bilayer C_2F/C_2OH systems intercalated with water. We started our modeling by replacing one water molecule with a single FA molecule in the already-simulated bulk and layered structures with an intercalated water layer and then carried out a full optimization of lattice constants and atomic positions. The results of our calculations demonstrate that the presence of a single FA molecule does not significantly affect the order of water molecules (compare Figure 5a with Figure 3a). In contrast to AC adsorption from air, the interaction of FA with the C_2F/C_2OH structure dissolved in water does not lead to a change in the band gap either in bulk or in two-layer C_2F/C_2OH (see Figure 5b,c). However, there are two distinct peaks near VBM and one distinct peak near CBM. These three peaks can be the sources of two additional features in the absorption spectra of C_2F/C_2OH in the near UV region of the spectrum, and C_2F/C_2OH can thus be used for simple optical detection of FA in water, by analogy with the previously discussed MOF-based method. [18]. Unlike the detection of molecules by detecting changes in the spectrum (color) of emitted fluorescent radiation, the detection of molecules by detecting changes in the absorption spectrum of the supporting active nanomaterial (matrix) requires more effort in sample preparation for its use in sensor devices in order to provide appropriate changes in the intensity of the light transmitted through a matrix with guest molecules.

The final step of our work was to compare the proposed material with previously reported carbon-based sensing systems. Graphene-based compounds are usually discussed as the best prospective materials for carbon-based sensing materials, mainly due to the changes in their conductivity and in the spectrum of electronic states during the sorption of external agents by means of van der Waals forces, because they have a zero gap and are essentially semimetals in the initial state. In contrast, sensors based on graphene oxide or reduced graphene oxide have a large nonzero gap (~2.2 eV or 1~1.7 eV, depending upon the degree of reduction). This makes it possible to use the fluorescence of these materials for the detection of agents that have the properties of luminescence quenchers [42]. Among the latter are the cations of a number of metals. The same applies to the grafting onto graphene oxide particles of sensitive fluorescent organic molecules that can effectively interact with the analyte and whose intrinsic luminescence is not suppressed by a wide-gap carrier particle [42]. Nanoscale graphene quantum dots also have a gap and exhibit photoluminescent properties that can be used to detect luminescence quenching agents in trace amounts. Graphene quantum dots are sometimes used in combination with other metal nanoparticles to study the analyte by measuring the optical absorption of the solution and detecting the spectral shifts associated with plasmon resonance and the presence of the analyte [43]. The main advantage of the described systems over the proposed C_2F/C_2OH system is the developed procedure of the production. The main disadvantage of other carbon-based materials is that the exact atomic and electronic structure of the materials is uncontrollable.

4. Conclusions

Based on the results of ab initio modeling, we conclude that layered compounds built from C_2F/C_2OH interfaces are stable in air and water. This paper also demonstrates the possibility of UV light conversion into visible light, as in the case of light-emitting carbon dots. The adsorption of acetone molecules from the air changes the spectral composition of emitted radiation from orange-yellow to red. On the contrary, intercalation with water leads to a blueshift in the luminescence to the violet region of the spectrum. These results demonstrate that C_2F/C_2OH-based thin-layer materials have a unique optical response selectivity for different types of analytes. In contrast to adsorption from the gas phase, the presence of a formaldehyde molecule in water on or between the layers of materials based on C_2F/C_2OH does not change the color of the emitted light, but changes their absorption spectra in the near-ultraviolet spectral range. Therefore, C_2F/C_2OH-based materials can also be proposed as materials for the detection of various simple molecules in both gaseous and liquid media.

Author Contributions: Conceptualization, D.W.B. and V.Y.O.; methodology, D.W.B.; software, D.W.B.; validation, D.W.B., V.Y.O.; formal analysis, D.W.B.; investigation, D.W.B.; resources, D.W.B.; data curation, V.Y.O.; writing—original draft preparation, D.W.B.; writing—review and editing, V.Y.O.; visualization, D.W.B.; funding acquisition, D.W.B. and V.Y.O. All authors have read and agreed to the published version of the manuscript.

Funding: D.W.B. was funded by Russian Science Foundation (project no. 21-12-00392). V.Yu.O. was funded by the Ioffe Institute (project no. 0040-20140013).

Institutional Review Board Statement: Not applicable.

Informed Consent Statement: Not applicable.

Data Availability Statement: Data available on request due to restrictions eg privacy or ethical.

Acknowledgments: D.W.B. acknowledges the support of the Russian Science Foundation (project no. 21-12-00392). V.Yu.O. acknowledges the support of the Ioffe Institute (project no. 0040-2014-0013).

Conflicts of Interest: The authors declare no conflict of interest.

References

1. Venkateswarulu, M.; Gaur, P.; Koner, R.R. Sensitive molecular optical material for signaling primary amine vapors. *Sens. Actuators B Chem.* **2015**, *210*, 144–148. [CrossRef]
2. Silva, A.P.; Fox, D.B.; Huxley, A.J.M.; Moody, T.S. Current developments in fluorescent PET (photoinduced electron transfer) sensors and switches. *Coord. Chem. Rev.* **2000**, *205*, 41–57. [CrossRef]
3. Mader, H.S.; Wolfbeis, O.S. Optical ammonia sensor based on upconverting luminescent nanoparticles. *Anal Chem.* **2010**, *82*, 5002–5004. [CrossRef] [PubMed]
4. Mukherjee, S.; Betal, S.; Chattopadhyay, A.P. Luminescence sensing, DFT, extraction and monitoring of Cr^{3+} and Al^{3+}via the application of first derivative fluorescence spectroscopy. *New J. Chem.* **2020**, *44*, 12692–12703. [CrossRef]
5. Chai, B.-L.; Yao, S.-L.; Xie, X.; Xu, H.; Zheng, T.-F.; Li, J.-Y.; Chen, J.-L.; Liu, S.-J.; Wen, H.-R. Luminescent Metal–Organic Framework-Based Fluorescence Turn-On and Red-Shift Sensor toward Al3+ and Ga3+: Experimental Study and DFT Calculation. *Cryst. Growth Des.* **2022**, *22*, 277–284. [CrossRef]
6. Vellingiri, K.; Boukhvalov, D.W.; Pandey, S.K.; Deep, A.; Kim, K.H. Luminescent metal-organic frameworks for the detection of nitrobenzene in aqueous media. *Sens. Actuators B Chem.* **2017**, *245*, 305–313. [CrossRef]
7. Aulsebrook, M.L.; Graham, B.; Grace, M.R.; Tuck, K.L. Lanthanide complexes for luminescence-based sensing of low molecular weight analytes. *Coord. Chem. Rev.* **2018**, *375*, 191–220. [CrossRef]

8. Timmer, B.; Olthuis, W.; van den Berg., A. Ammonia sensors and their applications—A review. *Sens. Actuators. B Chem.* **2005**, *107*, 666–677. [CrossRef]
9. Roy, S.; Chakraborty, A.; Maji, T.K. Lanthanide–organic frameworks for gas storage and as magneto-luminescent materials. *Coord. Chem. Rev.* **2014**, *273–274*, 139–164. [CrossRef]
10. Fujisaku, T.; Igarashi, R.; Shirakawa, M. Nanometre-scale visualization of chemical parameter changes by T1-weighted ODMR imaging using a fluorescent nanodiamond. *Chemosensors* **2020**, *8*, 68. [CrossRef]
11. Kočí, M.; Kromka, A.; Bouřa, A.; Szabó, O.; Husák, M. Hydrogen-terminated diamond surface as a gas sensor: A comparative study of its sensitivities. *Sensors* **2021**, *21*, 5390. [CrossRef]
12. Dettenrieder, C.; Türkmen, D.; Mattsson, A.; Österlund, L.; Karlsson, M.; Mizaikoff, B. Determination of Volatile Organic Compounds in Water by Attenuated Total Reflection Infrared Spectroscopy and Diamond-Like Carbon Coated Silicon Wafers. *Chemosensors* **2020**, *8*, 75. [CrossRef]
13. da Silva, Á.R.L.; de Araujo, D.M.; da Silva, E.B.S.; Vieira, D.S.; Monteiro, N.D.K.V.; Martinez-Huitle, C.A. Understanding the behavior of caffeine on a boron-doped diamond surface: Voltammetric, DFT, QTAIM and ELF studies. *New J. Chem.* **2017**, *41*, 7766–7774. [CrossRef]
14. Yuan, X.; Gao, N.; Gao, X.; Qiu, D.; Xu, R.; Sun, Z.; Jiang, Z.; Liu, J.; Li, H. Nanopyramid boron-doped diamond electrode realizing nanomolar detection limit of 4-nonylphenol. *Sens. Actuators B Chem.* **2019**, *281*, 830–836. [CrossRef]
15. Leenaerts, O.; Partoens, B.; Peeters, F.M. Hydrogenation of bilayer graphene and the formation of bilayer graphane from first principles. *Phys. Rev. B* **2009**, *80*, 245422. [CrossRef]
16. Barboza, A.P.M.; Guimaraes, M.H.D.; Massote, D.V.P.; Campos, L.C.; Barbosa Neto, N.M.; Cancado, L.G.; Lacerda, R.G.; Chacham, H.; Mazzoni, M.S.C.; Neves, B.R.A. Room-temperature compression-induced diamondization of few-layer graphene. *Adv. Mater.* **2011**, *23*, 3014. [CrossRef]
17. Bakharev, P.V.; Huang, M.; Saxena, M.; Lee, S.W.; Joo, S.H.; Park, S.O.; Dong, J.; Camacho-Mojica, D.C.; Jin, S.; Kwon, Y.; et al. Chemically induced transformation of chemical vapour deposition grown bilayer graphene into fluorinated single-layer diamond. *Nat. Nanotechnol.* **2020**, *15*, 59–66. [CrossRef]
18. Lavini, F.; Rejhon, M.; Reido, E. Two-dimensional diamonds from sp2-to-sp3 phase transitions. *Nat. Rev. Mater.* **2022**, *7*, 814–832. [CrossRef]
19. Martins, L.G.P.; Matos, M.J.S.; Paschoal, A.R.; Freire, P.T.C.; Andrade, N.F.; Aguiar, A.L.; Kong, J.; Neves, B.R.A.; de Oliveira, A.B.; Mazzoni, M.S.C.; et al. Raman evidence for pressure-induced formation of diamondene. *Nat. Comm.* **2017**, *8*, 96. [CrossRef]
20. BenMouss, A.; Soltani, A.; Schühle, U.; Haenen, K.; Chong, Y.M.; Zhang, W.J.; Hochedez, J.F. Recent developments of wide-bandgap semiconductor based UV sensors. *Diam. Rel. Mater.* **2009**, *18*, 860–864. [CrossRef]
21. Pearton, S.J.; Ren, F.; Yu-Lin Wang, Y.-L.; Chu, B.H.; Chen, K.H.; Chang, C.Y.; Lim, W.; Lin, J.; Norton, D.P. Recent advances in wide bandgap semiconductor biological and gas sensors. *Prog. Mater. Sci.* **2010**, *55*, 1–59. [CrossRef]
22. Saasa, V.; Malwela, T.; Beukes, M.; Mokgotho, M.; Liu, C.-P.; Mwakikunga, B. Sensing Technologies for Detection of Acetone in Human Breath for Diabetes Diagnosis and Monitoring. *Diagnostics* **2018**, *8*, 12. [CrossRef] [PubMed]
23. Vellingiri, K.; Deep, A.; Kim, K.-H.; Boukhvalov, D.W.; Kumar, P.; Yao, Q. The sensitive detection of formaldehyde in aqueous media using zirconium-based metal organic frameworks. *Sens. Actuators B: Chem.* **2017**, *241*, 938–948. [CrossRef]
24. Soler, J.M.; Artacho, E.; Gale, J.D.; Garsia, A.; Junquera, J.; Orejon, P.; Sanchez-Portal, D. The SIESTA Method for Ab-Initio Order-N Materials Simulation. *J. Phys. Condens. Matter.* **2002**, *14*, 2745. [CrossRef]
25. Perdew, J.P.; Burke, K.; Ernzerhof, M. Generalized Gradient Approximation Made Simple. *Phys. Rev. Lett.* **1996**, *77*, 3865. [CrossRef]
26. Dion, M.; Rydberg, H.; Schröder, E.; Langreth, D.C.; Lundqvist, B.I. Van der Waals density functional for general geometries. *Phys. Rev. Lett.* **2004**, *92*, 246401. [CrossRef]
27. Troullier, O.N.; Martins, J.L. Efficient Pseudopotentials for Plane-Wave Calculations. *Phys. Rev. B* **1991**, *43*, 1993. [CrossRef]
28. Mentel, M.; Baerends, E.J. Can the Counterpoise Correction for Basis Set Superposition Effect Be Justified? *J. Chem. Theory Comput.* **2014**, *10*, 252–267. [CrossRef]
29. Gutowski, M.; Grzegorz Chałasiński, G. Critical evaluation of some computational approaches to the problem of basis set superposition error. *J. Chem. Phys.* **1993**, *98*, 5540–5554. [CrossRef]
30. van Schilfgaarde, M.; Kotani, T.; Faleev, S. Quasiparticle Self-Consistent GW Theory. *Phys. Rev. Lett.* **2006**, *96*, 226402. [CrossRef]
31. Tian, L.; Chen, F.; Ding, H.; Li, X.; Li, X. The influence of inorganic electrolyte on the properties of carbon quantum dots in electrochemical exfoliation. *J. Electrochem. Soc.* **2020**, *878*, 114673. [CrossRef]
32. Ding, M.; Xiao, R.; Zhao, C.; Bukhvalov, D.; Chen, Z.; Xu, H.; Tang, H.; Xu, J.; Yang, X. Evidencing Interfacial Charge Transfer in 2D CdS/2D MXene Schottky Heterojunctions toward High-Efficiency Photocatalytic Hydrogen Production. *Solar Rrl* **2021**, *5*, 2000414. [CrossRef]
33. Boukhvalov, D.W.; Dreyer, D.R.; Bielawski, C.W.; Son, Y.-W. A Computational Investigation of the Catalytic Properties of Graphene Oxide: Exploring Mechanisms by using DFT Methods. *ChemCatChem* **2012**, *4*, 1844–1849. [CrossRef]
34. Perrin, C.L.; Nielson, J.B. "Strong" hydrogen bonds in chemistry and biology. *Ann. Rev. Phys. Chem.* **1997**, *48*, 511–544. [CrossRef] [PubMed]
35. Spanu, L.; Sorella, S.; Galli, G. Nature and strength of interlayer binding in graphite. *Phys. Rev. Lett.* **2009**, *103*, 196401. [CrossRef] [PubMed]

36. Kim, J.-S.; Choi, J.S.; Lee, M.J.; Park, B.H.; Boukhvalov, D.W.; Son, Y.-W.; Yoon, D.; Cheong, H.; Park, J.Y.; Salmeron, M. Between Scylla and Carbides: Hydrophobic Graphene-Guided Water Diffusion on Hydrophilic Substrates. *Sci. Rep.* **2013**, *3*, 2309. [CrossRef]
37. Zheng, Y.; Su, C.; Lu, J.; Loh, K.P. Room-Temperature Ice Growth on Graphite Seeded by Nano-Graphene Oxide. *Angew. Chem.* **2013**, *125*, 8870–8874. [CrossRef]
38. 30 Boukhvalov, D.W.; Katsnelson, M.I.; Son, Y.-W. Origin of anomalous water permeation through graphene oxide membrane. *Nano Lett.* **2013**, *13*, 3930–3935. [CrossRef]
39. Wakisaka, A.; Shimizu, Y.; Nishi, N.; Tokumaru, K.; Sakuragi, H. Interaction of hydrophobic molecules with water influenced by the clustering conditions of acetonitrile–water mixtures. *J. Chem. Soc. Faraday Trans.* **1992**, *88*, 1129–1135. [CrossRef]
40. NIST Chemistry Webbook. Available online: https://webbook.nist.gov/cgi/cbook.cgi?ID=C67641&Mask=4 (accessed on 29 October 2022).
41. Bondino, F.; Duman, S.; Nappini, S.; D'Olimpio, G.; Ghica, C.; Menteş, T.O.; Politano, A. Improving the efficiency of gallium telluride (GaTe) for photocatalysis, electrocatalysis, and chemical sensing through defects engineering and interfacing with its native oxide. *Adv. Funct. Mater.* **2022**, *32*, 2205923. [CrossRef]
42. Sivasubramanian, S.; David, C.I.; Prabhu, J.; Raju, N. Functionalized graphene oxide materials for fluorometric sensing of various analytes: A mini review. *Mater. Adv.* **2021**, *2*, 6197–6212. [CrossRef]
43. Babazadeh, S.; Moghaddam, P.A.; Keshipour, S.; Mollazade, K. Colorimetric sensing of imidacloprid in cucumber fruits using a graphene quantum dot/Au (III) chemosensor. *Sci. Rep.* **2020**, *10*, 14327. [CrossRef] [PubMed]

chemosensors MDPI

Article

Energy-Efficient Chemiresistive Sensor Array Based on SWCNT Networks, WO_3 Nanochannels and SWCNT-Pt Heterojunctions for NH_3 Detection against the Background Humidity

Alexey V. Romashkin [1,*], Andrey V. Lashkov [1], Victor V. Sysoev [2,*], Nikolay S. Struchkov [1], Evgeny V. Alexandrov [1] and Denis D. Levin [1]

[1] Center for Probe Microscopy and Nanotechnology, National Research University of Electronic Technology, Moscow 124498, Russia
[2] Department of Physics, Yuri Gagarin State Technical University of Saratov, Saratov 410054, Russia
* Correspondence: romaleval@gmail.com (A.V.R.); vsysoev@sstu.ru (V.V.S.)

Abstract: Recently, promising results have been achieved in improving the sensitivity to ammonia in gas sensors through the use of structures composed of heterojunctions or nanochannels. However, their sensitivity is highly dependent on the background humidity under air conditions. The sensor structures which could ensure selective ammonia detection with a low detection limit, despite interference from changing background humidity, remain highly demanded. In this work, we consider sensing units containing (i) nanochannels formed by a continuous tungsten oxide nanolayer to appear in contact between single-walled carbon nanotubes (SWCNTs) and a Pt sublayer and (ii) SWCNT-Pt junctions in frames of mass-scale microelectronic technologies. SWCNTs were deposited by spray-coating on a thin WO_3/Pt/W sublayer formed by a photolithographic pattern to be accompanied by satellite samples with just SWCNTs for reference purposes. We elucidate the specific differences that appeared in the response of sensors based on SWCNT-Pt junctions and WO_3 nanochannels relative to satellite SWCNT samples with a similar SWCNT network density. Particularly, while a similar response to NH_3 vapors mixed with dry air is observed for each sensor type, the response to NH_3 is reduced significantly in the presence of background humidity, of 45 rel.%, especially in the case of WO_3 nanochannel structures even at room temperature. A multisensor array based on the four various sensing structures involving SWCNT-Pt junctions, WO_3 nanochannels, and their satellite-only-SWCNT ones allowed us to determine a correct ammonia concentration via utilizing the linear discriminant analysis despite the presence of background air humidity. Thus, such an energy-efficient multisensor system can be used for environmental monitoring of ammonia content, health monitoring, and other applications.

Keywords: carbon nanotube; nanochannel; tungsten oxide; heterojunction; multisensor array; ammonia; gas sensor; spray deposition

Citation: Romashkin, A.V.; Lashkov, A.V.; Sysoev, V.V.; Struchkov, N.S.; Alexandrov, E.V.; Levin, D.D. Energy-Efficient Chemiresistive Sensor Array Based on SWCNT Networks, WO_3 Nanochannels and SWCNT-Pt Heterojunctions for NH_3 Detection against the Background Humidity. *Chemosensors* **2022**, *10*, 476. https://doi.org/10.3390/chemosensors10110476

Academic Editor: Pi-Guey Su

Received: 7 October 2022
Accepted: 10 November 2022
Published: 12 November 2022

1. Introduction

The ammonia vapors present a high demand for detection because of numerous applications linked to this analyte [1,2]. These vapors are extremely hazardous to heavily influence the eyes, lungs, and skin under direct contact causing nose and throat irritation combined with coughing and respiratory tract irritation [3]. Currently, the long-term, 8 h permissible concentration of NH_3 for workers is 25 ppm [4]. At this level, the vapors substantially inhibit the immune system of animals and reduce their ability to eliminate infections [5]. At higher concentrations, in the range of 15–28% by volume, ammonia vapors are flammable in the air. Still, humans feel its odor at the threshold of ca. 6 ppm [6], which allows us to distinguish this danger organoleptically. However, the technical units are obviously needed for continuous monitoring of this toxic gas at the spots of interest, such as an industry's emission, environmental monitoring, and human health. Therefore, to be effective for these applications, the sensors have to yield signals at lower detection limits combined with high stability of the response and small mass/volume parameters [4,7]. At the same time, the NH_3 vapors are the most common ones to appear in compost areas where their concentration reaches ca. 4000 ppm [8], which heavily exceeds the noted permissions and requires sensor-based units capable of working instantly for automatic monitoring tasks. Furthermore, ammonia can react in the atmosphere with other substances, contributing to the particle's aggregation, of which the most harmful are ones smaller than 2.5 microns in size. Thus, the detection threshold for particles, and hence ammonia, is reduced down to concentrations of ca. 30 ppb in the ambient air [7]. The other source of ammonia to be important for humans is meat or poultry, which is subject to metabolic spoiling under a microbe's growth, damaging the organic matter [9], and last but not least, ammonia appears as a biomarker produced by a human body under various metabolic activities: its excessive presence in the exhaled breath, at concentrations higher than 1 ppm [10], indicates diseases related primarily to dysfunctions of liver and kidneys [11,12].

While the ammonia vapors could be well detected by analytical instrumentation such as chromatography [13], these techniques do not suit the requirements to have a compact device that yields "here and now" the corresponding information/signal. Therefore, the NH_3 sensing units are intensively developed employing gas sensors of various principles.

Currently, sensors based on oxide materials with channels formed by continuous thin films are mostly wide-spread to allow one measuring NH_3 concentrations of several ppm as a detection limit [14], and only recently, notable progress has been made in increasing the sensitivity of sensor structures, reaching a detection limit of less than 1 ppm through, for instance, using tungsten oxide as thin layers [15] or nanoflakes [16]. However, these sensors have a number of features that limit their usage: (i) a low selectivity to specific gases, (ii) degradation of the sensitive layer over time upon exposure to high concentrations of various gases and volatile organic compounds (VOCs), and (iii) advanced operating temperature of the sensor, 200 °C and higher, which require an extensive power consumption. Therefore, novel chemiresistive sensors based on nanomaterials matured from carbon-related structures [17], such as carbon nanotubes (CNTs), are intensively developed to operate at room temperature (RT) [18,19]. In addition, these structures have low electrical noise and compatibility with the microprocessor's processing of the sensor signals in a real-time scale [20] to meet entirely all the challenges related, among others, to the Internet-of-Things paradigm [21].

In order to advance the gas-sensing performance of carbon structures, they are complemented with foreign additives for designing heterojunctions [22–24]. However, one of the negative properties of such sensors is a strong dependence on the sensor's response to NH_3 under humidity interference. In some cases, enhancing relative humidity (RH) from 25 rel.% to 64 rel.% might modify the sensor response by six times at the same ammonia concentration [18].

The most significant results in terms of the sensor response magnitude were achieved for structures containing heterojunctions based on materials that have a significant difference in the positions of energy levels in the band structure [18,19] though not all the heterojunctions contribute to advancing the sensor response [25]. From this viewpoint, the efficient approach is combining CNTs with metal oxides. For example, SWCNTs and WO_3 differ by the type of conductivity: the edge of the SWCNT valence band, E_v, is located near the edge of the oxide conduction band, E_c, but with a fairly large difference between these levels. Therefore, this junction might significantly change the nature of the response of such a structure to various gases. CNT-SnO_2 heterostructures are also shown to obtain enhanced sensor response [19]. However, tungsten oxide has significantly more predictable sensing performance yielding, for instance, a linear chemiresistive response to gaseous analytes both at low and high levels of background humidity [26] and higher responses to water vapors [27,28]. Therefore, in this work, we consider namely the combination of WO_3 with SWCNTs. Thus far, reported data in the literature highlight that responses of sensors based on WO_3/CNTs exceed those based on just WO_3 at certain optimal ratios between the components in the composite. Furthermore, there are strong changes in sensor response in response to even minor deviations in optimal component ratios of such composite sensor layers. In addition, CNTs themselves also exhibit differences in sensor response depending on background humidity [29]. Thus, there is a necessity to fabricate heterostructures and nanochannels in a controlled manner on a nanoscale for obtaining sensors with specified characteristics.

One of the solutions can be using photolithography to deposit thin layers properly in specified areas on the substrate to provide the designed topology at the micro level. Together with utilizing nanostructures such as SWCNTs, this makes it possible to design nanosized junctions and channels already at the nanoscale level and fabricate sensor layers in a reproducible and controllable manner. Furthermore, this technology allows one easier placing a number of sensors as a multisensor array [30] on a single substrate with controlled parameters to advance a selectivity while detecting gases and their mixtures [31,32]. Herein, we propose sensors based on SWCNTs in combination with WO_3 and Pt nanolayers, as well as their fabrication technology. The control of the stages of the technological process makes it possible to obtain structures that differ from each other in the gas sensitivity mechanisms. The manufacture of sensors with specified characteristics has a positive effect on their joint application as a part of a multisensor gas-analytical system, where significant differences in the behavior of sensor responses to gases are welcome to advance their selectivity to analytes. The developed multisensor array has been tested regarding the recognition of NH_3 vapors against the background of humidity.

2. Materials and Methods

2.1. Fabrication of Structures under Study

Using magnetron sputtering and lift-off photolithography on glass substrates, 12 × 12 mm, the metallization regions, 20 × 20 μm, were formed as an array with a 30 μm period (Figure 1). The metallization areas consisted of successively deposited W/Pt/W layers, each 10 nm thick. The upper layer of tungsten (W) was oxidized to grow a tungsten oxide (WO_3) by annealing in air at the temperature of ca. 380 °C in oven (Mikroterm-70, Spark-Don, Volgodonsk, Russia). This temperature has been found sufficient for the oxidation of tungsten, accounting for numerous literature data (see, for example, [33,34]), and has been further confirmed in the experiment. Another sample type was obtained in a similar way but with the top tungsten layer covering the entire substrate surface.

The use of structures with island metallization areas made it possible to target forming the nanojunctions of CNT-Pt and CNT-WO_3, which is difficult to implement in composites due to the heterogeneity of the distribution of materials regarding each other, even at the micro level. The continuous WO_3 layers, without making the Pt-sublayer periodic structure with island metallization areas, would not allow one yielding the WO_3 nanochannels to form between the SWCNT and Pt layers. The vertically aligned WO_3 nanochannels, whose length has been targeted by thickness of pristine continuous WO_3 layer, are located between Pt sublayer and top SWCNT layer. The width of these WO_3 nanochannels is controlled by SWCNT diameter (Figure 1, cross-section of SWCNT40-WO_3 sample). Still, top non-continuous networked SWCNT layer does not inhibit access of NH_3 or H_2O molecules to the nanochannel. These nanochannels allowed us to reduce the operating temperature of the sensor down to RT due to the lower resistance of such an oxide nanochannel structure when compared to rather long microchannels between metal electrodes to be conventionally employed in the sensors.

Then, SWCNTs (P3-SWNT, Carbon Solutions, Riverside, CA, USA) were deposited by spray coating with a low aerosol flow density on the surface of six samples in four separate processes (Figure 1) employing own-designed pneumatic spray-coating system [35] from the N-methylpyrrolidone: water dispersion, of 3:7 ratio, with various volumes chosen for each process purposes. The dispersion's output going from the spray nozzle was governed by managing the effective throughput via shifting a locking needle. The deposition was performed over the substrate heated up to 120 °C to be followed by treating the samples with formic acid of analytical grade purity in order to remove the residual solvent from the as-prepared SWCNT layer. For each of the structure types, the deposition process was individual. Together with the WO_3/Pt/W layer-based structure under study, we have employed a pure glass substrate in the chamber as a reference during SWCNT's deposition to prepare a satellite sample. In a separate process, to compare the effect of SWCNT network density on sensor response, two additional samples were fabricated on glass substrates, with a similar (called hereafter, rSWCNT54) and higher (called hereafter, rSWCNT174) network density than the above-described structures.

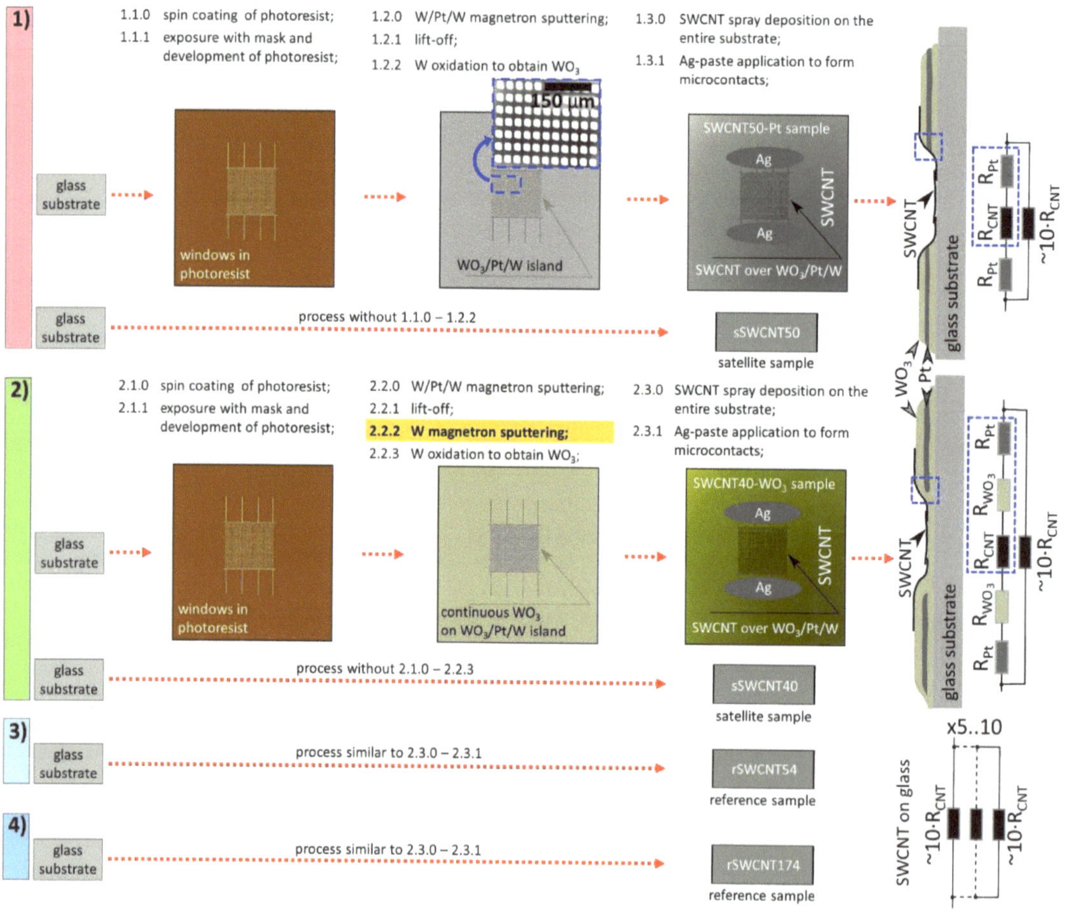

Figure 1. The scheme and technology routs to produce the gas-sensing structures under the study via employing SWCNT layer deposited over WO_3/Pt/W sublayer formed with lift-off photolithography and over the pure glass substrates. In the right, the cross-section and equivalent circuit schemes of the structures are given.

Contacts to sensors were fabricated with Ag paste followed by its drying under infrared irradiation at 85–100 °C with the formation of wire leads at negligibly small, relative to the resistance of the sensor channel, contact resistance.

2.2. Material Characterization Methods

The layer non-uniformity and the SWCNT network density were estimated by atomic force microscopy (AFM) (Solver-Pro, NT-MDT, Zelenograd, Moscow, Russia) and via analyzing the SWCNT G-peak intensity, I_G, in Raman maps. The Raman spectroscopy maps were acquired with the spectrometer (Centaur-UHR, Nano Scan Technology, Dolgoprudny, Russia) equipped with a laser, 532 nm wavelength [36]. The layer non-uniformity was estimated as a ratio of standard deviation of I_G in the Raman map to the median value of I_G.

Thus, six gas-sensing structures employing SWCNTs were prepared altogether for the study: (1) with a WO_3 layer located only on Pt/W to form the SWCNT-Pt junction, called hereafter as SWCNT50-Pt, (2) with a continuous layer of WO_3 formed over the entire surface of the substrate to form SWCNT-WO_3 heterojunctions with WO_3 nanochannels, called hereafter as SWCNT40-WO_3; (3) two corresponding satellite samples with SWCNTs, called hereafter as sSWCNT50 and sSWCNT40, where the network density has been equal to 50 a.u. and 40 a.u., respectively, to be manufactured by spray deposition according to routes numbered by 1 and 2 in Figure 1; (4) two additional reference samples with varied SWCNT network densities, rSWCNT54 and rSWCNT174, where the network density is equal to 54 a.u. and 174 a.u., respectively, according to routes numbered by 3 and 4 in Figure 1. These sensors were measured simultaneously under all the test measurements as an array. The temperature coefficient of resistance (TCR) was evaluated for the fabricated structures as the averaged changes in the resistivity of each sample with a stepwise change of temperature by 10 °C in the range of 50–100 °C under dry airflow conditions.

2.3. Gas-Sensing Measurements

In order to measure the gas sensor response, the NH_3 vapors were obtained by evaporating a measured micro-volume of 25% aqueous ammonia solution, of analytical grade, in a bottle, of 26 l volume to yield the basic concentrations of about 18 ppm, 40 ppm, 70 ppm, and 135 ppm. We have utilized dry or humid, 45 rel.%, air as the background carrier gas. The humidity level of 45% was chosen as rather typical one for ambient air humidity, close to the average value of 50–60% RH found in Europe [37]. At the same time, the former research showed that quite significant changes in the sensor response are observed for SWCNTs and composite structures of CNTs/WO_3 at this range of background RH when compared to one in dry air [18,29]. The temperature of 30 °C, which is somewhat higher than RT of about 23–28 °C, was chosen to keep the sensors stable in all the measurements and to ensure its independence from environment temperature.

The NH_3 exposures were independently monitored/registered with a commercial semiconductor sensor (TGS826 sensor , Figaro Engineering Inc., Minoo city, Osaka, Japan). Vapors were let into a chamber containing all the sensors with a 0.2 L/min flow rate to be managed by the gas-flow controller (MKS Instruments Inc., Burlington, MA USA).

The sensor recovery was carried out by passing an analyte-free air flow at 2.5 L/min flow rate. The resistance of the sensors was measured with a 16-channel multiplexing system equipped with computer software, which provides data storage and a display as a graph on a real-time scale (IPS-16, Praktik-NTs, Zelenograd, Moscow, Russia). The sensor resistances were read in a sequential mode with a rate of 1 s per sensor. The measuring setup is shown in Figure 2. In order to deliver the prepared ammonia concentration, a closed supply cycle was used. Here, NH_3 vapors were forwarded from the evaporation bottle to the measuring chamber containing the sensors and then returned by the pump (Figure 2, elements grouped by a red dotted line).

Prior the measuring, the samples were annealed at 100 °C in a dry air flow for 2 h to stabilize the properties and to remove any residuals following the sensor's preparation.

2.4. Processing of Sensor Responses

Linear discriminant analysis (LDA) was applied to the resistance data sets from the multisensor array as a recognition algorithm in order to classify these vector data into clusters at the artificial coordinate system in accordance with the test gas mixtures as classes to be recognized [38]. We have considered six gas mixtures, namely 135 ppm and 40 ppm NH_3 in dry air, 135 ppm NH_3 in mixture with wet air, 45 rel.%, and air with RH = 0%, 13%, and 45%. For LDA processing, we have considered taking vector data sampling of

50 points measured under quasi-stationary conditions of the sensors upon exposure to each tested analyte at temperature of ca. 30 °C. The gas response of the sensor structures has been estimated as a relative change of resistance in percent, $S = \Delta R/R_0$, where ΔR is a resistance change upon sensor exposure to analytes at each measurement cycle, R_0 is the resistance in the background (air) before each measuring cycle with a new analyte's or its changed concentration exposure.

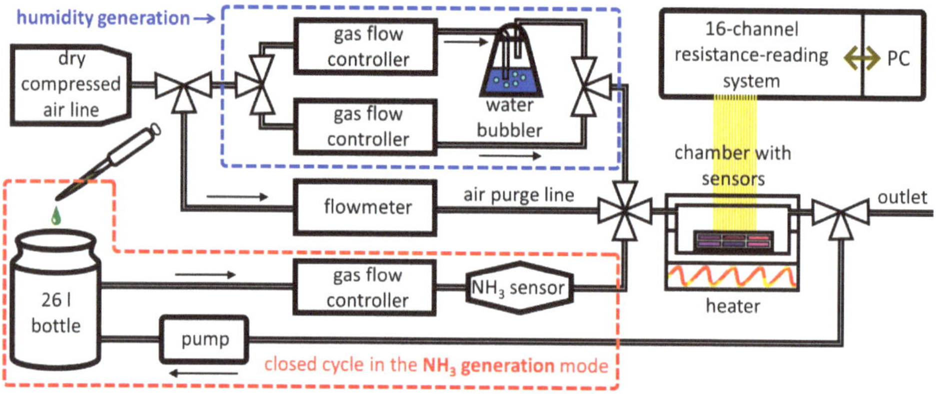

Figure 2. The gas-supply setup to feed the multisensor array under study for exposing to H_2O an NH_3 vapors of various concentrations. The sealed chamber containing the sensors is equipped with built-in heater. Data collection is carried out by a resistance measurement unit equipped with multiple relays to be managed by a personal computer (PC).

The limit of detection (LOD) was estimated by extrapolating the experimental data, which could be well fitted with a power function describing the dependence of the sensor response (%), S, on the ammonia concentration in dry air (ppm), C, as $S = k{\cdot}C^n$, where k and n are the coefficients to be estimated by fitting.

While going to the S(C) range of low analyte' concentration, the LOD value was derived at the point where the noise magnitude is equalized with the sensor response taken from the S(C) fitting function. At the same time, the noise magnitude was evaluated as 5 times of standard deviation, taken over a sampling of 50 resistance measurement points under dry background air according to methodology proposed previously for SWCNT-based sensors [39]. The approach proved to be viable and described the behavior of the sensor response quite well for sub-ppm concentrations down to 30 ppb as described elsewhere [29].

3. Results and Discussion

3.1. Characterization of the Sensor Structures Layers Parameters

Due to the oxidation of the top W layer, the overall height of the final WO_3/Pt/W layer in SWCNT50-Pt and SWCNT40-WO_3 sensors increased from ca. 35 nm to more than ca. 50 nm (Figure 3).

The appearance of WO_3 with crystalline structure was indexed by measuring the Raman spectra via recording the corresponding peaks at 265 cm^{-1}, 705 cm^{-1}, and 805 cm^{-1} [40,41], as shown in Figure 4a,b. Figure 4a gives an optical image of the sensor layer taken with ×100 microscope objective where the marked area has been considered for taking the Raman map. Still, it is worth noting that the peaks were observed at the thin WO_3 layer located on top of the Pt layer and were not registered at the thin WO_3 layer located on the glass surface. This is due to enhancing Raman scattering from the oxide located over Pt islands in frames of so-called the surface-enhanced Raman scattering (SERS) effect [42]. The island-like structure of the Pt layer is characterized by the specific resistivity value of ca. 0.5 kΩ/Sq., which is almost 30 times higher than the known data for the bulk Pt and significantly more, 3–5 times than one observed for the layers with a thickness of 3–5 nm [43]. This effect appears at thicknesses being noticeably smaller than that of the thin Pt layer having an island-like structure [43] and smaller than ones in our structures, equal to 10 nm. After the oxidation process, the WO_3 thickness was 20–25 nm (Figure 3c).

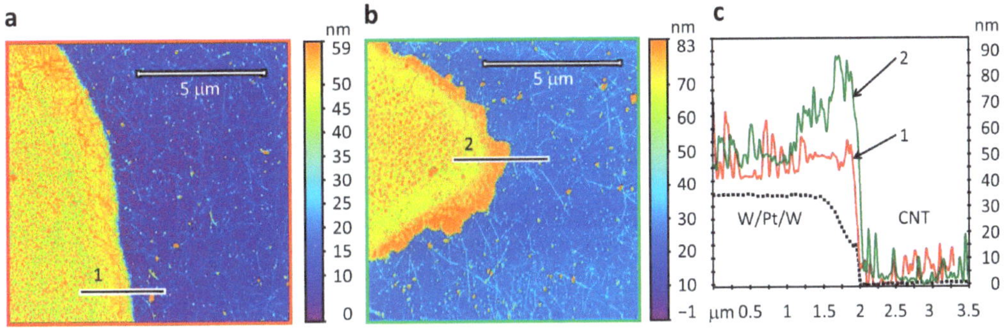

Figure 3. The exploring of sensor structures under study with atomic force microscopy: (**a**) AFM image of SWCNT-Pt structure (SWCNT50-Pt); (**b**) AFM image of SWCNT-WO_3 structure (SWCNT40-WO_3); (**c**) cross-sections of edge of initial metal W/Pt/W area and of "1", "2" lines marked at 1a,b corresponding to profiles of SWCNT50-Pt and SWCNT40-WO_3 structures.

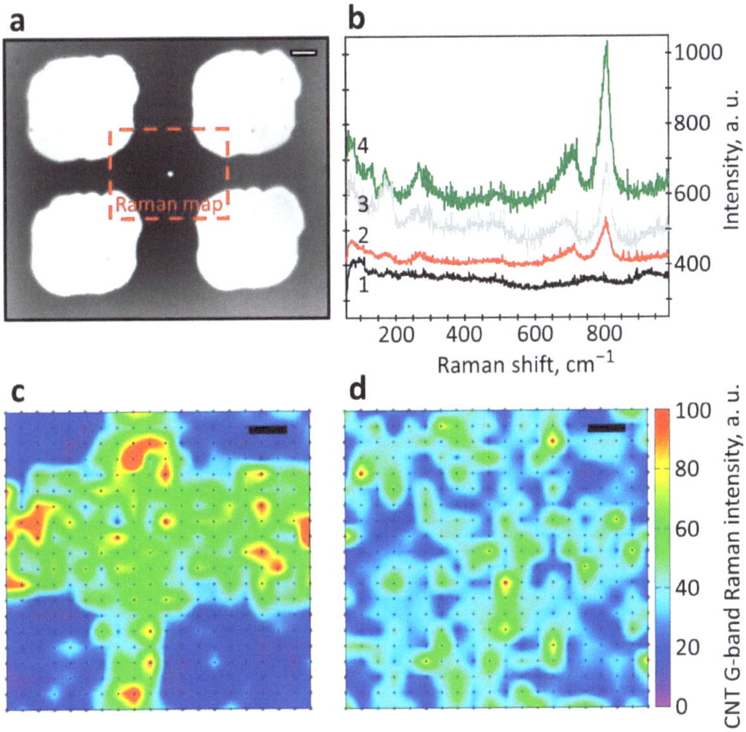

Figure 4. The characterization of the sensor structures under study with Raman spectroscopy: (**a**) optical image of WO_3/Pt/W areas on glass substrate of SWCNT50-Pt and SWCNT40-WO_3 structures; (**b**) Raman spectra related to thin WO_3 layer, of 7–15 nm, on glass area (1), thin WO_3 on Pt/W, SWCNT50-Pt (2), WO_3 continuous film, ca. ~25 nm, on glass, SWCNT40-WO_3 (3), WO_3 continuous film on Pt/W, SWCNT40-WO_3 (4); (**c**,**d**) G-peak Raman maps for SWCNT50-Pt (**c**) and SWCNT40-WO_3 (**d**) structures. The scale bar is 5 µm.

The resistance-to-temperature data have shown that TCR in SWCNT50-Pt structure is about −0.26%/K, while for the TCR of its satellite, sSWCNT50, the sample is about −0.39%/K, as plotted in Figure 5. In Figure 5, the values along the horizontal axis are the I_G median values characterizing the CNT network density of the corresponding structure. This mean value of SWCNT network density can be visually evaluated by the prevailing color in the recorded color map, which displays the distribution of the G-peak intensity in Raman maps for two types of structures, as shown in Figure 4c,d. These values indicate that the SWCNT-Pt junction and the Pt sublayer have a significant influence on the resistance upon heating. TCR modulus of the SWCN50-Pt structure is lower than one of satellite sSWCNT50, which indicates the influence of both the SWCNT-WO_3 junction and WO_3, is negligible in the overall conducting channel for that structure type. Still, the TCR of WO_3 is about −2.4%/K [44]. In the case of composite layers with a WO_3 content of about 30–50%, as is partially the case for the SWCNT40-WO_3 sample (Figure 1), where a part of the conductive channels is caused not only by oxide but also another material, the TCR modulus significantly reduces down to 1.5%/K [45]). The effect of this WO_3 conducting channel on the overall conductance in the structure should enhance the TCR modulus value over that of pure SWCNTs. In addition, it should be noted that the resistivity of the SWCNT50-Pt structure was only 180 kΩ/Sq. against 1.8 MΩ/Sq. of the satellite, sSWCNT50, sample.

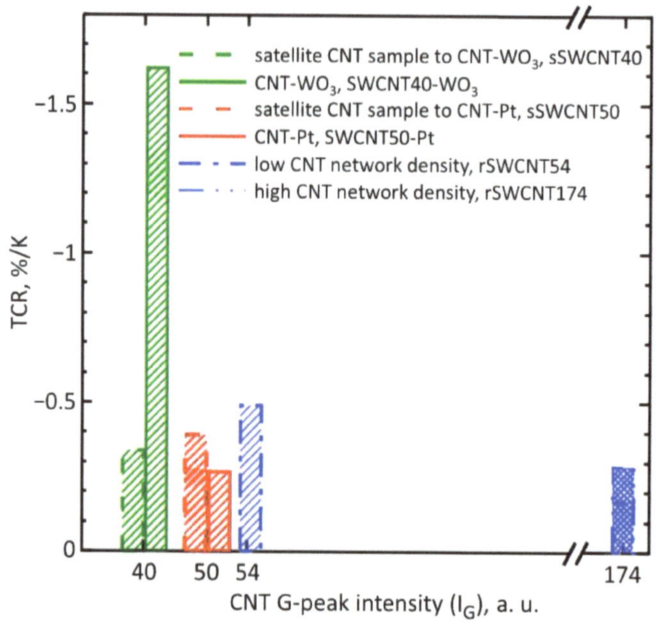

Figure 5. Temperature coefficient of resistance (TCR) for all the SWCNT-based sensor structures under study.

Thanks to the TCR measurement, it is possible to confirm the contribution of the SWCNT-Pt or SWCNT-WO_3 junctions and, therefore, WO_3 nanochannels to the resistance of the sensor layer. Despite the fact that the parameters of the SWCNT network, density and uniformity, for the pairs of structures, SWCNT50-Pt, sSWCNT50 and SWCNT40-WO_3, sSWCNT40, are the same, the TCR of such structures significantly differs.

3.2. The Characterization of Gas-Sensing Performance of the Sensor Structures

Despite the differences in TCR and resistance, SWCNT50-Pt and sSWCNT50 structures were found to exhibit almost the same sensor response to 135 ppm of NH_3 at RT, about 30%

and 25%, respectively (Figure 6a). From the literature data, the semiconducting SWCNTs make a major contribution to the sensor response [46]. Thus, we may suggest that the SWCNT50-Pt structure with a lower value of TCR modulus and, consequently, a smaller effect on the resulting resistance and TCR from semiconducting SWCNTs would also have a lower sensor response relative to the satellite, sSWCNT50, sample.

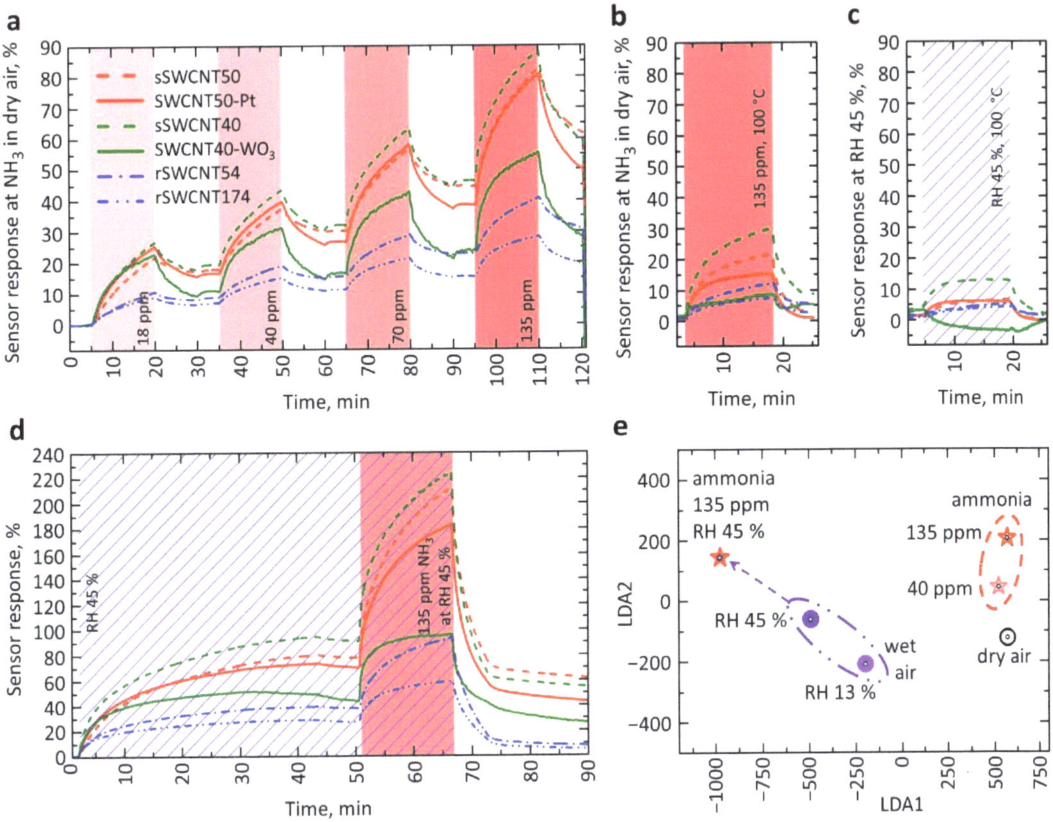

Figure 6. Sensor responses of SWCNT-based structures under study to NH_3: (**a**) dry air background, RT; (**b**) dry air background, 100 °C; (**c**) wet air, RH = 45%, 100 °C; (**d**) NH_3 at wet air background, RH = 45%, RT; (**e**) LDA processing of the vector signal from the multisensor array composed of the four sensor structures: SWCNT50-Pt, SWCNT40-WO_3, their satellite samples of sSWCNT50, and sSWCNT40; the 2D cross-section related to two primary components of 5D LDA space is shown, points represent the vector signals.

This might be explained via a partial shunting of the nanotubes with other conductance channels formed by the Pt layer with SWCNT-Pt junctions. Because a reduction of the sensor response is not observed, it can be concluded that the junction between SWCNTs and Pt has a significant effect on the sensor response. Still, the SWCNT-Pt junction is known to improve significantly, up to ca. 6–10% upon 150 ppm of NH_3 exposure, the sensor response even when the metal-type CNTs are used [47,48]. This also explains that the rSWCNT174 sample with a high-density SWCNT network, but with a similar TCR, of ca. −0.27%/K and resistivity of ca. 80 kΩ/Sq. yields a lower sensor response to 135 ppm of NH_3, equal to ~11%, than that of SWCNT-Pt structure.

Despite minor differences in the TCR, the sensor responses of satellite samples and reference samples (rSWCNT54) with similar network densities and with different SWCNT network densities (rSWCNT174) are significantly varied. With a decrease in the SWCNT network density, we have observed enhancing the sensor response. It seems this matures from a significant contribution of semiconducting SWCNTs [49] whose effective length, between SWCNT junctions, is increased at the overall conducting channel. It is worth noting that the effective length of SWCNTs is enlarged as (i) the network density and/or (ii) the non-uniformity of the SWCNT layer are reduced. The network's non-uniformity of satellite SWCNT samples, being estimated equal to 0.42 for I_G = 50 a.u., is lower than one of the additional samples, rSWCNT54, where the value was 0.65 for I_G = 54 a.u., at the similar network density.

At the same time, the structures with a lower non-uniformity of the SWCNT network, even at a similar SWCNT network density, exhibit a significantly greater sensor response, which can be seen by comparing the responses of the sSWCNT50 and rSWCNT54 samples. The obtained responses to ammonia in dry air are significantly higher than in other studies with sensors based on an SWCNT network [29,46] both for satellite samples containing only SWCNTs sSWCNT40 and sSWCNT50 and for the SWCNT40-WO_3 and SWCNT50-Pt structures, which are drawn in Figure S1. For comparison purposes, Figure S1 shows how the sensor response varies s with the NH_3 concentration for all the studied sensors and the literature data. The deposition of SWCNTs for the SWCNT-Pt structure occurs with some non-uniformity: a lower SWCNT network density was found in the WO_3/Pt/W regions, in contrast to uniformity over the entire surface for the SWCNT50-WO_3 sample. Nevertheless, the non-uniformity of the SWCNT network itself over glass regions for both structures of the SWCNT-Pt and the SWCNT-WO_3 is similar to 0.46. Thus, some differences in SWCNT deposition for SWCNT50-Pt and SWCNT40-WO_3 structures cannot explain the differences observed in their gas-sensing responses.

The sensor response of the SWCNT-Pt structure measured in forward to 135 ppm of NH_3, as exemplary exposure, has significantly decreased to 14.5% when the operating temperature has raised from RT up to 100 °C (Figure 6b). At the same time, the satellite sample, sSWCNT50, demonstrated under the heating the sensor response to be only slightly lower, equal to 21%, than one at 30 °C. The sensor response of the rSWCNT54 sample, having a low network density, I_G = 54 a.u., resistivity of ~1 MΩ/Sq. and TCR of −0.49%/K has also slightly reduced from 14% at 30 °C to 11% at 100 °C. This indicates that the contribution of SWCNT-Pt junctions to the sensor response decreases with increasing temperature, which is similar to the behavior of other heterojunctions [19] because the effect of the Schottky barrier in the SWCNT-Pt junction is diminished. On the other hand, the SWCNT40-WO_3 structure does not exhibit significant differences in sensor response to NH_3 relative to its satellite sample, sSWCNT40, at 30 °C (Figure 6a). At the same time, the SWCNT-WO_3 structure has a higher TCR modulus (Figure 5) and higher resistivity than the satellite sample, to be 8.5 MΩ/Sq. versus 1.2 MΩ/Sq.

It is rather well known that WO_3 does not yield a noticeable sensor response to NH_3 at RT but being heated to 100–150 °C the WO_3 nanostructures could exhibit a significant sensor response as a decrease in resistance, for instance, yielding the response to 120 ppm NH_3 at 150 °C equal to 75%, and at 50 °C, −25% [40]. However, our measurements demonstrated a decrease in the sensor response of the SWCNT-WO_3 structure to NH_3 vapors at a temperature of 100 °C. We guess the lower sensor response, 6.5%, at 100 °C is due to the opposite reaction of SWCNTs and WO_3 materials to the analyte. The sensor response of the satellite, sSWCNT40, the sample under similar conditions is 28.6%. Thus, it can be assumed that the negative sensor response, as a resistance reduction, of the WO_3 nanochannel might be about 22% (6.5% minus 28.6%). In the absence of doping, the estimated response is sufficiently high for WO_3 [40,50]. This is due to the small width, length, and thickness of the channel, which is equal to the film thickness of ca. 25 nm. In the structures where the channel is given as a continuous oxide film, the response is lower at 100 °C by a few orders [51]. Although the estimate of the WO_3 response in the SWCNT40-WO_3 structure under study is still lower than the best response known from the literature for this material, the fabricated structure exhibits a substantially lower response time, being of ca. 10 min, when compared, for instance, to WO_3 nanoflakes [16]) where the characteristic time has been observed equally to ca. 100 min.

The SWCNT50-Pt and SWCNT40-WO_3 structures exhibit a higher initial sensor response rate compared to the satellite samples and saturate earlier when exposed to NH_3. This effect is more pronounced when exposed to NH_3 vapor against the background of air with RH = 45% (Figure 6a,d). The reason for such behavior seems to come from the significant influence of small contact areas on the sensor response. As a result, sensors based on SWCNT-Pt and SWCNT-WO_3 junctions exhibit not only increasing the speed of the sensor response but also yield some selectivity to a particular gas mixture. The chemiresistive response of the WO_3 layer to humidity is known to be quite high even at RT; as reported elsewhere [28], there might be observed a drop in resistance by 50%. At the same time, the sensor response of satellite samples and additional samples with SWCNTs to wet air of 45 rel.% is characterized by increasing the resistance (called frequently as a positive sensor response) both under 30 °C and 100 °C.

Thus, the chemiresistive effect in the SWCNT-WO_3 structure is contributed by two oppositely-directed responses of SWCNTs and WO_3. Based on the difference in the responses of the SWCNT-WO_3 structure and the satellite sample, it can be concluded that WO_3 changes its resistance in humidity, of 45 rel.%, by 40% at 30 °C (85–45%, Figure 6d) and by 15% at 100 °C (10%–(−5%), Figure 6c). The sensor response of the satellite SWCNT sample also goes down from 85% to 10% under the heating (Figure 6c,d). It is worth noting that the sensor response of SWCNTs decreases more than one of WO_3; therefore, the effect of the WO_3 channel on the overall resistance is more pronounced when the sample is heated. As a result, the response of the SWCNT-WO_3 structure to H_2O vapors is lower than that of the satellite sample at 30 °C. For comparison, the sensor response's dependence on H_2O vapor concentration is drawn in Figure S2a for all the SWCNT-based structures under study at RT compared to known literature data. The typical sensor response transients, response, and recovery, recorded upon exposure to H_2O at various concentrations, 13%, 30%, 45%, and 55% RH, in the dry air background under RT are displayed in Figure S2b. At 100 °C, the contribution of the WO_3 channel to the sensor layer conductance becomes decisive, and the resistance of the SWCNT-WO_3 structure in water vapors reduces (Figure 6c).

3.3. The Analysis of the Gas-Selectivity of Manufactured Multisensor Array

It should be recognized that the structures with heterojunctions studied in this work do not demonstrate a significant enhancement of the sensor response when compared to other published results [18,19]. At the same time, significant differences were observed in the behavior of the sensor response of these structures, even at RT. For example, the sensor responses of SWCNT50-Pt and SWCNT40-WO_3 structures to NH_3 vapors in dry air are the same, but the sensor responses of these structures recorded in wet air, of 45 rel.% differ significantly in comparison to their satellite SWCNT samples. Therefore, combining the sensors under study into a multisensor array makes it possible to selectively recognize the presence of NH_3 against the air under varied humidity interference (Figure 6a,d).

Following the LDA processing of the multisensor vector signals collected from the sensors, we have built the five-dimensional LDA space accounting for the classes related to testing analytes. The two-dimensional cross-section of this space is displayed in Figure 6e. As one can see, there is a reliable separation of multisensor vector signals versus gas mixtures into clusters, even when taking only four sensor structures, SWCNT-Pt, SWCNT-WO_3, and their satellite samples, when combining them into the multisensory array. This manifests itself in the form of large distances between centers of gravity characterizing all the clusters, much larger than the radius of the cluster corresponding to each mixture of gases in Figure 6e, which was about 3 a.u., is marked on the graph with small black circles. These circles are developed under the assumption of normal distribution of signals, at 0.9 confidence, within each class. This allows us to conclude that it is possible to reliably recognize both the humidity and different concentrations of ammonia even in the presence of background air humidity.

Thus, the specific behavior of these sensors upon exposure to the vapors was sufficient to recognize different concentrations of NH_3 against the background of dry and wet, 45 rel.%, air and to recognize the wet air containing various levels of humidity, RH = 0%, 13%, and 45%. The mutual arrangement of clusters corresponding to various gas mixtures eliminates the possibility of incorrect recognition during a "transition" from clusters related to one vapor concentration to another one, from air to ammonia in dry air, and to ammonia in air containing water vapors. Compared to other works, where the difference between the sensing structures composing the on-chip multisensor array is designed via changing, for instance, a surface filter of varied thickness [52], or varied sensor layer thickness [38], the proposed multisensor array provides higher selectivity, which could be valued via distance between gravity centers of the analyte-related clusters in the LDA space, to NH_3 in air, even containing various content of humidity. It should be noted that the multisensor array, which includes structures with just SWCNTs and SWCNT-Pt while leaving out the SWCNT-WO_3 structure, provides a lower selective signal though it still ensures the recognition of the gas mixtures. At the same time, the array composed of four sensors that do not include SWCNT-WO_3 or SWCNT-Pt structures could not allow us to obtain a selectivity since it contains only pure SWCNT-based samples which do not deliver a specific behavior of the sensor response to the analytes. In addition, based on the measurements carried out at 30 °C and 100 °C, in order to further increase the selectivity of the multisensor array, different temperature operation modes implemented by the temperature gradient on the substrate can be set for the sensors [53], which can also advance the response speed of the sensors and reduce their recovery time.

3.4. Evaluation of Sensor Detection Limits

As described above, the LOD value was evaluated taking the methodology proposed previously for SWCNT-based sensors [39], where LOD was estimated to be 3 ppb. With this purpose, the observed experimental data (Figure 6a) have been fit by a power function to describe the dependence of the sensor response (Figure S1). These data, as well as the results of the LOD evaluation, are shown in Table 1. It should be noted that the NH_3-sensing results obtained for the SWCNT50-Pt, sSWCNT50 structures, and even for the SWCNT40-WO_3 structure, given that it has a significantly higher channel resistance and TCR, are better than the data known from the literature for sensors based on SWCNTs.

Table 1. The sensing characteristics of the sensor structures under study.

Structure	Noise, 5·σ, kΩ [1]	R_0, kΩ [2]	Power Law of S(C)	LOD Estimation, Ppb
SWCNT40-WO_3	6.563	8240	$5.2 \cdot C^{0.48}$	4.8
sSWCNT40	3.352	1160	$5.1 \cdot C^{0.58}$	7.1
SWCNT50-Pt	0.138	181	$5.1 \cdot C^{0.56}$	0.6
sSWCNT50	0.752	1840	$3.5 \cdot C^{0.64}$	0.9
rSWCNT54	0.763	957	$2.0 \cdot C^{0.61}$	5.1
rSWCNT174	0.115	87	$2.6 \cdot C^{0.49}$	2.4

[1] σ is standard deviation of resistance, sampling of 50 points, in air prior exposing to NH_3, 40 ppm concentration.
[2] R_0 is the resistance of samples in air before NH_3 exposure.

This can be explained by the rather high sensitivity of sensor structures based on low-density SWCNT networks in combination with the low noise that these structures exhibit. In general, for all types of sensors under this study, the estimated LOD is at an acceptable level for the analysis of ammonia in ambient air (less than 30 ppb, [7]). This allows us to conclude that the studied set of gas sensors, together with the fact that it also provides a sufficiently reliable recognition of ammonia against the background of air humidity, could be used in electronic devices for environmental monitoring.

4. Conclusions

In summary, the lift-off photolithography of the W/Pt/W sublayer, subsequent oxidation of the upper tungsten layer, and controlled spray deposition of SWCNTs make it possible to form gas-sensor structures based on SWCNT-Pt junctions. We have observed the SERS effect while characterizing WO_3 on the island sublayer of Pt. The TCR modulus of the SWCNT-Pt structure is found to be significantly lower than that of the satellite SWCNT one, but the sensor response to NH_3 vapors in dry air differs very slightly for these sensors at RT. It seems that the influence of the SWCNT-Pt junction on the sensor response is equally important as the effect of the SWCNT network and semiconducting SWCNTs in it. At the same time, the sensor response of the SWCNT-Pt structure to NH_3 is less than one of the satellite SWCNT samples at 100 °C. This behavior is associated with a decrease in the influence of the heterojunctions on the resulting response of the SWCNT-Pt structure under heating. Furthermore, the response of the SWCNT-Pt structure to NH_3 vapors in wet air, of RH = 45%, is lower than that of its satellite SWCNT sample, even at 30 °C. The deposition of SWCNTs onto a WO_3 layer placed over the entire surface of a glass substrate with Pt/W regions makes it possible to form SWCNT-WO_3 junctions and, thereby, WO_3 nanochannels. The TCR modulus of such a structure is much higher than that of the satellite SWCNT sample due to the contribution of WO_3 nanochannels to the overall conductance. The sensor response of the SWCNT-WO_3 structure to ammonia is comparable to that of the satellite structure at 30 °C. When the sensors are heated to 100 °C, the SWCNT40-WO_3 structure has a lower response relative to the satellite sSWCNT40 sample as a result of increasing the effect of the WO_3 nanochannels that reduces the sensor resistance in NH_3. The response of an SWCNT-WO_3 structure to ammonia in wet air of 45 rel.% is significantly lower than that of a satellite SWCNT sample. This is only partially explained by the significant response of WO_3, which, unlike SWCNTs, reduces its resistance in water vapors even at RT. In other words, the effect cannot be reduced only to a simple algebraic sum of multidirectional responses of CNTs and WO_3 for this mixture of gases, but it can also be partially explained by a change in the properties of the substrate surface at which SWCNT networks located: from hydrophilic glass to more hydrophobic WO_3.

As a result, a specific difference in the sensor response for SWCNT-Pt and even more pronounced for the SWCNT-WO_3 structure with respect to SWCNTs was realized even at RT. This makes it possible to manufacture a multisensor array based on a set of just four sensors (SWCNT-Pt, SWCNT-WO_3, and their satellite samples only with SWCNTs) which allows one using LDA to recognize different concentrations of NH_3 against the background of humidity at various levels. Still, a multisensor array based only on SWCNTs samples does not allow this. Such a multisensor array with an estimated detection limit at the ppb range can be used in devices for NH_3 monitoring in the environment, health monitoring, and other applications.

Supplementary Materials: The following supporting information can be downloaded at: https://www.mdpi.com/article/10.3390/chemosensors10110476/s1, Figure S1: The dependencies of sensor responses on NH_3 concentration for SWCNT-based structures under study at dry air background at RT, in comparison to the reference literature data. The experimental results are indicated by dots, the lines are built under fitting with a power function to the experimental data. The parameters of the fitting functions for sensors under study are given in Table 1; Figure S2: The responses of sensor structures to H_2O vapors. (a) The dependencies of sensor responses on relative humidity content in air for SWCNT-based structures under study at RT, in comparison to the reference literature data. The experimental results are indicated by dots, the lines designate fitting with a power function to experimental data. (b) Changes in the sensor response magnitudes in time at exposing to water vapors of various concentrations, 13%, 30%, 45%, 55% RH, at RT, at dry air background and in the process of sensors recovering in a dry air flow. The shaded areas show the analyte's exposure time intervals. References [29,46] are cited in the supplementary materials.

Author Contributions: Conceptualization, A.V.R. and A.V.L.; data curation, A.V.R. and A.V.L.; formal analysis, A.V.R., A.V.L. and N.S.S.; funding acquisition, A.V.R., A.V.L. and V.V.S.; investigation, A.V.R. and A.V.L.; methodology, A.V.R., N.S.S., E.V.A. and D.D.L.; project administration, A.V.R. and A.V.L.; resources, A.V.L., V.V.S. and D.D.L.; software, A.V.L., supervision, V.V.S.; writing—original draft, A.V.R., A.V.L. and V.V.S.; writing—review and editing, A.V.R., A.V.L. and V.V.S. All authors have read and agreed to the published version of the manuscript.

Funding: This work was supported by RFBR, project number 19-38-60034 (A.V.L.), to metal-oxide sub-layer formation by photolithography, sensor response study, LDA; and was carried out with the financial assistance of the Ministry of Education and Science in the framework of state task FSMR-2020-0017 (A.V.R., N.S.S., E.V.A., D.D.L.), to development, carbon nanotube spray-deposition, Raman spectroscopy, AFM and TCR measurements. V.V.S. thanks Russian Science Foundation, grant no. 22-29-00793, for a partial support.

Institutional Review Board Statement: Not applicable.

Informed Consent Statement: Not applicable.

Data Availability Statement: All relevant data are contained within the article; further data are available from the corresponding author under request.

Conflicts of Interest: The authors declare no conflict of interest.

References

1. Timmer, B.; Olthuis, W.; Van Den Berg, A. Ammonia sensors and their applications—A review. *Sens. Actuators B Chem.* **2005**, *107*, 666–677. [CrossRef]
2. Kwak, D.; Lei, Y.; Maric, R. Ammonia gas sensors: A comprehensive review. *Talanta* **2019**, *204*, 713–730. [CrossRef] [PubMed]
3. Goswami, P.; Gupta, G. Recent progress of flexible NO_2 and NH_3 gas sensors based on transition metal dichalcogenides for room temperature sensing. *Mater. Today Chem.* **2022**, *23*, 100726. [CrossRef]
4. Kalita, A.; Hussain, S.; Malik, A.H.; Subbarao, N.V.V.; Iyer, P.K. Vapor phase sensing of ammonia at the sub-ppm level using a perylene diimide thin film device. *J. Mater. Chem. C* **2015**, *3*, 10767–10774. [CrossRef]
5. Targowski, S.P.; Klucinski, W.; Babiker, S.; Nonnecke, B.J. Effect of ammonia on in vivo and in vitro immune responses. *Infect. Immun.* **1984**, *43*, 289–293. [CrossRef]
6. Devos, M.; Patte, F.; Rouault, J.; Laffort, P.; Van Gemert, L.J. *Standardized Human Olfactory Thresholds*; Oxford University Press: Oxford, UK, 1990.
7. Petrus, M.; Popa, C.; Bratu, A.M. Ammonia concentration in ambient air in a Peri-urban area using a laser photoacoustic spectroscopy detector. *Materials* **2022**, *15*, 3182. [CrossRef]
8. Wang, Y.-C.; Han, M.-F.; Jia, T.-P.; Hu, X.-R.; Zhu, H.-Q.; Tong, Z.; Lin, Y.-T.; Wang, C.; Liu, D.-Z.; Peng, Y.-Z.; et al. Emissions, measurement, and control of odor in livestock farms: A review. *Sci. Total Environ.* **2021**, *776*, 145735. [CrossRef]
9. Shaalan, N.M.; Ahmed, F.; Saber, O.; Kumar, S. Gases in food production and monitoring: Recent advances in target chemiresistive gas sensors. *Chemosensors* **2022**, *10*, 338. [CrossRef]
10. Tang, X.; Debliquy, M.; Lahem, D.; Yan, Y.; Raskin, J.-P. A review on functionalized graphene sensors for detection of ammonia. *Sensors* **2021**, *21*, 1443. [CrossRef]
11. Bevc, S.; Mohorko, E.; Kolar, M.; Brglez, P.; Holobar, A.; Kniepeiss, D.; Podbregar, M.; Piko, N.; Hojs, N.; Knehtl, M.; et al. Measurement of breath ammonia for detection of patients with chronic kidney disease. *Clin. Nephrol.* **2017**, *88*, S14–S17. [CrossRef]
12. Das, S.; Pal, S.; Mitra, M. Significance of exhaled breath test in clinical diagnosis: A special focus on the detection of diabetes mellitus. *J. Med. Biol. Eng.* **2016**, *36*, 605–624. [CrossRef] [PubMed]
13. Michalski, R.; Pecyna-Utylska, P.; Kernert, J. Determination of ammonium and biogenic amines by ion chromatography. A review. *J. Chromatogr. A* **2021**, *1651*, 462319. [CrossRef] [PubMed]
14. Gupta, P.; Maurya, S.; Pandey, K.N.; Verma, V. Metal-oxide based ammonia gas sensors: A review. *Nanosci. Nanotechnol.-Asia* **2021**, *11*, 270–289. [CrossRef]
15. Chou, T.C.; Chang, C.H.; Lee, C.; Liu, W.C. Ammonia sensing characteristics of a tungsten trioxide thin-film-based sensor. *IEEE Trans. Electron Devices* **2018**, *66*, 696–701. [CrossRef]
16. Büyükköse, S. Highly selective and sensitive WO_3 nanoflakes based ammonia sensor. *Mater. Sci. Semicond. Process.* **2020**, *110*, 104969. [CrossRef]
17. Luo, W.; Zhao, T.; Li, Y.; Wei, J.; Xu, P.; Li, X.; Wang, Y.; Zhang, W.; Elzatahry, A.A.; Alghamdi, A.; et al. A micelle fusion-aggregation assembly approach to mesoporous carbon materials with rich active sites for ultrasensitive ammonia sensing. *J. Am. Chem. Soc.* **2016**, *138*, 12586–12595. [CrossRef]
18. Duong, V.; Nguyen, C.; Luong, H.; Nguyen, D.; Nguyen, H. Ultralow-detection limit ammonia gas sensors at room temperature based on MWCNT/WO_3 nanocomposite and effect of humidity. *Solid State Sci.* **2021**, *113*, 106534. [CrossRef]

19. Nguyet, Q.T.M.; Van Duy, N.; Manh Hung, C.; Hoa, N.D.; Van Hieu, N. Ultrasensitive NO_2 gas sensors using hybrid heterojunctions of multi-walled carbon nanotubes and on-chip grown SnO_2 nanowires. *Appl. Phys. Lett.* **2018**, *112*, 153110. [CrossRef]
20. Chiou, J.-C.; Wu, C.-C. A wearable and wireless gas-sensing system using flexible polymer/multi-walled carbon nanotube composite films. *Polymers* **2017**, *9*, 457. [CrossRef]
21. Kumar, N.; Prajesh, R. Selectivity enhancement for metal oxide (MOX) based gas sensor using thermally modulated datasets coupled with golden section optimization and chemometric techniques. *Rev. Sci. Instrum.* **2022**, *93*, 064702. [CrossRef]
22. Norizan, M.N.; Abdullah, N.; Halim, N.A.; Demon, S.Z.N.; Mohamad, I.S. Heterojunctions of rGO/metal oxide nanocomposites as promising gas-sensing materials—A review. *Nanomaterials* **2022**, *12*, 2278. [CrossRef] [PubMed]
23. Bannov, A.G.; Popov, M.V.; Brester, A.E.; Kurmashov, P.B. Recent advances in ammonia gas sensors based on carbon nanomaterials. *Micromachines* **2021**, *12*, 186. [CrossRef] [PubMed]
24. Liu, Y.; Xiao, S.; Du, K. Chemiresistive gas sensors based on hollow heterojunction: A review. *Adv. Mater. Interfaces* **2021**, *8*, 2002122. [CrossRef]
25. Lei, G.; Lou, C.; Liu, X.; Xie, J.; Li, Z.; Zheng, W.; Zhang, J. Thin films of tungsten oxide materials for advanced gas sensors. *Sens. Actuators B Chem.* **2021**, *341*, 129996. [CrossRef]
26. Ma, N.; Suematsu, K.; Yuasa, M.; Kida, T.; Shimanoe, K. Effect of water vapor on Pd-loaded SnO_2 nanoparticles gas sensor. *ACS Appl. Mater. Interfaces* **2015**, *7*, 5863–5869. [CrossRef] [PubMed]
27. Tischner, A.; Maier, T.; Stepper, C.; Köck, A. Ultrathin SnO_2 gas sensors fabricated by spray pyrolysis for the detection of humidity and carbon monoxide. *Sens. Actuators B Chem.* **2008**, *134*, 796–802. [CrossRef]
28. Qian, J.; Peng, Z.; Shen, Z.; Zhao, Z.; Zhang, G.; Fu, X. Positive impedance humidity sensors via single-component materials. *Sci. Rep.* **2016**, *6*, 25574. [CrossRef]
29. Rigoni, F.; Freddi, S.; Pagliara, S.; Drera, G.; Sangaletti, L.; Suisse, J.-M.; Bouvet, M.; Malovichko, A.M.; Emelianov, A.V.; Bobrinetskiy, I.I. Humidity-enhanced sub-ppm sensitivity to ammonia of covalently functionalized single-wall carbon nanotube bundle layers. *Nanotechnology* **2017**, *28*, 255502. [CrossRef]
30. Sysoev, V.V.; Strelcov, E.; Kolmakov, A. Multisensor micro-arrays based on metal oxide nanowires for electronic nose applications. In *Metal Oxide Nanomaterials for Chemical Sensors*; Integrated Analytical Systems; Carpenter, M., Mathur, S., Kolmakov, A., Eds.; Springer: New York, NY, USA, 2013; pp. 465–502. [CrossRef]
31. Potyrailo, R.A. Multivariable sensors for ubiquitous monitoring of gases in the era of internet of things and industrial internet. *Chem. Rev.* **2016**, *116*, 11877–11923. [CrossRef]
32. Jeong, S.-Y.; Kim, J.-S.; Lee, J.-H. Rational design of semiconductor-based chemiresistors and their libraries for next-generation artificial olfaction. *Adv. Mater.* **2020**, *32*, 2002075. [CrossRef]
33. Gulbransen, E.A.; Andrew, K.F. Kinetics of the oxidation of pure tungsten from 500 to 1300 C. *J. Electrochem. Soc.* **1960**, *107*, 619. [CrossRef]
34. Manciu, F.S.; Enriquez, J.L.; Durrer, W.G.; Yun, Y.; Ramana, C.V.; Gullapalli, S.K. Spectroscopic analysis of tungsten oxide thin films. *J. Mater. Res.* **2010**, *25*, 2401–2406. [CrossRef]
35. Polikarpov, Y.A.; Romashkin, A.V.; Struchkov, N.S.; Levin, D.D. High uniform carbon nanotube thin films spray deposition on substrates with patterned structures having height difference. In Proceedings of the 2019 IEEE Conference of Russian Young Researchers in Electrical and Electronic Engineering, St. Petersburg, Russia, 28–31 January 2019; pp. 1980–1985. [CrossRef]
36. Romashkin, A.V.; Polikarpov, Y.A.; Silakov, G.O.; Alexandrov, E.V. Spray deposited thin uniform NiO/Spiro-OMeTAD composite hole transport layer with top carbon nanotube electrode. *J. Phys. Conf. Ser.* **2021**, *2086*, 012097. [CrossRef]
37. Frick, C.; Steiner, H.; Mazurkiewicz, A.; Riediger, U.; Rauthe, M.; Reich, T.; Gratzki, A. Central European high-resolution gridded daily data sets (HYRAS): Mean temperature and relative humidity. *Meteorol. Z.* **2014**, *23*, 15–32. [CrossRef]
38. Rabchinskii, M.K.; Varezhnikov, A.S.; Sysoev, V.V.; Solomatin, M.A.; Ryzhkov, S.A.; Baidakova, M.V.; Stolyarova, D.Y.; Shnitov, V.V.; Pavlov, S.S.; Kirilenko, D.A.; et al. Hole-matrixed carbonylated graphene: Synthesis, properties, and highly-selective ammonia gas sensing. *Carbon* **2021**, *172*, 236–247. [CrossRef]
39. Rigoni, F.; Tognolini, S.; Borghetti, P.; Drera, G.; Pagliara, S.; Goldoni, A.; Sangaletti, L. Enhancing the sensitivity of chemiresistor gas sensors based on pristine carbon nanotubes to detect low-ppb ammonia concentrations in the environment. *Analyst* **2013**, *138*, 7392–7399. [CrossRef]
40. Kolhe, P.S.; Mutadak, P.; Maiti, N.; Sonawane, K.M. Synthesis of WO_3 nanoflakes by hydrothermal route and its gas sensing application. *Sens. Actuators A Phys.* **2020**, *304*, 111877. [CrossRef]
41. Diaz-Reyes, J.; Delgado-Macuil, R.J.; Dorantes-García, V.; Perez-Benitez, A.; Balderas-Lopez, J.A.; Ariza-Ortega, J.A. Physical properties characterization of WO_3 films grown by hot-filament metal oxide deposition. *Mater. Sci. Eng. B* **2010**, *174*, 182–186. [CrossRef]
42. Lee, C.; Robertson, C.S.; Nguyen, A.H.; Kahraman, M.; Wachsmann-Hogiu, S. Thickness of a metallic film, in addition to its roughness, plays a significant role in SERS activity. *Sci. Rep.* **2015**, *5*, 1–10. [CrossRef]
43. Avrekh, M.; Monteiro, O.R.; Brown, I.G. Electrical resistivity of vacuum-arc-deposited platinum thin films. *Appl. Surf. Sci.* **2000**, *158*, 217–222. [CrossRef]
44. Wang, Z.; Su, J.; Qi, H.; Pan, P.; Jiang, M. Porous nanocrystalline WO_3 thin films: Fabrication, electrical and optical properties. *Surf. Innov.* **2021**, *9*, 214–221. [CrossRef]

45. Sastry, D.N.; Revanasiddappa, M.; Basavaraja, C.; Suresh, T.; Raghavendra, S.C. DC conductivity studies of doped polyaniline tungsten oxide nanocomposites. *Indian J. Eng. Mater. Sci.* **2013**, *20*, 435–442.
46. Abdellah, A.; Abdelhalim, A.; Horn, M.; Scarpa, G.; Lugli, P. Scalable spray deposition process for high-performance carbon nanotube gas sensors. *IEEE Trans. Nanotechnol.* **2013**, *12*, 174–181. [CrossRef]
47. Penza, M.; Rossi, R.; Alvisi, M.; Cassano, G.; Signore, M.A.; Serra, E.; Giorgi, R. Pt- and Pd-nanoclusters functionalized carbon nanotubes networked films for sub-ppm gas sensors. *Sens. Actuators B Chem.* **2008**, *135*, 289–297. [CrossRef]
48. Vu, T.D.; Cong, T.N.; Huu, B.L.; Duc, C.N.; Huu, L.N. Surface-modified carbon nanotubes for enhanced ammonia gas sensitivity at room temperature. *J. Nanosci. Nanotechnol.* **2019**, *19*, 7447–7451. [CrossRef]
49. Abdelhalim, A.; Winkler, M.; Loghin, F.; Zeiser, C.; Lugli, P.; Abdellah, A. Highly sensitive and selective carbon nanotube-based gas sensor arrays functionalized with different metallic nanoparticles. *Sens. Actuators B Chem.* **2015**, *220*, 1288–1296. [CrossRef]
50. Xiong, Y.; Zhu, Z.; Guo, T.; Li, H.; Xue, Q. Synthesis of nanowire bundle-like WO_3-$W_{18}O_{49}$ heterostructures for highly sensitive NH_3 sensor application. *J. Hazard. Mater.* **2018**, *353*, 290–299. [CrossRef]
51. Ani, A.; Poornesh, P.; Antony, A.; Shchetinin, I.V.; Nagaraja, K.K.; Chattopadhyay, S.; Vinayakumar, K.B. Impact of Ag on the limit of detection towards NH_3-sensing in spray-coated WO_3 thin-films. *Sensors* **2022**, *22*, 2033. [CrossRef]
52. Sysoev, V.V.; Kiselev, I.; Trouillet, V.; Bruns, M. Enhancing the gas selectivity of single-crystal SnO_2:Pt thin film chemiresistor microarray by SiO_2 membrane coating. *Sens. Actuators B Chem.* **2013**, *185*, 59–69. [CrossRef]
53. Sysoev, V.V.; Kiselev, I.; Frietsch, M.; Goschnick, J. Temperature gradient effect on gas discrimination power of a metal-oxide thin-film sensor microarray. *Sensors* **2004**, *4*, 37–46. [CrossRef]

chemosensors

MDPI

Article

Gas Sensors Based on Exfoliated g-C_3N_4 for CO_2 Detection

Ahmed Kotbi [1,2], Manal Benyoussef [1], El Mostafa Ressami [2], Michael Lejeune [1], Brahim Lakssir [2] and Mustapha Jouiad [1,*]

1 Laboratory of Physics of Condensed Mater, University of Picardie Jules Verne, 80039 Amiens, France
2 Moroccan Foundation for Advanced Science, Innovation and Research, Rabat 10100, Morocco
* Correspondence: mustapha.jouiad@u-picardie.fr

Abstract: We report on the investigation of graphitic carbon nitride (g-C_3N_4) for carbon dioxide (CO_2) sensor applications. g-C_3N_4 is prepared by the thermal polycondensation of thiourea and sprayed onto a substrate with interdigitated electrodes. The resulting sensor device exhibited a high sensitivity to CO_2 molecules of ~200 ppm, a high responsivity of ~730 ms at 40 °C and a full recovery time of 36 s. Furthermore, a set of various characterization measurements demonstrated the excellent stability of both the g-C_3N_4 nanosheets and the fabricated gas sensor device. Meanwhile, density functional theory (DFT) calculations for the bulk and monolayer models, based on tri-s-triazine, revealed the optoelectronic properties of g-C_3N_4 and the interaction energy with CO_2, which is evaluated at −0.59 eV. This value indicates the very good affinity of g-C_3N_4 nanosheets to CO_2 molecules. Our findings shed light on the potential for g-C_3N_4 to be used for the development of high-performing gas sensor devices.

Keywords: graphitic carbon nitride (g-C_3N_4); density functional theory (DFT); tri-s-triazine; optoelectronic properties; CO_2 gas sensors

Citation: Kotbi, A.; Benyoussef, M.; Ressami, E.M.; Lejeune, M.; Lakssir, B.; Jouiad, M. Gas Sensors Based on Exfoliated g-C_3N_4 for CO_2 Detection. *Chemosensors* **2022**, *10*, 470. https://doi.org/10.3390/chemosensors10110470

Academic Editors: Kai Xu and Zhong Li

Received: 24 October 2022
Accepted: 8 November 2022
Published: 10 November 2022

1. Introduction

Natural and industrial air pollutions due to toxic gas emissions constitute one of the major threats to the environment, causing serious hazards to human health such as respiratory and cardiovascular diseases [1]. Hence, there is an urgent need to reduce these emissions by issuing strict government regulations [2] while developing materials that can detect these gases before they reach the atmosphere. For gas detection, two-dimensional (2D) materials appear among the emerging solutions for gas sensing [3,4]. Indeed, due to their high surface area and good electronic and optical properties, 2D materials have attracted great attention in several domains such as photocatalysis, photovoltaics and gas sensing [5]. Among 2D materials, graphitic carbon nitride g-C_3N_4 (gCN) has shown several advantages for the development of gas sensors at a low processing cost and with low toxicity and high stability [6,7]. Indeed, gCN exhibits a planar structure similar to graphene, where C and N are covalently bonded, and adjacent layers are linked together by Van Der Waals forces. gCN has a tri-s-triazine cyclic structure, and the strong covalent coupling between C and N atoms offers this material very a high chemical and thermal stability under ambient conditions [8]. In addition, gCN can be used as a non-toxic organic semiconductor material possessing a bandgap energy of ~2.7 eV [9,10], making it a suitable catalyst for hydrogen evolution reaction, energy storage, contaminant degradation, CO_2 reduction and gas sensing [11]. In gas sensing applications, the performance of single-layer gCN remains attractive for two main reasons: (1) its intrinsic vacancies provide gCN with natural pathways for gas diffusion, unlike the artificial defects introduced in the other 2D materials [12,13]; (2) its large specific surface area offers gCN a high gas adsorption capacity [14]. gCN is easily synthesized by the thermal polycondensation process from affordable precursors such as thiourea (CH_4N_2S) [15], melamine ($C_3H_6N_6$) [16], dicyandiamide ($C_2H_4N_4$) [17] and urea (CH_4N_2O) [18], suggesting that it is one of the most promising candidates for industrial scaling up for various applications.

In the present work, a step-by-step fabrication and characterization of gCN are carried out to explore the routes of obtaining pure gCN to be used for gas sensing applications and especially CO_2 detection. Moreover, first-principles calculations are also performed to determine the optoelectronic properties of gCN responsible for its outstanding CO_2 detection.

2. Materials and Methods

Thiourea (CH_4N_2S, 99% purity, Alfa Aesar A12828.0E) and chloroform ($CHCl_3$, 67-66-3) were obtained from Fisher Scientific (Waltham, MA, USA). A crucible of 100 mL (Cole Parmer, Vernon Hills, IL, USA) and a furnace (Carbolite Gero, AAF 1100, Hope, UK) were used to synthesize gCN. Fourier transform infrared (FTIR) spectroscopy was conducted using KBr discs on an ABB Bomem FTLA2000 spectrophotometer. Thermogravimetric analysis (TGA) was carried out under an air atmosphere on a TA Instrument Q500 thermal analyzer. The crystal structure of thiourea was determined using X-ray powder diffraction (XRD) recorded on a Bruker-D4 diffractometer operating with Cu Kα irradiation. The vibration modes were recorded using a micro-Raman Renishaw spectrometer using UV light excitation (325 nm). A scanning electron microscope (SEM) Quanta 200 FEG and Transmission Electron Microscope (TEM) Tecnai from Thermo Fisher Sientific were used to examine the microstructure of the samples. A Retsch Milling Sieving Assisting machine was used for powder grinding, and a Palmsens 4 electrical measurements station was used for electrical measurements. A micro-probe system from the Nextron company (Basel, Switzerland) was used for gas detection tests (Figure 1). This system allows in-situ measurements of electrical properties under different environmental conditions: vacuum, temperature and controlled gas flow. The full sensing measurement set up was fully controlled by integrated software from Nextron. All detection tests were carried out in a gray room with a controlled relative humidity of 47%.

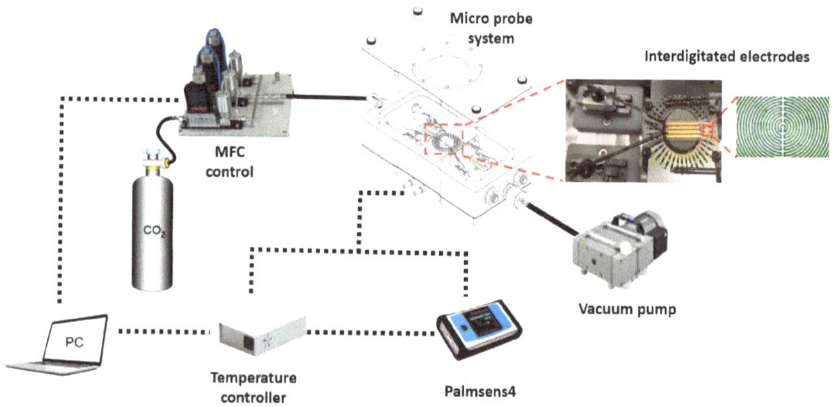

Figure 1. Sensing measurement set up used in this work.

The gCN nanosheets were elaborated by the thermal polycondensation of thiourea (CH_4N_2S). For this, 45 g of thiourea was placed in a closed alumina crucible and heated at 550 °C in a muffle furnace for 2 h at a heating rate of 5 °C/min. After natural cooling to room temperature, the obtained compound in the form of yellow powder was subsequently crashed using a ball milling machine. The powder was then exfoliated using ultrasound for 3 h in chloroform solution. The resulting solution was sprayed to form a thin film onto a substrate with interdigitated electrodes using the spray coating technique. The mixture of chloroform and gCN powder at 5.3×10^{-2} g/L concentration was sprayed at a 1 mL.mn^{-1} flow rate for 40 min while heating the substrate at 400 °C. During the whole spraying process, the air pressure was maintained at 2 bar, and the spraying nozzle was placed 20 cm from the substrate.

3. Computational Method

The computed gCN has a hexagonal crystal structure, "space group P$\bar{6}$m2 (No. 187)". The valence electron configurations selected for C and N atoms are $2s^2$ $2p^2$ and $2s^2$ $2p^3$, respectively. The values of the relaxed lattice constants for bulk gCN are a = 7.135 Å and c = 7.040 Å. Figure 2a shows the 2 × 2 × 2 k-points used to obtain the band structure along the path H-K-Γ-M-L-H in the Brillouin zone. For the gCN monolayer (Figure 2b), 2 × 2 × 1 k-points were used, and a 15 Å vacuum space along the z-axis was added to isolate the gCN and prevent interactions between the adjacent layers.

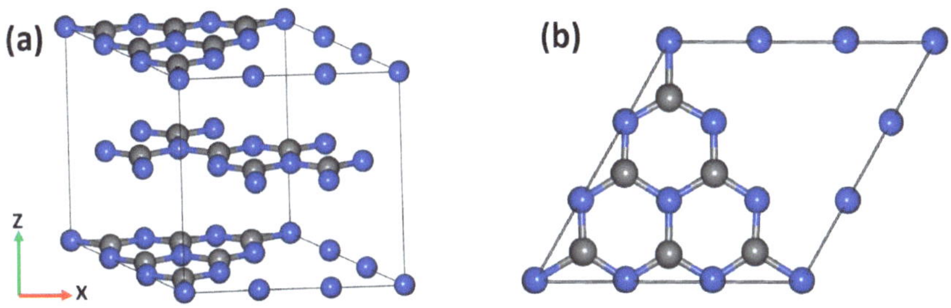

Figure 2. Geometric structure of heptazine-based gCN: (**a**) bulk; (**b**) monolayer. (C and N atoms are in black and blue respectively).

In the structure of gCN, the C and N atoms are represented, respectively, in black and in blue. The gCN monolayer model is obtained by removing other layers from the unit cell of the bulk gCN along the c direction. The structure of gCN includes eight N atoms and six C atoms for each layer. The energy cutoff of the plane basis was set to 500 eV. The calculations were performed using the Perdew–Burke–Ernzerhof (PBE) exchange-correlation function with generalized gradient approximation (GGA) and the Heyd–Scuseria–Ernzerhof (HSE06) hybrid function, implemented in Cambridge Sequential Total Energy Package (CASTEP) simulation software (22.1.1 version, CDG, Cambridge, UK).

4. Results and Discussion

4.1. Material Characterization

Figure 3 shows the TGA weight loss of thiourea. The first degradation is observed at ~200 °C with two endothermic peaks caused by the sublimation of thiourea. In the temperature range of 225–400 °C, there is a slight weight loss due to the thermal condensation of thiourea into triazine (melamine) and tri-s-triazine (heptazine). Tri-s-triazine could form by the intermolecular rearrangement of triazine. A second degradation is observed around ~450 °C, which corresponds to the onset of polymerization of gCN. The polycondensation of thiourea gives several products before the complete formation of gCN, such as triazine and tri-s-triazine [19].

Figure 4a shows the XRD diagram of gCN. The main peaks are located at 27.25° and 12.99°, indicating the formation of pure gCN. The XRD peak shapes confirm the graphitic stacking structure, highlighted by the presence of two peaks characteristic of gCN bulk and gCN nanosheets. The intense diffraction peak at 27.25° corresponds to the (002) reflection of the graphitic interlayer stacking structure of gCN, and the weak diffraction peak at 12.99° is attributed to the (100) plane of the planar repeating units in gCN. The crystallite size d is estimated using Scherrer's equation—$d = k.\lambda/\beta.\cos\theta$—to be ~9.28 nm and ~3.12 nm, respectively, for gCN bulk and nanosheets, where k is the shape constant (0.94), λ is the wavelength, θ is the Bragg angle and β is the peak full width at half-maximum. The XRD pattern of the polymeric gCN nanosheets indicates a graphitic structure with an interplanar stacking distance of aromatic units of ~0.33 nm.

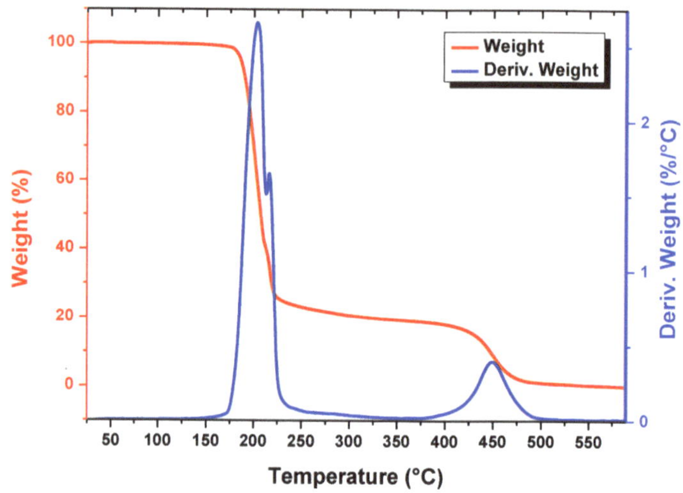

Figure 3. Weight loss of thiourea versus temperature and gCN formation.

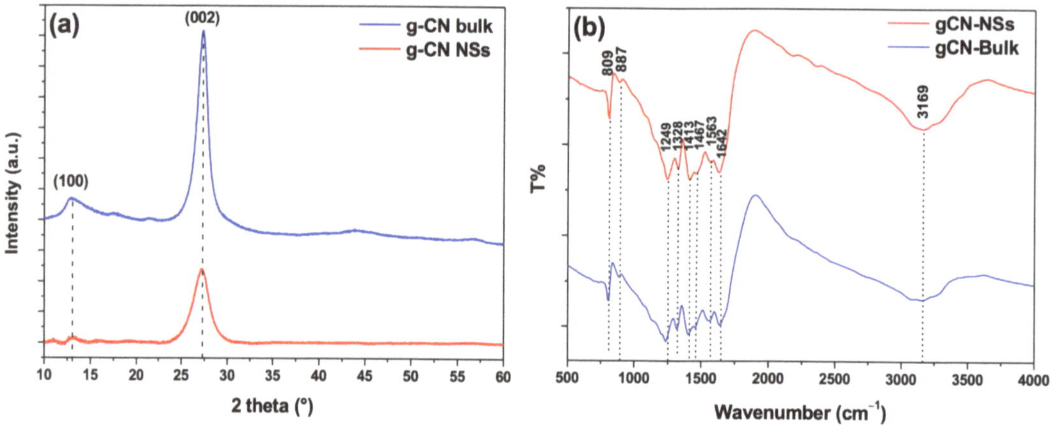

Figure 4. (**a**) X-ray diffraction diagram and (**b**) FTIR spectra of gCN bulk and nanosheets.

The FTIR spectra of the elaborated gCN are provided in Figure 4b. Two main peaks at 809 cm^{-1} and 887 cm^{-1} show the presence of s-triazine units, caused by the out-of-plane bending vibrations of the triazine ring. The peaks appearing at 1249 cm^{-1}, 1328 cm^{-1}, 1413 cm^{-1}, 1467 cm^{-1} and 1563 cm^{-1} are ascribed to the stretching vibrations of the C–N bonds in tri-s-triazine. The peak at 1642 cm^{-1} is due to the stretching vibration of the C=N bond in the heptazine units. The broadband absorption peak at 3169 cm^{-1} shows the stretching vibration of the N–H bond of NH and NH_2 groups, presumably caused by the incomplete polycondensation of gCN. As illustrated in Figure 4b, the FTIR spectra of gCN nanosheets and bulk gCN have the same chemical structure.

Note that, under visible light excitation, the Raman spectra of gCN present a single broad response, which is due to the strong fluorescence and photoluminescence effects occurring in the graphitic gCN phase when excited by visible light [20,21]. Hence, to better apprehend the Raman response, Raman analyses are conducted using UV laser excitation. Figure 5 depicts the temperature-dependent Raman spectra of the gCN sample collected in the frequency range of 300–1800 cm^{-1}. The Raman response at room temperature of gCN is in good agreement with the reported results [21–23]. Moreover, one can note the presence of high intensity Raman modes at ~705 cm^{-1}, ~764 cm^{-1}, ~977 cm^{-1}, ~1113 cm^{-1}, ~1400 cm^{-1}, ~1483 cm^{-1}, ~1576 cm^{-1} and ~1620 cm^{-1}.

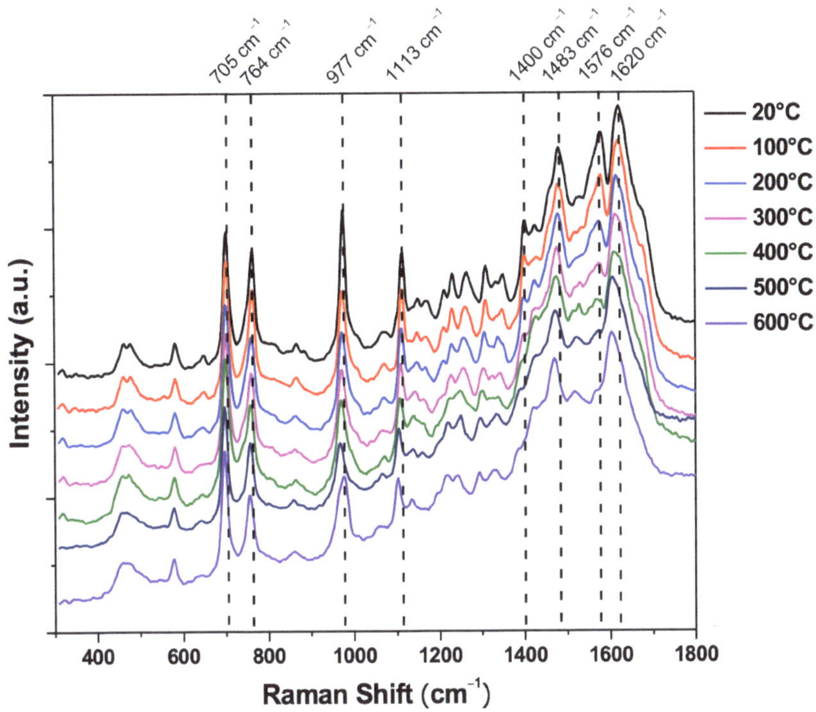

Figure 5. Temperature-dependent vibration modes of gCN excited with a UV laser, 325 nm.

The strong Raman vibration modes observed at 705 and 764 cm^{-1} are related to the intraplanar bending vibrations of the C−N=C linked heptazine units. Moreover, the mode at ~977 cm^{-1} was reported to be due to the symmetric N-breathing mode of the tri-s-triazine units [24]. These last modes are observed to present a blue shift as the temperature increases. Besides, the modes assigned to the D (1400 cm^{-1}) and G (1576 cm^{-1}) bands for the C sp^2 sites are also visible. The D band has an A_{1g} symmetry attributed to the breathing modes of C-sp^2 atoms in rings, and it is linked to structural defects, whereas the G band has E_{2g} symmetry involving the in-plane bond stretching motion of all C-sp^2 atoms and is linked to the degree of gCN graphitization [25,26]. As can also be seen, the temperature increase induced a weakening of the D and G Raman intensities ID and IG, respectively. To investigate the influence of the temperature change on the gCN structural distortions, the D and G bands' intensity ratio ID/IG was monitored. Indeed, our findings show that when the temperature increased from ambient to 600 °C, the ID/IG ratio also increased. This is the signature of a high degree of structural distortion in the gCN. In addition, all vibrational modes are seen to downshift at lower frequencies with increasing temperature. Of particular interest, these results concur with a high thermal stability of our elaborated gCN nanosheets for temperatures as high as 600 °C since no gCN degradation was registered.

The microstructures of elaborated gCN were analyzed by SEM and TEM, and the respective micrographs are depicted in Figure 6.

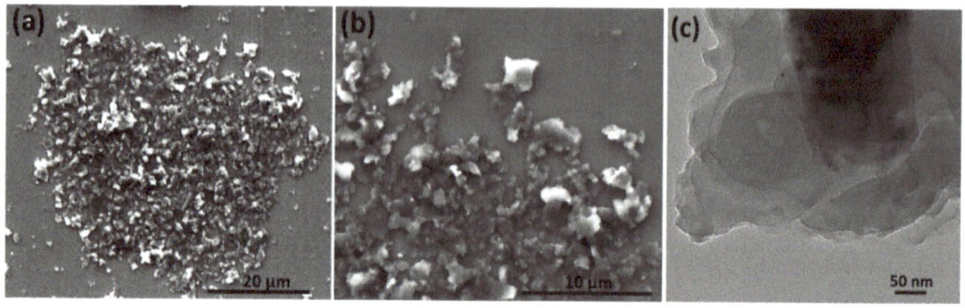

Figure 6. (**a**,**b**) Low and high magnification SEM images of gCN. (**c**) TEM image of layered gCN nanosheets.

The low and high magnification SEM images of dispersed nanosheets of gCN are given in Figure 6a,b, respectively. The nanosheet form is noticeable in these images, exhibiting a lateral size ranging from a few nanometers to a few micrometers. The TEM image shown in Figure 6c highlights the layered nanosheet morphology of the elaborated gCN. The size of the nanosheets is in the range of hundreds of nanometers. The stacking layers are evidenced; one can count four to five layers in Figure 6c.

4.2. Optical Properties

The band gap energy (E_g) of elaborated gCN is extracted from the absorption coefficient according to the model proposed by Tauc [27]:

$$\alpha h\nu = A(h\nu - E_g)^n, \quad (1)$$

where $h\nu$ is the photon energy, E_g is the band gap energy, A is a constant, and n is an exponent indicating the optical absorption nature. Plotting $(\alpha h\nu)^{1/2}$ against photon energy provides the indirect optical gap value. The band gap energy results are extracted here from the optical transmittance data of Figure 7a for the gCN bulk and monolayer and inserted in Figure 7b.

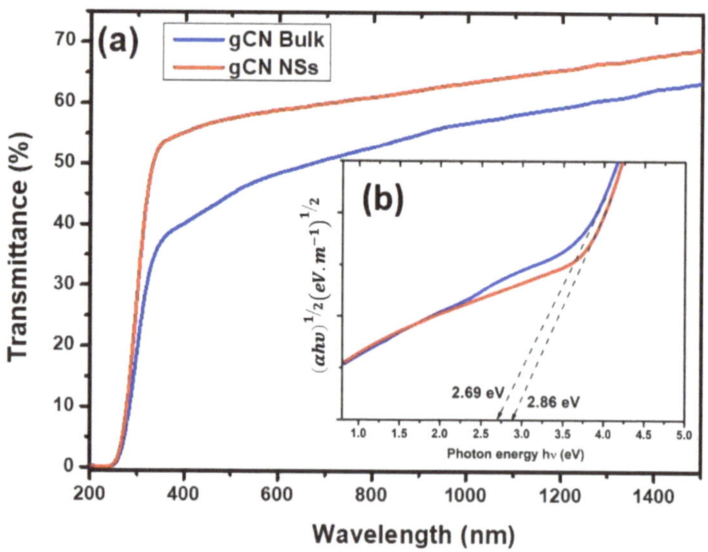

Figure 7. (**a**) Optical transmittance of both bulk and monolayer gCN, (**b**) $(\alpha h\nu)^{1/2}$ vs. photon energy for gCN bulk and monolayer and their respective band gap energy.

The gCN nanosheets exhibit higher transmittance in the full analyzed wavelength range of 300–1500 nm. The obtained band gap energy is 2.69 eV and 2.86 eV for the bulk and the nanosheets, respectively. These values are comparable to reported values [28–30].

4.3. Optoelectronic Properties

The band structures of gCN bulk and nanosheets, determined with different approximations, show that gCN has an indirect band gap with a minimum of the conduction band at point M of the Brillouin zone and the maximum of the valence band at point G of the Brillouin zone (e.g., Figure 8).

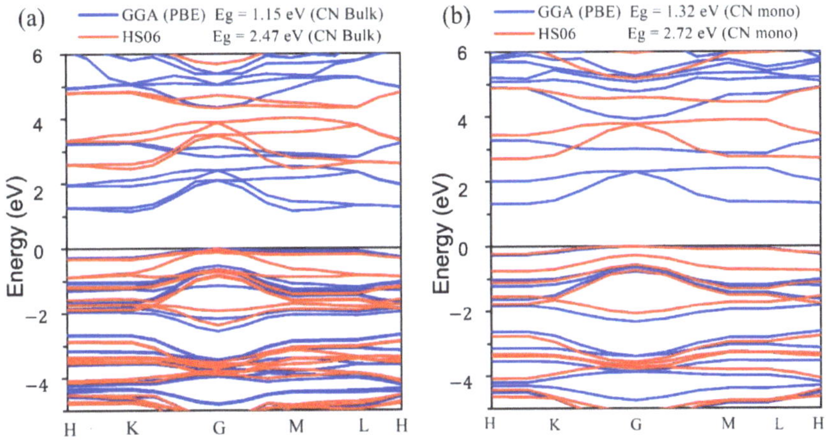

Figure 8. Band structure of gCN bulk (**a**) and gCN monolayer (**b**) obtained by DFT calculations using GGA-PBE (blue) and HS06 (red) approximations.

The computed band gap energies using GGA-PBE approximation show very low values of 1.15 eV and 1.32 eV for the gCN bulk and monolayer, respectively, which are surprisingly highly underestimated. However, using HSE06 approximation, the obtained band gap energies values are 2.47 eV and 2.72 eV for the gCN bulk abd gCN monolayer, respectively. These values are in agreement with the experimental results. We have gathered in the following Table 1 various reported band gap energies of the gCN bulk and monolayer computed using both approximations to be compared to experimental values.

Table 1. gCN band gap energy comparison—experimental and theoretical.

System	Theoretical "DFT" E_g (eV)	Approximation	Experimental E_g (eV)
gCN bulk	1.36 [31]; 1.41 [32]; 1.15 [this work]	GGA-PBE	2.51 [33]; 2.59 eV [34]
	2.73 [31]; 2.71 [6]; [35], 2.68 [36]; 2.80 [37]; 2.47 [this work]	HSE06	2.69 eV [this work]
gCN monolayer	1.59 [38] 1.32 [this work]	GGA-PBE	2.70 [39], 2.71 [40]
	2.76 [41], 2.93 [42], 2.96 [43], 2.97 [38], 3.20 [44], 3.22 [45], 4.02 [46] 2.72 [this work]	HSE06	2.86 eV [this work]

From this table, it is clear that the GGA-PBE approximation provides underestimated band gap energy values, as reported in various work, while the HSE06 approximation shows a very good agreement with the experimental results.

The total and partial densities of states (TDOS and PDOS) of the gCN bulk and monolayer are then computed using the HSE06 approximation and given in Figure 9a,b, respectively.

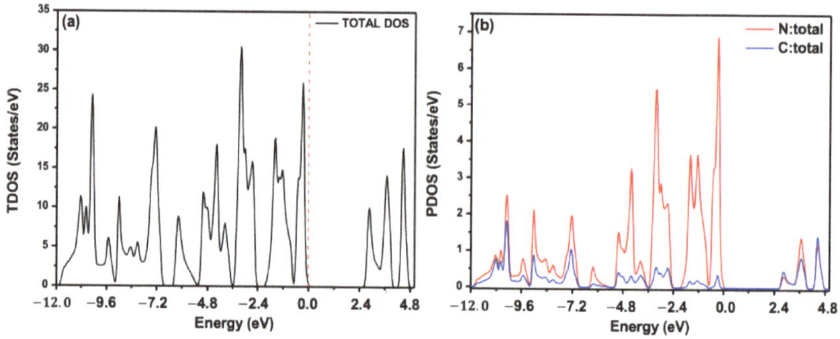

Figure 9. (**a**) Total density of states (TDOS) and (**b**) partial density of states (PDOS) plots of gCN bulk and monolayer.

TDOS provides the overall states of both N and C atoms, whereas PDOS permits us to segregate both constituents. Indeed, our PDOS results show that the energy band states near the Fermi energy in the valence band are mainly occupied by the electrons of the N atoms, and only a small contribution of the electrons is derived from C atoms. In contrast, the conduction band is mainly constituted of atomic orbitals of both N and C with a large contribution from the C atom. This band is dominated by a strong hybridization of the electrons of the N and C atoms. Hence, both atoms could participate in gCN interactions with volatile gas, providing more possible sites for gas detection.

Moreover, we have determined the adsorption energy (E_{ads}) between the CO_2 gas molecule and the gCN monolayer using GGA-PBE approximation [19]. The obtained adsorption energy of the system is calculated using the following relationship [19]:

$$E_{ads} = E_{gCN-CO2} - \left(E_{gCN} + E_{gas}\right), \tag{2}$$

where $E_{gCN-CO2}$ is the total energy of the adsorbed CO_2 molecules by the gCN surface, and E_{gCN} and E_{CO2} are, respectively, the total energies of gCN and a CO_2 molecule. A positive adsorption energy value indicates that the reaction is endothermic and therefore unstable and energetically unfavorable, while a negative value concurs with energetically favorable adsorption as the reaction is exothermic and thus stable. Table 2 shows the optimized structure of gCN while performing the adsorption calculations, including the distance between the CO_2 molecule and the gCN edge and the bond length.

Table 2. Structural comparisons after adsorption of CO_2 at the surface of gCN and their adsorption energy.

	Before Optimization	Optimized Structure
gCN-CO_2 configuration	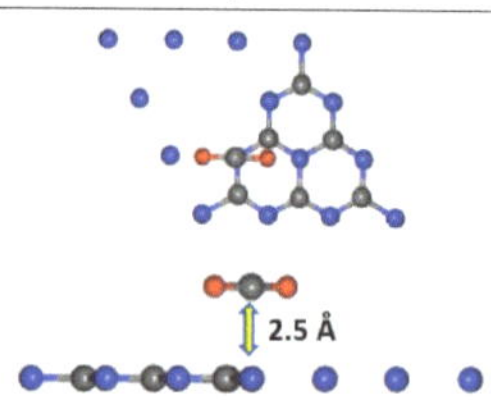	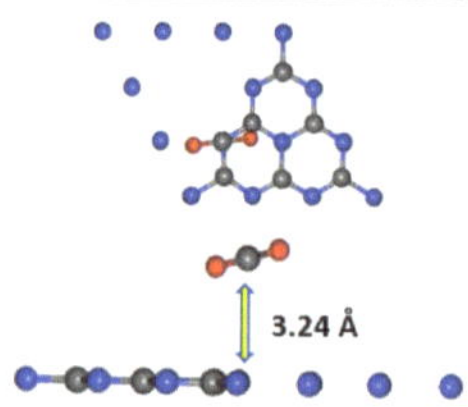
Angle (degree)	180	179.42
Bond length (Å)	1.181	1.172/1.170
Distance gCN–CO_2 (Å)	2.5	3.24
E_{ads} (eV)	−0.59	

Our finding revealed an adsorption energy value $E_{ad} = -0.59$ eV. This result indicates that the reaction is exothermic and therefore stable and energetically favorable. This is consistent with the high surface stability of gCN against CO_2 molecules, which could lead to its high reactivity and is beneficial to enhancing gCN sensing performances. From the above, one can notice that the elaborated gCN nanosheets are highly suitable for CO_2 sensing.

4.4. Gas Sensing Tests

In this work, we have used a resistive gas sensor based on a charge transfer process, in which the sensor acts as a charge donor [47]. Once the sensor device is exposed to CO_2 gas, it is assumed that the charge transfer reaction occurs in different directions between the sensor and the adsorbed gases. This results in changes in the resistance of the sensor [48]. Adsorption-based sensors use the change in resistance of the g-CN nanosheet caused by the adsorption of CO_2. A physical adsorption takes place at the g-CN–CO_2 interfaces, leading to a reduction of CO_2 to CO_2^- ; thus, electrons exchange, modifying the electrical performances of the material as per the following reaction [49]:

$$CO_{2\ (gas)} + e^- \leftrightarrow CO^-_{2(ads)},$$

To investigate the device recovery time dependent on temperature, we heated the gCN device to fully restore its base strength; then, we conducted CO_2 detection tests at 30 °C, 35 °C and 40 °C while monitoring the recovery time as shown in Figure 10.

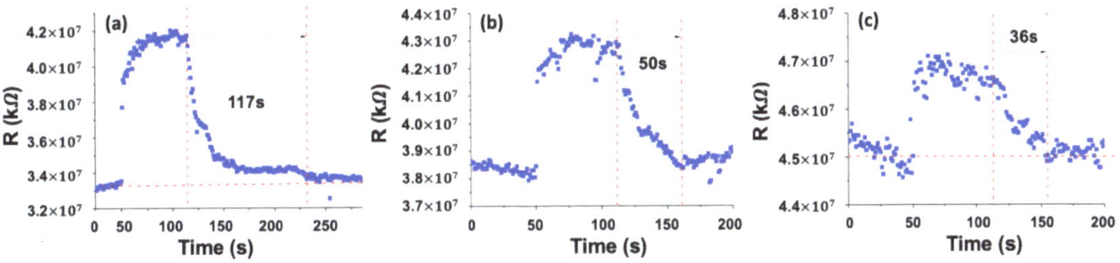

Figure 10. Exposure of gCN sensor device at 1000 ppm of CO_2 at (**a**) 30 °C, (**b**) 35 °C and (**c**) 40 °C.

From the plots of Figure 10, the recovery time as a function of temperature indicates the shortest recovery time of ~36 s achieved at 40 °C, while higher recovery times were registered at 30 °C and 35 °C, at ~117 s and ~50 s, respectively.

Hence, in the following, all our sensing tests were carried out with a 120 s dwell interval, which corresponds to a recovery time of 36 s achieved at 40 °C. A good recovery time was already reported to be considered as an important indicator of a high-performing gas sensor [50–53]. In this sense, the gCN nanosheets were tested for CO_2 detection to evaluate their stability, sensitivity and response. The gCN device was then exposed cyclically to three CO_2 concentrations at 40 °C—200, 600 and 1000 ppm—each for 60 s and maintained at 120 s as dwell time. To monitor the gCN response, a voltage sweep of 2 V at a 0.1 V step was conducted, and the change of the gCN resistance during the CO_2 exposition was collected. The result is illustrated in Figure 11.

As can be seen in Figure 11, for all the three concentrations, the gCN device showed a very fast response of about 730 ms once in contact with CO_2 and a very good recovery time of about 62 s, especially at the lower CO_2 concentration of 200 ppm. These results indicate the high CO_2 reactivity of the elaborated gCN device, as predicted by the adsorption energy calculations.

Moreover, to examine the sensitivity, we exposed (two cycles) the gCN device to CO_2, successively using the following concentrations: 200 ppm, 600 ppm and 1000 ppm. The sensor response is presented in Figure 12.

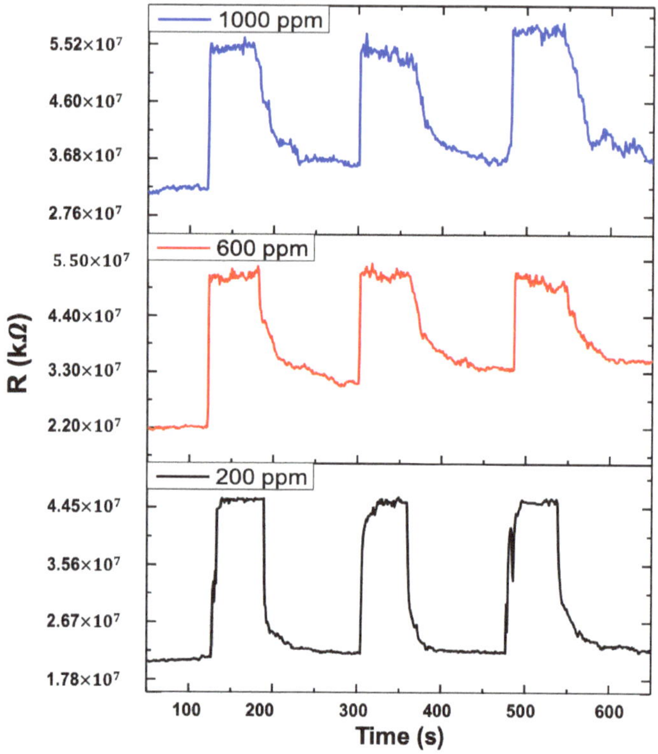

Figure 11. Relative responsiveness of the gCN sensor cyclic exposition to different CO_2 concentrations at 40 °C.

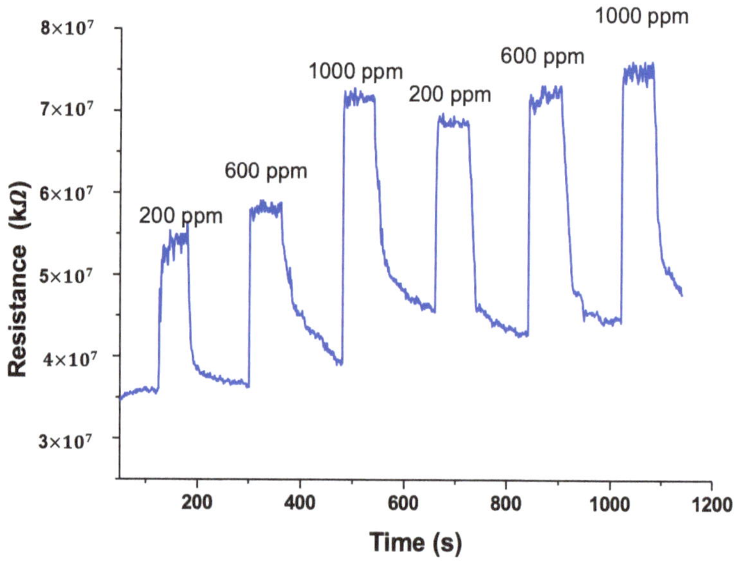

Figure 12. Two cycles of exposure of gCN sensor to CO_2 at 200, 600 and 1000 ppm concentrations at 40 °C.

In this test, the device was repeatedly exposed to gas pulses for 60 s at each concentration, separated by periods of 120 s where the CO_2 stream was cut. The resulting resistance of the device gradually increased with increasing CO_2 gas concentrations. For instance, 5.26×10^7, 5.85×10^7 and 7.17×10^7 kΩ corresponded to 200, 600 and 1000 ppm CO_2 in the first cycle and 6.86×10^7, 7.19×10^7 and 7.50×10^7 kΩ in the second cycle, respectively. This increase in the resistance demonstrates that the gCN device exhibited a very good sensitivity to CO_2 concentration changes.

5. Conclusions

gCN nanosheets were successfully fabricated by the thermal polycondensation of thiourea. The experimental and theoretical investigations have demonstrated an indirect band gap of approximately 2.47 eV and 2.72 eV for bulk and monolayer gCN, respectively. The elaborated gCN was then deposited by spray coating and tested for CO_2 gas sensing. We have evaluated the adsorption of CO_2 into the gas sensor at −0.59 eV, highlighting the very good affinity of gCN nanosheets to CO_2 molecules. The sensing tests revealed that the gCN-based gas sensor has a high CO_2 detection capability of a few hundreds of ppm, high sensitivity and a fast recovery time of 36 s recorded at 40 °C. These findings are expected to provide insights for the low-cost incorporation of gCN to develop very promising, scalable and high-performance gas sensor devices.

Author Contributions: A.K. and M.J. conceived the study; A.K., M.B., E.M.R. and M.L. carried out the experimental and theoretical investigations; A.K., M.B., B.L. and M.J. analyzed the data. All authors contributed equally to writing, editing and reviewing the manuscript. All authors have read and agreed to the published version of the manuscript.

Funding: This research received no external funding.

Data Availability Statement: Data would be made available upon request to the corresponding author.

Acknowledgments: This work is a research collaboration between the University of Picardie Jules Verne (UPJV) and Moroccan Foundation for Advanced Science, Innovation and Research (MAScIR).

Conflicts of Interest: The authors declare no competing interest.

References

1. Yadav, P.; Usha, K.; Singh, B. *Chapter 10—Air Pollution Mitigation and Global Dimming: A Challenge to Agriculture under Changing Climate*; Shanker, A.K., Shanker, C., Anand, A., Maheswari, M., Eds.; Academic Press: Cambridge, MA, USA, 2022; pp. 271–298, ISBN 978-0-12-816091-6.
2. Sikorska, P.E. The need for legal regulation of global emissions from the aviation industry in the context of emerging aerospace vehicles. *Int. Comp. Jurisprud.* **2015**, *1*, 133–142. [CrossRef]
3. Donarelli, M.; Ottaviano, L. 2d materials for gas sensing applications: A review on graphene oxide, mos2, ws2 and phosphorene. *Sensors* **2018**, *18*, 3638. [CrossRef] [PubMed]
4. Liu, X.; Ma, T.; Pinna, N.; Zhang, J. Two-Dimensional Nanostructured Materials for Gas Sensing. *Adv. Funct. Mater.* **2017**, *27*, 1702168. [CrossRef]
5. Mouloua, D.; Kotbi, A.; Deokar, G.; Kaja, K.; El Marssi, M.; EL Khakani, M.A.; Jouiad, M. Recent Progress in the Synthesis of MoS2 Thin Films for Sensing, Photovoltaic and Plasmonic Applications: A Review. *Materials* **2021**, *14*, 3283. [CrossRef] [PubMed]
6. Zhu, B.; Zhang, L.; Xu, D.; Cheng, B.; Yu, J. Adsorption investigation of CO_2 on g-C_3N_4 surface by DFT calculation. *J. CO2 Util.* **2017**, *21*, 327–335. [CrossRef]
7. Hang, N.T.; Zhang, S.; Yang, W. Efficient exfoliation of g-C_3N_4 and NO_2 sensing behavior of graphene/g-C_3N_4 nanocomposite. *Sens. Actuators B Chem.* **2017**, *248*, 940–948. [CrossRef]
8. Zuo, H.W.; Lu, C.H.; Ren, Y.R.; Li, Y.; Zhang, Y.F.; Chen, W.K. Pt_4 clusters supported on monolayer graphitic carbon nitride sheets for oxygen adsorption: A first-principles study. *Wuli Huaxue Xuebao/Acta Phys. Chim. Sin.* **2016**, *32*, 1183–1190. [CrossRef]
9. Song, H.; Liu, L.; Wang, H.; Feng, B.; Xiao, M.; Tang, Y.; Qu, X.; Gai, H.; Huang, T. Adjustment of the band gap of co-doped KCl/NH_4Cl/g-C_3N_4 for enhanced photocatalytic performance under visible light. *Mater. Sci. Semicond. Process.* **2021**, *128*, 105757. [CrossRef]
10. Iqbal, W.; Qiu, B.; Lei, J.; Wang, L.; Zhang, J.; Anpo, M. One-step large-scale highly active g-C_3N_4 nanosheets for efficient sunlight-driven photocatalytic hydrogen production. *Dalt. Trans.* **2017**, *46*, 10678–10684. [CrossRef]

11. Basharnavaz, H.; Habibi-Yangjeh, A.; Kamali, S.H. A first-principle investigation of NO_2 adsorption behavior on Co, Rh, and Ir-embedded graphitic carbon nitride: Looking for highly sensitive gas sensor. *Phys. Lett. Sect. A Gen. At. Solid State Phys.* **2020**, *384*, 126057. [CrossRef]
12. Liu, X.; Kang, W.; Zeng, W.; Zhang, Y.; Qi, L.; Ling, F.; Fang, L.; Chen, Q.; Zhou, M. Structural, electronic and photocatalytic properties of g-C_3N_4 with intrinsic defects: A first-principles hybrid functional investigation. *Appl. Surf. Sci.* **2020**, *499*, 143994. [CrossRef]
13. Ji, Y.; Dong, H.; Lin, H.; Zhang, L.; Hou, T.; Li, Y. Heptazine-based graphitic carbon nitride as an effective hydrogen purification membrane. *RSC Adv.* **2016**, *6*, 52377–52383. [CrossRef]
14. Wu, H.Z.; Liu, L.M.; Zhao, S.J. The effect of water on the structural, electronic and photocatalytic properties of graphitic carbon nitride. *Phys. Chem. Chem. Phys.* **2014**, *16*, 3299–3304. [CrossRef] [PubMed]
15. Jia, L.; Zhang, H.; Wu, P.; Liu, Q.; Yang, W.; He, J.; Liu, C.; Jiang, W. Graphite-like C3N4-coated transparent superhydrophilic glass with controllable superwettability and high stability. *Appl. Surf. Sci.* **2020**, *532*, 147309. [CrossRef]
16. Mohammadi, I.; Zeraatpisheh, F.; Ashiri, E.; Abdi, K. Solvothermal synthesis of g-C_3N_4 and ZnO nanoparticles on TiO_2 nanotube as photoanode in DSSC. *Int. J. Hydrogen Energy* **2020**, *45*, 18831–18839. [CrossRef]
17. Ravichandran, K.; Kalpana, K.; Mohamed Ibrahim, M.; Shantha Seelan, K. Effect of source material of g-C_3N_4 on the photocatalytic activity of ZnO/g-C_3N_4 thin film coated on stainless steel mesh substrate. *Mater. Today Proc.* **2020**, *48*, 207–215. [CrossRef]
18. Mohamed, N.A.; Safaei, J.; Ismail, A.F.; Jailani, M.F.A.M.; Khalid, M.N.; Noh, M.F.M.; Aadenan, A.; Nasir, S.N.S.; Sagu, J.S.; Teridi, M.A.M. The influences of post-annealing temperatures on fabrication graphitic carbon nitride, (g-C_3N_4) thin film. *Appl. Surf. Sci.* **2019**, *489*, 92–100. [CrossRef]
19. Kotbi, A.; Imran, M.; Kaja, K.; Rahaman, A.; Ressami, E.M.; Lejeune, M.; Lakssir, B.; Jouiad, M. Graphene and g-C_3N_4-Based Gas Sensors. *J. Nanotechnol.* **2022**, *2022*, 9671619. [CrossRef]
20. Zinin, P.V.; Ming, L.-C.; Sharma, S.K.; Khabashesku, V.N.; Liu, X.; Hong, S.; Endo, S.; Acosta, T. Ultraviolet and near-infrared Raman spectroscopy of graphitic C_3N_4 phase. *Chem. Phys. Lett.* **2009**, *472*, 69–73. [CrossRef]
21. Bormett, R.W.; Asher, S.A.; Witowski, R.E.; Partlow, W.D.; Lizewski, R.; Pettit, F. Ultraviolet Raman spectroscopy characterizes chemical vapor deposition diamond film growth and oxidation. *J. Appl. Phys.* **1995**, *77*, 5916–5923. [CrossRef]
22. Lin, L.; Ou, H.; Zhang, Y.; Wang, X. Tri-s-triazine-Based Crystalline Graphitic Carbon Nitrides for Highly Efficient Hydrogen Evolution Photocatalysis. *ACS Catal.* **2016**, *6*, 3921–3931. [CrossRef]
23. Wu, J.; Ji, X.; Yuan, X.; Zhao, Z.; Li, Y.; Wen, B.; Zhang, H.; Yu, D.; Zhao, Y.; Tian, Y. Regulating Polymerization in Graphitic Carbon Nitride To Improve Photocatalytic Activity. *Chem. Mater.* **2019**, *31*, 9188–9199. [CrossRef]
24. Li, H.; Jin, C.; Wang, Z.; Liu, Y.; Wang, P.; Zheng, Z.; Whangbo, M.H.; Kou, L.; Li, Y.; Dai, Y.; et al. Effect of the intra- and inter-triazine N-vacancies on the photocatalytic hydrogen evolution of graphitic carbon nitride. *Chem. Eng. J.* **2019**, *369*, 263–271. [CrossRef]
25. Ferrari, A.C.; Robertson, J.; Ferrari, A.C.; Robertson, J. Raman spectroscopy of amorphous, nanostructured, diamond-like carbon, and nanodiamond. *Philos. Trans. R. Soc. Lond. Ser. A Math. Phys. Eng. Sci.* **2004**, *362*, 2477–2512. [CrossRef] [PubMed]
26. Ferrari, A.C.; Robertson, J. Interpretation of Raman spectra of disordered and amorphous carbon. *Phys. Rev. B* **2000**, *61*, 14095–14107. [CrossRef]
27. Kotbi, A.; Hartiti, B.; Ridah, A.; Thevenin, P. Characteristics of CuInS2 thin films synthesizes by chemical spray pyrolysis. *Opt. Quantum Electron.* **2016**, *48*, 75. [CrossRef]
28. Wang, H.; Zhang, X.; Xie, J.; Zhang, J.; Ma, P.; Pan, B.; Xie, Y. Structural distortion in graphitic-C3N4 realizing an efficient photoreactivity. *Nanoscale* **2015**, *7*, 5152–5156. [CrossRef]
29. Pareek, S.; Quamara, J.K. Dielectric and optical properties of graphitic carbon nitride–titanium dioxide nanocomposite with enhanced charge seperation. *J. Mater. Sci.* **2018**, *53*, 604–612. [CrossRef]
30. Dong, F.; Zhao, Z.; Xiong, T.; Ni, Z.; Zhang, W.; Sun, Y.; Ho, W. In Situ Construction of g-C_3N_4/g-C_3N_4 Metal-Free Heterojunction for Enhanced Visible-Light Photocatalysis. *ACS Appl. Mater. Interfaces* **2013**, *5*, 11392–11401. [CrossRef]
31. Liu, J. Effect of phosphorus doping on electronic structure and photocatalytic performance of g-C_3N_4: Insights from hybrid density functional calculation. *J. Alloys Compd.* **2016**, *672*, 271–276. [CrossRef]
32. Yu, W.; Xu, D.; Peng, T. Enhanced photocatalytic activity of g-C_3N_4 for selective CO_2 reduction to CH_3OH via facile coupling of ZnO: A direct Z-scheme mechanism. *J. Mater. Chem. A* **2015**, *3*, 19936–19947. [CrossRef]
33. Li, L.; Huang, Z.; Li, Z.; Li, H.; Wang, A. Defect-rich porous g-C_3N_4 nanosheets photocatalyst with enhanced photocatalytic activity. *J. Mater. Sci. Mater. Electron.* **2021**, *32*, 6465–6474. [CrossRef]
34. Liang, Q.; Li, Z.; Huang, Z.H.; Kang, F.; Yang, Q.H. Holey Graphitic Carbon Nitride Nanosheets with Carbon Vacancies for Highly Improved Photocatalytic Hydrogen Production. *Adv. Funct. Mater.* **2015**, *25*, 6885–6892. [CrossRef]
35. Azofra, L.M.; MacFarlane, D.R.; Sun, C. A DFT study of planar: Vs. corrugated graphene-like carbon nitride (g-C_3N_4) and its role in the catalytic performance of CO_2 conversion. *Phys. Chem. Chem. Phys.* **2016**, *18*, 18507–18514. [CrossRef] [PubMed]
36. Ding, K.; Wen, L.; Huang, M.; Zhang, Y.; Lu, Y.; Chen, Z. How does the B,F-monodoping and B/F-codoping affect the photocatalytic water-splitting performance of g-C_3N_4? *Phys. Chem. Chem. Phys.* **2016**, *18*, 19217–19226. [CrossRef]
37. Wang, Y.; Tian, Y.; Yan, L.; Su, Z. DFT Study on Sulfur-Doped g-C_3N_4 Nanosheets as a Photocatalyst for CO_2 Reduction Reaction. *J. Phys. Chem. C* **2018**, *122*, 7712–7719. [CrossRef]

38. Yao, Q.; Lu, M.; Du, Y.; Wu, F.; Deng, K.; Kan, E. Designing half-metallic ferromagnetism by a new strategy: An example of superhalogen modified graphitic C_3N_4. *J. Mater. Chem. C* **2018**, *6*, 1709–1714. [CrossRef]
39. Zhang, X.; Xie, X.; Wang, H.; Zhang, J.; Pan, B.; Xie, Y. Enhanced photoresponsive ultrathin graphitic-phase C_3N_4 nanosheets for bioimaging. *J. Am. Chem. Soc.* **2013**, *135*, 18–21. [CrossRef]
40. Meng, J.; Pei, J.; He, Z.; Wu, S.; Lin, Q.; Wei, X.; Li, J.; Zhang, Z. Facile synthesis of g-C_3N_4 nanosheets loaded with WO_3 nanoparticles with enhanced photocatalytic performance under visible light irradiation. *RSC Adv.* **2017**, *7*, 24097–24104. [CrossRef]
41. Liu, J.; Hua, E. High Photocatalytic Activity of Heptazine-Based g-C_3N_4/SnS_2 Heterojunction and Its Origin: Insights from Hybrid DFT. *J. Phys. Chem. C* **2017**, *121*, 25827–25835. [CrossRef]
42. Opoku, F.; Govender, K.K.; van Sittert, C.G.C.E.; Govender, P.P. Insights into the photocatalytic mechanism of mediator-free direct Z-scheme g-C_3N_4/Bi_2MoO_6(010) and g-C_3N_4/Bi_2WO_6(010) heterostructures: A hybrid density functional theory study. *Appl. Surf. Sci.* **2018**, *427*, 487–498. [CrossRef]
43. Xu, L.; Huang, W.Q.; Wang, L.L.; Tian, Z.A.; Hu, W.; Ma, Y.; Wang, X.; Pan, A.; Huang, G.F. Insights into enhanced visible-light photocatalytic hydrogen evolution of g-C_3N_4 and highly reduced graphene oxide composite: The role of oxygen. *Chem. Mater.* **2015**, *27*, 1612–1621. [CrossRef]
44. Liu, J.; Cheng, B.; Yu, J. A new understanding of the photocatalytic mechanism of the direct Z-scheme g-C_3N_4/TiO_2 heterostructure. *Phys. Chem. Chem. Phys.* **2016**, *18*, 31175–31183. [CrossRef] [PubMed]
45. Chen, X.; Zhou, C.; Zhao, W.; He, H.; Li, R.; Zou, W. Hybrid triazine-based graphitic carbon nitride and molybdenum disulfide bilayer with and without Li/Mg intercalation: Structural, electronic and optical properties. *Comput. Mater. Sci.* **2017**, *134*, 84–92. [CrossRef]
46. Ruan, L.; Xu, G.; Gu, L.; Li, C.; Zhu, Y.; Lu, Y. The physical properties of Li-doped g-C_3N_4 monolayer sheet investigated by the first-principles. *Mater. Res. Bull.* **2015**, *66*, 156–162. [CrossRef]
47. Deokar, G.; Casanova-Cháfer, J.; Rajput, N.S.; Aubry, C.; Llobet, E.; Jouiad, M.; Costa, P.M.F.J. Wafer-scale few-layer graphene growth on Cu/Ni films for gas sensing applications. *Sens. Actuators B Chem.* **2020**, *305*, 127458. [CrossRef]
48. Zhang, Y.H.; Chen, Y.B.; Zhou, K.G.; Liu, C.H.; Zeng, J.; Zhang, H.L.; Peng, Y. Improving gas sensing properties of graphene by introducing dopants and defects: A first-principles study. *Nanotechnology* **2009**, *20*, 185504. [CrossRef]
49. Dervieux, E.; Théron, M.; Uhring, W. Carbon Dioxide Sensing-Biomedical Applications to Human Subjects. *Sensors* **2022**, *22*, 188. [CrossRef]
50. Li, W.; Zhang, Y.; Long, X.; Cao, J.; Xin, X.; Guan, X.; Peng, J.; Zheng, X. Gas Sensors Based on Mechanically Exfoliated MoS_2 Nanosheets for Room-Temperature NO_2 Detection. *Sensors* **2019**, *19*, 2123. [CrossRef]
51. Presmanes, L.; Thimont, Y.; Chapelle, A.; Blanc, F.; Talhi, C.; Bonningue, C.; Barnabé, A.; Menini, P.; Tailhades, P. Highly sensitive sputtered zno:Ga thin films integrated by a simple stencil mask process on microsensor platforms for sub-ppm acetaldehyde detection. *Sensors* **2017**, *17*, 1055. [CrossRef]
52. Presmanes, L.; Thimont, Y.; El Younsi, I.; Chapelle, A.; Blanc, F.; Talhi, C.; Bonningue, C.; Barnabé, A.; Menini, P.; Tailhades, P. Integration of P-Cuo thin sputtered layers onto microsensor platforms for gas sensing. *Sensors* **2017**, *17*, 1409. [CrossRef] [PubMed]
53. Zhang, X.; Wang, Y.; Luo, G.; Xing, M. Two-Dimensional Graphene Family Material. *Sensors* **2019**, *19*, 2966. [CrossRef] [PubMed]

chemosensors

Article

Photoactivated In_2O_3-GaN Gas Sensors for Monitoring NO_2 with High Sensitivity and Ultralow Operating Power at Room Temperature

Jafetra Rambeloson [1], Dimitris E. Ioannou [1], Parameswari Raju [1], Xiao Wang [1], Abhishek Motayed [2], Hyeong Jin Yun [2] and Qiliang Li [1,*]

1 Department of Electrical and Computer Engineering, George Mason University, Fairfax, VA 22030, USA
2 N5 Sensors, Inc., Rockville, MD 20850, USA
* Correspondence: qli6@gmu.edu

Abstract: Photoactivated gallium nitride (GaN) nanowire-based gas sensors, functionalized with either bare In_2O_3 or In_2O_3 coated with a nanolayer of evaporated Au (Au/In_2O_3), were designed and fabricated for high-sensitivity sensing of NO_2 and low-power operation. The sensors were tested at room temperature under 265 nm and 365 nm ultraviolet illumination at several power levels and in relative humidity ranging from over 20% to 80%. Under all conditions, photoconductivity was lower in the Au/In_2O_3-functionalized sensors compared to that of sensors functionalized with bare In_2O_3. However, when tested in the presence of NO_2, the Au/In_2O_3 sensors consistently outperformed In_2O_3 sensors, the measured sensitivity being greater at 265 nm compared to 365 nm. The results show significant power reduction (×12) when photoactivating at (265 nm, 5 mW) compared to (365 nm, 60 mW). Maximum sensitivities of 27% and 42% were demonstrated with the Au/In_2O_3 sensors under illumination at (265 nm, 5 mW) for 1 ppm and 10 ppm concentration, respectively.

Keywords: gallium nitride (GaN); Au functionalization; nanowire; low power; gas sensing; photoactivation

Citation: Rambeloson, J.; Ioannou, D.E.; Raju, P.; Wang, X.; Motayed, A.; Yun, H.J.; Li, Q. Photoactivated In_2O_3-GaN Gas Sensors for Monitoring NO_2 with High Sensitivity and Ultralow Operating Power at Room Temperature. *Chemosensors* **2022**, *10*, 405. https://doi.org/10.3390/chemosensors10100405

Academic Editors: Kai Xu and Zhong Li

Received: 1 September 2022
Accepted: 4 October 2022
Published: 9 October 2022

1. Introduction

With the rise of interest in environmental protection, health concerns, air quality monitoring, toxic gasses and industrial waste detection and control, the development of gas sensors has been gaining increasing interest over the last several decades [1]. Among the various sensor technologies, those based on semiconducting metal oxides have important advantages, such as a simple principle of operation, low cost, and portability, and have been studied extensively [2]. Because their operation critically depends on the reaction of the analyte with the sensor surface, improvements have been achieved by forming bare metal oxide nanowires, nanoribbons, hollow porous structures, etc., which increase the specific surface area. Further improvements have been attained by combining bare metal oxide with suitable metal and semiconducting materials and by exploiting the catalytic and heterojunction properties therein. Functionalizing the sensing layer with noble metals, such as Au, Pt, Pd, or Ag, can improve sensitivity and tailor selectivity [3–5]. This route can be a promising solution to solve the inherent selectivity issues of semiconductor gas sensors. To achieve the required performance, however, these sensors need to operate at elevated temperatures, which increase their operating power. The recent rapid development of the Internet of Things (IoT) necessitates the availability of compact and portable gas sensors for integration into personal smart devices such as phones, watches, etc., which must, therefore, be designed for low power operation. The need for sensors which operate at room temperature has, therefore, become apparent and led to the emergence of "photoactivated" gas sensors—i.e., sensors where the energy needed to activate and promote analyte/sensor interaction is supplied by illumination with the light of appropriate wavelength and irradiance. Interest in photoactivated sensors is increasing rapidly, and several excellent reviews have recently been published [6–8]. According to these reviews and the references therein, most published works assume that the light-generated electron–hole pairs react with the gas molecules and facilitate their chemisorption and desorption. This simple view, however, cannot explain all the experimental results, where the best outcome is obtained, at times, with wavelengths close to the energy gap and other times, with wavelengths much shorter than that, and where, sometimes, higher irradiance leads to worse performance. These inconsistencies have since been resolved by Geng et al. [9], who proposed and proved that, as well as generating electron–hole pairs, light also interacts and excites the analyte molecules.

Due to the harmful effect of NO_2 on human health and the environment, there is a great deal of interest in NO_2 gas sensors, and their state-of-the-art characteristics have been thoroughly discussed in a comprehensive review by M. Setka et al. [10]. ZnO is among the most popular oxides for NO_2 gas sensors and has extensively been used to fabricate NO_2 gas sensors with promising results [11]. Excellent NO_2 gas sensors have also been made from bare In_2O_3, which is characterized by a wide band gap (direct/indirect 3.5 eV/2.8 eV), low resistivity, and high chemical stability [12]. Zhang et al. [13] developed a coral-like In_2O_3 sensor able to achieve sensitivity of 132% with an optimal operating temperature of 130 °C. Similarly, Xiao et al. [14] achieved a very high sensitivity of 350% to 1 ppm NO_2 with sensors made from In_2O_3 nanospheres, which, however, needed to be operated at a temperature of 120 °C. As mentioned above, an especially interesting family of gas sensors is based on nanostructures such as nanowires (NWs), nanotubes, and nanorods, which have seen significant progress in recent years [15–20] due to their large surface-to-volume ratio which considerably increases sensitivity and improves the time response of the analyte gases. Gallium nitride (GaN) nanostructure-based gas sensors have recently become very popular due to their unique physical features such as direct bandgap, excellent carrier mobility, high heat capacity, and high breakdown voltage [21,22]. These properties make GaN an excellent candidate for portable gas sensors. Furthermore, the surface engineering of GaN nanostructures significantly improves sensor performance. Recently, Shi et al. [23] developed a high-performance photoactivated GaN nanowire (GaN NW) sensor functionalized with TiO_2, which exhibits a 25% response to 500 ppm NO_2 concentration under 365 nm UV illumination. Khan et al. [24] developed multiple sensors based on oxide/GaN NW combinations with various metal oxides, such as ZnO, WO_3, and SnO_2, for NO_2 sensing. An optimum response of 7.1% for 10 ppm NO_2 was achieved using ZnO oxide at room temperature under 365 nm UV illumination.

The present study combines photoactivation with the advantages of In_2O_3 and GaN nanowires to design and fabricate NO_2 gas sensors operating at room temperature with excellent sensitivity and low-power operation. The oxide layer was either bare In_2O_3 or coated with an evaporated Au nanolayer. Two UV LEDs (265 nm and 365 nm) and several power levels were investigated to achieve the optimal solution for lowering the operating power of the finished sensor without sacrificing the analyte gas sensitivity. The sensors were tested at two different NO_2 concentrations (1 ppm and 10 ppm) and over a range of relative humidity (RH) levels, from 20% to 80% at room temperature.

2. Materials and Methods

2.1. Device Fabrication & Characterization

The sensors were fabricated following steps similar to those by Khan et al. [24]. An AlGaN buffer layer was first deposited on the Si substrate to minimize lattice mismatch and improve adhesion between Si and GaN NW. Then, the GaN NWs were patterned by stepper-lithography-assisted dry etching, followed by induced coupled plasma etching, using a metal hard mask to protect the GaN NW. The nanowire width target ranges between 200 and 400 nm. Subsequently, electrodes composed of a Ti/Al/Ti/Au metal stack were deposited on top of the GaN layer to form ohmic contacts. Following this, RF magnetron sputtering was used to deposit a thin layer of In_2O_3 (in the range of 5–10 nm) on top of the exposed GaN NW, which, for a subset of sensors, was followed by the deposition of 1 nm evaporated Au nanolayer on top of the In_2O_3 layer. For brevity, these two groups of sensors will be referred to as In_2O_3 and Au/In_2O_3, respectively, in the figure below. Finally, rapid thermal annealing (RTA) at 700 °C was performed to crystalize the receptor layers and improve ohmic contacts. The devices were characterized by Zeiss Auriga-Field Emission Scanning Electron Microscope (SEM) with energy dispersive spectroscopy (SEM w/EDS). Figure 1 shows schematics and SEM micrographs of the finished based sensors. Several sensors from each group were wire-bonded and mounted onto an array board chamber for testing.

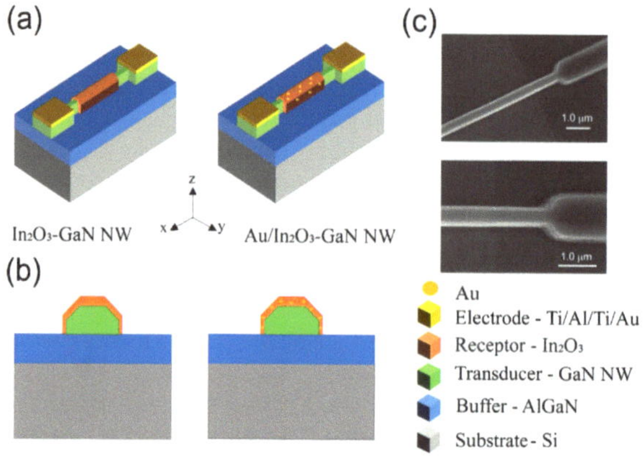

Figure 1. Schematics of proposed In_2O_3-GaN-NW-based sensors: (**a**) device structure, (**b**) cross-section at the center along y/z-axis, and (**c**) top view of SEM images of GaN NW.

EDS analysis was performed to characterize the composition of In_2O_3 and Au/In_2O_3 sensors, as observed in Figure 2a,b.

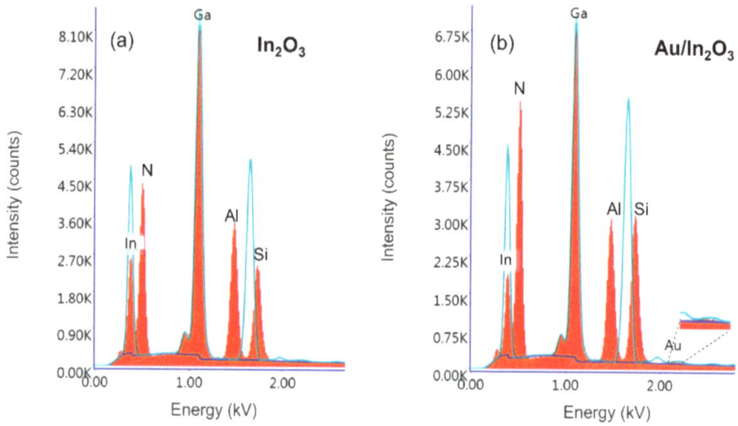

Figure 2. EDS spectrum of (**a**) In_2O_3 and (**b**) Au/In_2O_3 sensors. Small peak at 2.1 kV for Au/In_2O_3 sensors reveals the presence of deposited Au.

Figure 2a,b both reveal the presence of In, N, O, Ga, Al, and Si with peaks located at 0.365 kV, 0.392 kV, 0.525 kV, 1.098 kV, and 1.486 kV, respectively. The presence of a small peak of Au at 2.1 kV is observed in the inset of Figure 2b. A corresponding quantity of 0.11 measured in weight % confirms the deposition of a small layer of Au on the Au/In_2O_3 sensors. Details of EDS spectra values quantified in atomic % and weight % for Ga, N, and Au elements are listed in Table 1.

Table 1. EDS weight and atomic ratio analysis for In_2O_3 and Au/In_2O_3 sensors.

	In_2O_3			Au/In_2O_3		
Element	**Weight %**	**Atomic %**	**Error %**	**Weight %**	**Atomic %**	**Error %**
N	31.84	69.93	8.27	33.73	71.73	8.28
Ga	68.16	30.07	2.47	66.15	28.26	2.52
Au	0	0	3.4	0.11	0.02	3.52

2.2. Gas Sensor Measurement

During testing, the devices were biased at 5V DC. UV LEDs were mounted within the chamber 10 mm above the sensors, providing constant illumination for the duration of testing. Two UV light wavelengths were consecutively used: 265 nm and 365 nm. Both 265 nm and 365 nm were selected as their photon energies (4.6 eV and 3.4 eV, respectively) are both larger than the GaN bandgap (3.4 eV), a factor that is necessary for electron–hole pair generation. For each wavelength, testing was performed at three power levels: 5 mW, 30 mW, and 60 mW. Optical power was measured using a Network-Power Meter-Model 1928-C. Photoconductivity measurements, humidity testing, and NO_2 gas testing (1 ppm and 10 ppm concentration) under various humidity conditions were also conducted. Resistance responses were collected using an Arduino Mega 2560 controlled by NI-LabView software. A gaseous mixture of analyte and carrier gas (breathing air), controlled by mass flow controllers (MFC), was first introduced into a Bronkhorst CEM Evaporator W-101A humidifier prior to flowing into the chamber containing the sensors. The gas mixture was maintained at a constant flow rate of 0.6 slpm and a constant pressure of 5 psi. Figure 3 illustrates the experiment setup for sensor testing.

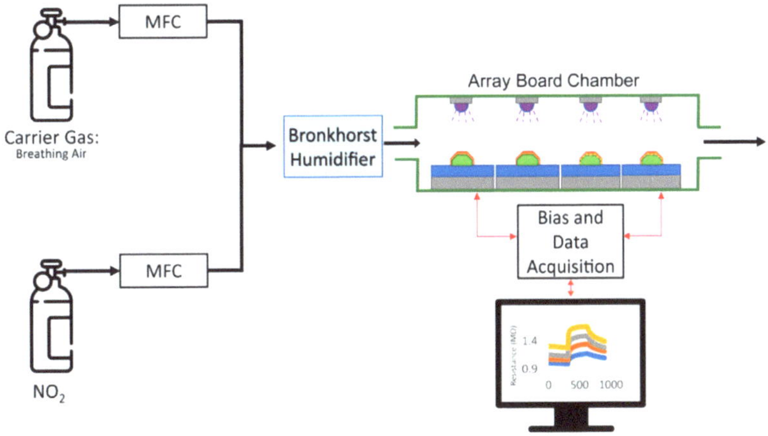

Figure 3. Experiment setup for sensor testing.

3. Results and Discussions

3.1. Photoconductivity

Photoconductivity and humidity testing of the sensors were performed under two illumination conditions: 265 nm at 5 mW (3.3 $\mu W/cm^2$ irradiance) and 365 nm at 60 mW (1462 $\mu W/cm^2$ irradiance). Figure 4 shows the I–V characteristics of both In_2O_3 and Au/In_2O_3 sensors in the dark and under illumination.

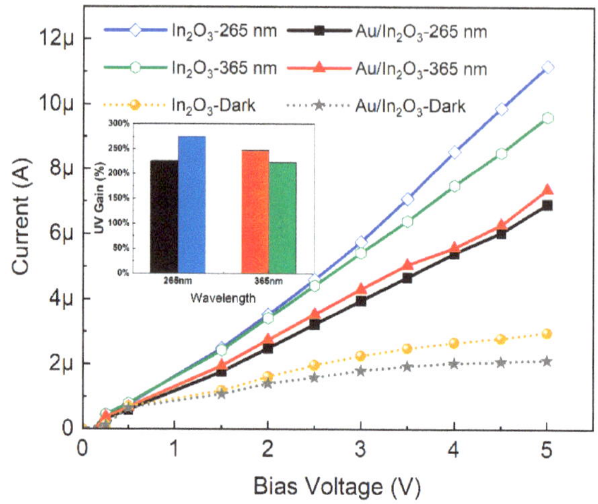

Figure 4. I–V Characteristics under UV illumination and dark condition; (inset) photoconductivity Gain at 265 nm and 365 nm.

In the dark condition, current levels are rather low for both sensor groups, similar to those observed by Shi et al. [23]. Electron–hole pairs are generated under UV illumination. The In_2O_3 layer absorbs a small part of the UV light to generate photocarriers, as it is quite thin. Consequently, the density of electrons in the GaN conducting channel increases, thereby explaining the improvement of conductivity between the dark and illuminated conditions observed in Figure 4. Chemisorption of environmental O_2 at the sensors surface can also explain the difference in conductivity seen in Figure 4, as described by the following equation [10]:

$$O_2 + e^- = O_2^-(ads) \tag{1}$$

Oxygen anions will deplete the electrons from the conduction channel of GaN. The amount of generated oxygen anions is small, while the amount of photogenerated electrons in the conducting channel of GaN is large and dominates the conductivity. It should be noted that there are limited active sites for O_2 adsorption. Functionalizing In_2O_3 with Au provides additional active sites for gas adsorption [25]. More O_2 molecules can be adsorbed at the Au interface, thus further depleting electrons from the channel. Consequently, the conductivity of Au/In_2O_3 sensors is lower than that of bare In_2O_3, as observed in Figure 4. The highest current levels were observed in bare In_2O_3 sensors, 11.2 µA and 9.6 µA for 265 nm and 365 nm, respectively, while the current of Au/In_2O_3 sensors diminished (6.9 µA and 7.2 µA). The inset in Figure 4 shows the UV response gain, defined as:

$$UV_{Gain} = \frac{I_{UV} - I_{dark}}{I_{dark}} \tag{2}$$

where I_{UV} represents the sensor current under UV illumination and I_{dark} the current under the dark condition. It is interesting to note that the obtained UV_{Gain} (250%) under low power illumination at (265 nm, 5 mW) is comparable to that obtained under much higher power illumination at (365 nm, 60 mW). Photoconductivity results indicate that using the 265 nm LED makes it possible for the sensor to be operated at 12 times less power than when using the 365 nm LED, without sacrificing any UV_{Gain}. Power reduction is an essential aspect when designing portable devices that rely on batteries.

3.2. Humidity Response

The performance of the sensors under the illuminated condition was evaluated for relative humidity (RH) in the range from 20% RH to 80% RH at room temperature (20 °C). The "Humidity Response" HR was defined by the relationship:

$$HR = \frac{R_{actual} - R_{20\%}}{R_{20\%}} \tag{3}$$

where R_{actual} is the resistance at the measured humidity, and $R_{20\%}$ represents the baseline resistance at 20% RH. Each sensor was sequentially tested for 900 s at RH of 20%, 40%, 60%, and 80%. Figure 5a,b show the results of the HR experiments.

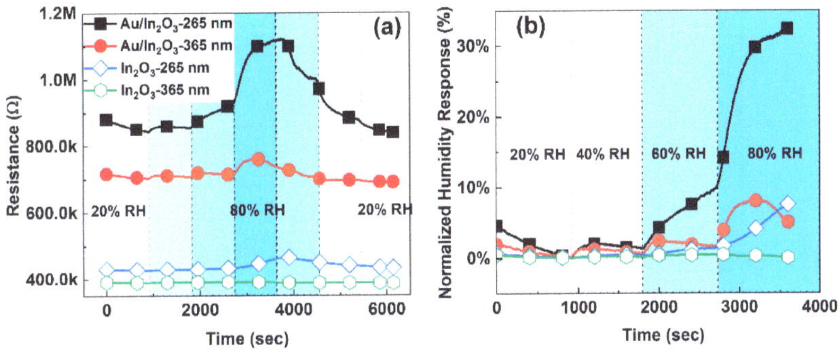

Figure 5. (**a**) Transient response of the various sensors to humidity change. (**b**) Normalized response from 20% RH to 80% RH.

At low humidity (<40% RH), all devices displayed a relatively low response to water adsorption under both illumination conditions, with the response deviating less than 5% from the baseline. Mostly, this small change in resistivity is due to chemosorbed water molecules with the metal oxide layer and photogenerated carriers [26]. At medium-to-high humidity (40–80% RH), resistance levels started to increase. Water layers accumulated on top of the chemosorbed layer, promoting proton (H^+) hopping [26]. Protons transfer from one molecule to the next by the exchange of a hydrogen bond which generates a change in conductivity. At elevated humidity (80% RH), more water layers accrued, providing additional paths for proton hopping and causing a drastic increase in device resistance. Moreover, water can condensate due to the capillary effect and infiltrate the nanopores, offering further conduction channel paths. This is known as the Grotthuss mechanism [27]. In addition, OH^- can be generated from H_2O under UV irradiation. The OH^- will form hydroxyl bond OH on the surface and deplete the electrons of n-type GaN, leading to an increase in resistance [28]. Au/In_2O_3 sensors produced a more pronounced response at elevated humidity compared to bare In_2O_3 sensors. As seen in Figure 5b, the humidity response of Au/In_2O_3 sensors sharply increased to 34%, whereas a relatively low response of 7% is observed for In_2O_3 under 265-nm illumination. Owing to its high catalytic proprieties, Au facilitates the dissociation of water molecules, and these separated species contribute to the increase in resistance. Interestingly, bare In_2O_3 devices illuminated under 365 nm did not respond to the change in humidity. An important observation from Figure 5 is that sensors illuminated at 265 nm showed a greater response to humidity increase compared to those illuminated at 365 nm. The LED power level may be playing an important role here on the adsorption/desorption rate, as well as on water molecules dissociation. Since (365 nm, 60 mW) irradiates at higher power, higher numbers of photogenerated carriers are available to react with dissociated water, accelerating its desorption rate and, hence, leading to stable response compared to (265 nm, 5 mW). Robust and stable sensors are more desired for environmental monitoring, as outdoor humidity often fluctuates.

3.3. NO_2 Response under Various LED Power and Wavelengths

After photoconductivity and humidity testing, the evaluation of NO_2 detection was performed. In the presence of NO_2, three possible surface reactions can occur, leading to a change in resistivity, measured by the sensitivity S:

$$S = \frac{R_{NO_2} - R_{air}}{R_{air}} \tag{4}$$

where R_{NO_2} is the sensor's resistance in the presence of NO_2, and R_{air} represents the baseline resistance under ambient air environment. In the first reaction, NO_2 interacts with the pre-adsorbed oxygen ion forming a NO_2^- ion and an oxygen gas molecule. Secondly, two molecules of NO_2 interact with the pre-adsorbed oxygen ion and photogenerated electron at the surface, producing two NO_3^- ions. Finally, in the third reaction, NO_2 directly reacts with a photogenerated electron at the surface, producing NO_2^- [6–10]. Figure 6a,b show the transient response of the Au/In_2O_3 sensors for 1 ppm NO_2 detection at the fixed humidity level of 40% and under three LED power levels (5 mW, 30 mW, 60 mW).

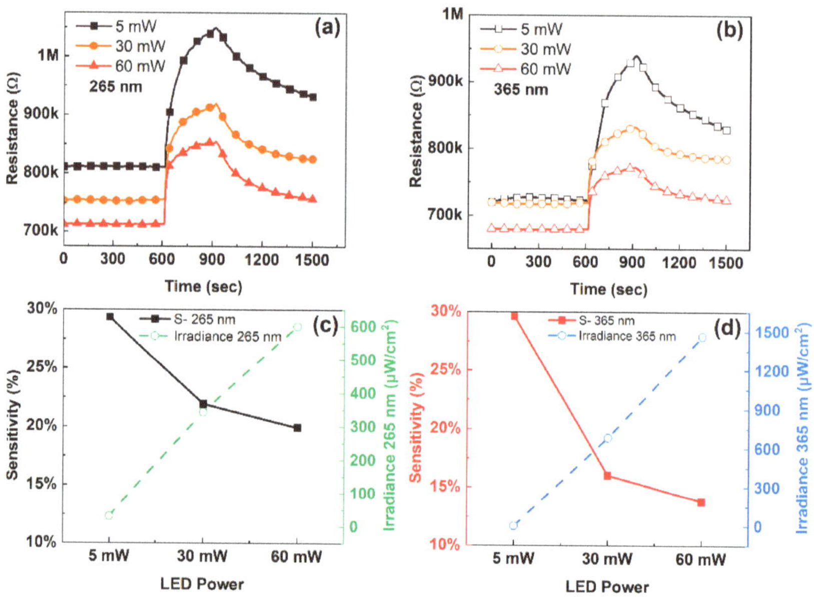

Figure 6. Transient response of 1 ppm NO_2 for Au/In_2O_3 sensors at fixed 40% RH under (**a**) 265 nm and (**b**) 365 nm irradiance with various LED power. Sensitivity under 265 nm (**c**) and 365 nm (**d**) degrades with increasing LED power.

An increase in resistance upon exposure to NO_2 is observed, confirming the oxidation type of NO_2 gas. NO_2 and the pre-adsorbed oxygen attract electrons at the surface, leading to an increase in the depletion layer width, causing a large increase in the sensor resistance. Figure 6c,d show NO_2 sensitivity at different LED power levels, along with the respective irradiances measured at the sensor surface. For the same power level, sensitivity is lower at a higher wavelength. For example, at 30 mW, the sensitivity was 23% under 265 nm and 15% at 365 nm. The reason behind this remains a complicated question which, despite many previous efforts to answer it, still remains open [6,29]. A notable observation is that the highest sensitivity (29%) was obtained at the lowest power (5 mW) for both wavelengths and that sensitivity degrades with increasing power. For example, sensitivity is reduced from 29% to 20% when the 265 nm LED power is raised from 5 mW (corresponding to an irradiance of 1.1 μW/cm^2) to 60 mW (600 μW/cm^2). A similar trend is observed for 365 nm illumination. This behavior agrees with Zhao's work [30].

Although sensitivity increases with the decreasing power of UV LED, Figure 7c,d show that response and recovery times decrease upon exposure to 1 ppm NO_2 with increasing power for both wavelengths. For instance, response time under 365 nm illumination decreased from 233.3 s at 5 mW to 159.6 s at 60 mW power. Interestingly, recovery time drastically decreases at higher LED power for 265 nm: from 572.9 s to 257.4 s. However, sensors under 365 nm illumination have a slow recovery time (i.e., more than 600 s). It is not completely understood why this is, but it can perhaps be explained by considering the newly proposed role of light-activated NO_2 [9] and absorption depth of light where the electrons/holes are generated in the semiconductor. Our devices Au/In_2O_3 demonstrate good short-term repeatability over 6 cycles of exposure to 1 ppm NO_2, as observed in Figure 7a,b. However, a slight degradation in sensor sensitivity is observable after 2 months, as evidenced by Figure S1a. This could be caused by the oxidation of our sensors. Finally, our Au/In_2O_3 sensors have excellent selectivity toward NO_2 compared to other gases, such as SO_2, H_2S, or HCN. Note that the Au/In_2O_3 did not respond to CO nor NH_3, as seen in Figure S1b. Our results indicate that LED power can be optimized for different applications needs.

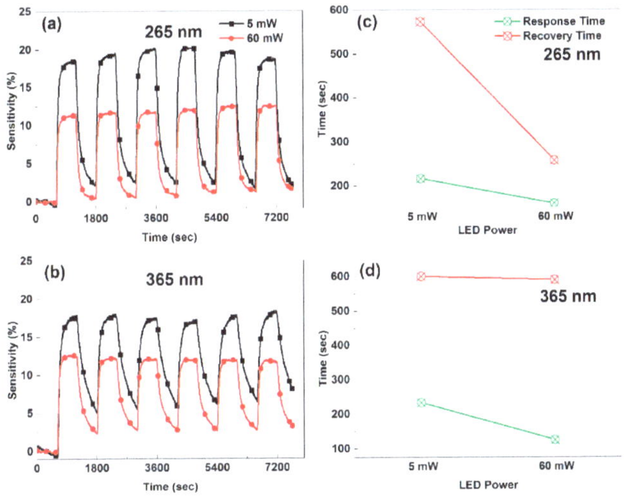

Figure 7. Short-term repeatability response at fixed 20% RH of 1 ppm NO_2 for Au/In_2O_3 sensors under (**a**) 265 nm and (**b**) 365 nm with 5 mW and 60 mW LED power. Response time and recovery time improve as LED power increases under 265 nm (**c**) and 365 nm (**d**).

3.4. NO_2 Response at Various Relative Humidity Levels

Figure 8 shows the sensitivity to 1 ppm and 10 ppm NO_2 detection at various relative humidity levels. In this experiment, two LED conditions were used: (265 nm, 5 mW) and (365 nm, 60 mW).

It was consistently observed that the Au/In_2O_3 sensors outperformed the bare In_2O_3 sensors. At 1 ppm, bare In_2O_3 sensors suffered from a low response rate (<5%). The gas sensing enhancement of Au/In_2O_3 sensors can be explained by the following two mechanisms. Firstly, the depletion layer modification caused by the presence of Au (work function 5.1 eV compared to 4.3 eV for GaN) causes electrons from the In_2O_3 to migrate toward the Au nanolayer and react with NO_2 [31–37]. Secondly, due to its high catalytic proprieties, Au allows for an easier dissociation of NO_2 and water molecules, known as spillover effect [38]. The separated species transfer toward the In_2O_3 layer, thereby altering the sensor resistance due to the concentration increase in the Au-induced active adsorption sites. For instance, at (1 ppm, 40% RH, 265 nm), Au/In_2O_3 exhibited a sensitivity of 27%, while the In_2O_3 sensors only showed 2% sensitivity. At (10 ppm, 40% RH, 265 nm), response of the In_2O_3 sensors was 16%, which increased to 33% for the Au/In_2O_3 sensors. Maximum sensitivities of 32% and 42% were achieved with the Au/In_2O_3 sensors under illumination at (265 nm, 5 mW) for 1 ppm and 10 ppm concentration, respectively. Remarkably, the NO_2 response signals did not degrade with increasing humidity. At higher concentrations, sensitivity increased linearly with increasing humidity. One possible explanation might be that water molecules introduce an additional electron cloud that is favorable to NO_2 adsorption due to its electrophilic nature.

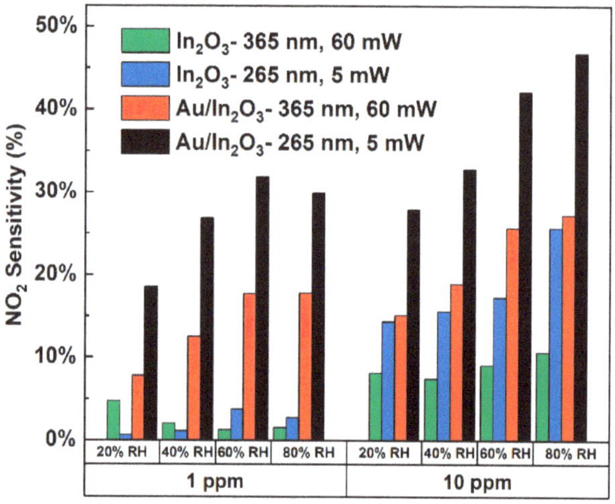

Figure 8. Sensitivities to 1 ppm and 10 ppm at various relative humidity.

Table 2 provides a comparison between the present work and the state of the art for NO_2 sensing. For instance, Redeppa et al. investigated the functionalizing of a GaN thin film layer with deoxyribonucleic acid (DNA-CTMA/GaN). The sensor's response was enhanced with UV illumination at 254 nm and 364 nm, similar to our work. At 10 ppm, the resulting sensitivity was around 10% for both wavelengths, a percentage which is lower compared to our sensor's response of 42% [39]. Recently, NO_2 sensing has also been demonstrated using a heterostructure of graphitic carbon nitride on GaN nanorods (g-C_3N_4/GaN NR) illuminated at 392 nm at room temperature. Despite having a faster response/recovery time and lower limit of detection (500 ppb) than our sensors, the g-C_3N_4/GaN NR sensor's response at 1 ppm was 10% lower than that of our Au/In_2O_3 sensor, which had a 27% response rate [40]. Sun et al. reported a response of 1.1% to a 1 ppm concentration of an AlGaN/GaN HMET sensor with a built-in microheater [41]. The microheater produces 200 mW, which is about 40 times more than the minimum LED power required for the present sensors. Shin et al. [42] demonstrated that a GaN NW sensor functionalized with graphene (G/GaN NW) could reach a response of 8% to 10 ppm NO_2 under 40% RH using LED with wavelengths ranging from 265–350 nm, whereas, under the same conditions, the present sensors exhibited a response of 37%. Recently, several studies have reported excellent sensitivity to NO_2 using In_2O_3 gas sensors with different morphologies. Ueda [34] reported an Au-loaded porous In_2O_3 sensitivity of 900% to 1 ppm NO_2, which is far greater than our sensor's response. Chen et al. [43] demonstrated that an In_2O_3/Zn nanofiber sensor could achieve a 130% sensitivity to 10 ppm. However, those In_2O_3-based sensors operated at a relatively high temperature (100 °C), which increases power consumption. Our sensors achieved reasonable sensitivity with a low power consumption.

Table 2. Comparison of the NO_2 sensitivity of the sensors presented in this study with those reported in the literature.

Sensing Material	Concentrations (ppm)	λ (nm)	Operating Temp. (°C)	Power Consump. (mW)	Sensitivity (%)	Reference
DNA-CTMA/GaN	10	264, 364	RT	-	10	[39]
g-C_3N_4/GaN NR	1	392	RT	-	10	[40]
AlGaN/GaN HMET	1	-	300	200	1.1	[41]
G/GaN	10	265–350	RT	-	8	[42]
In_2O_3/Zn nanofibers	1, 10	-	50	-	2.38, 130	[43]
Au-loaded Porous In_2O_3	1	-	100	-	900	[34]
Au/In_2O_3 GaN NW	1, 10	265	RT	5	27, 42	Present Study

4. Conclusions

In summary, GaN NW-based sensors functionalized with In_2O_3 and Au/In_2O_3 sensors were presented for NO_2 sensing. Photoconductivity results exhibited lower current levels in Au/In_2O_3-functionalized GaN NW sensors compared to In_2O_3-functionalized ones. However, the Au/In_2O_3 sensors consistently demonstrated a superior sensitivity to NO_2 than the In_2O_3 sensors at all tested humidity conditions. The highest sensitivities of 27% and 42% were achieved using the Au/In_2O_3 sensors under low-power illumination at (265 nm, 5 mW) for 1 ppm and 10 ppm concentrations, respectively. More importantly, the results show the achievement of significant power reduction (×12) when using (265 nm, 5 mW) UV illumination rather than (365 nm, 60 mW), without sacrificing sensitivity. This combination of power reduction and high NO_2 sensitivity makes our Au/In_2O_3-GaN NW sensors an excellent candidate for portable and smart sensor integration.

Supplementary Materials: The following supporting information can be downloaded at: https://www.mdpi.com/article/10.3390/chemosensors10100405/s1, Figure S1: (a) Long-term stability of Au/In_2O_3 at 5 mW and 60 mW LED power. (b) Selectivity of In_2O_3 and Au/In_2O_3 sensors under (265 nm, 5 mW) illumination.

Author Contributions: Conceptualization, A.M. and Q.L.; methodology, Q.L.; software, H.J.Y.; validation, J.R., P.R. and X.W.; formal analysis, J.R. and X.W.; investigation, J.R.; resources, A.M., H.J.Y. and Q.L.; data curation, J.R., P.R. and X.W.; writing—original draft preparation, J.R.; writing—review and editing, Q.L. and D.E.I.; visualization, J.R.; supervision, Q.L. and A.M.; project administration, Q.L. and A.M.; funding acquisition, A.M. and Q.L. All authors have read and agreed to the published version of the manuscript.

Funding: This research was funded by NASA research grant on hybrid gas sensors and Virginia Microelectronics Consortium (VMEC) research grant.

Institutional Review Board Statement: Not applicable.

Informed Consent Statement: Not applicable.

Data Availability Statement: Not applicable.

Acknowledgments: The authors would like to acknowledge the support of NASA research grant on hybrid gas sensors and Virginia Microelectronics Consortium (VMEC) research grant.

Conflicts of Interest: The authors declare no conflict of interest.

References

1. Arshak, K.; Moore, E.; Lyons, G.M.; Harris, J.; Clifford, S. A Review of Gas Sensors Employed in Electronic Nose Applications. *Sens. Rev.* **2004**, *24*, 181–198. [CrossRef]
2. Fine, G.F.; Cavanagh, L.M.; Afonja, A.; Binions, R. Metal Oxide Semi-Conductor Gas Sensors in Environmental Monitoring. *Sensors* **2010**, *10*, 5469–5502. [CrossRef] [PubMed]
3. Penza, M.; Cassano, G.; Rossi, R.; Alvisi, M.; Rizzo, A.; Signore, M.A.; Dikonimos, T.; Serra, E.; Giorgi, R. Enhancement of Sensitivity in Gas Chemiresistors Based on Carbon Nanotube Surface Functionalized with Noble Metal (Au, Pt) Nanoclusters. *Appl. Phys. Lett.* **2007**, *90*, 173123. [CrossRef]

4. Singhal, A.V.; Charaya, H.; Lahiri, I. Noble Metal Decorated Graphene-Based Gas Sensors and Their Fabrication: A Review. *Crit. Rev. Solid State Mater. Sci.* **2017**, *42*, 499–526. [CrossRef]
5. Lee, J.-H.; Mirzaei, A.; Kim, J.-Y.; Kim, J.-H.; Kim, H.W.; Kim, S.S. Optimization of the Surface Coverage of Metal Nanoparticles on Nanowires Gas Sensors to Achieve the Optimal Sensing Performance. *Sens. Actuators B Chem.* **2020**, *302*, 127196. [CrossRef]
6. Espid, E.; Taghipour, F. UV-LED Photo-Activated Chemical Gas Sensors: A Review. *Crit. Rev. Solid State Mater. Sci.* **2016**, *42*, 416–432. [CrossRef]
7. Xu, F.; HO, H.-P. Light-Activated Metal Oxide Gas Sensors: A Review. *Micromachines* **2017**, *8*, 333. [CrossRef]
8. Kumar, R.; Liu, X.; Zhang, J.; Kumar, M. Room-Temperature Gas Sensors under Photoactivation: From Metal Oxides to 2D Materials. *Nano-Micro Lett.* **2020**, *12*, 164. [CrossRef]
9. Geng, X.; Liu, X.; Mawella-Vithanage, L.; Hewa-Rahinduwage, C.C.; Zhang, L.; Brock, S.L.; Tan, T.; Luo, L. Photoexcited NO_2 Enables Accelerated Response and Recovery Kinetics in Light-Activated NO_2 Gas Sensing. *ACS Sens.* **2021**, *6*, 4389–4397. [CrossRef]
10. Šetka, M.; Claros, M.; Chmela, O.; Vallejos, S. Photoactivated Materials and Sensors for NO_2 Monitoring. *J. Mater. Chem. C* **2021**, *9*, 16804–16827. [CrossRef]
11. Wang, H.; Zhou, L.; Liu, Y.; Liu, F.; Liang, X.; Liu, F.; Gao, Y.; Yan, X.; Lu, G. UV-Activated Ultrasensitive and Fast Reversible ppb NO_2 Sensing Based on ZnO Nanorod Modified by Constructing Interfacial Electric Field with In_2O_3 Nanoparticles. *Sens. Actuators B Chem.* **2020**, *305*, 127498. [CrossRef]
12. Weiher, R.L.; Ley, R.P. Optical Properties of Indium Oxide. *J. Appl. Phys.* **1966**, *37*, 299–302. [CrossRef]
13. Zhang, H.; Xu, X.; Zhu, Y.; Bao, K.; Lu, Z.; Sun, P.; Sun, Y.; Lu, G. Synthesis and NO_2 Gas-Sensing Properties of Coral-like Indium Oxide via a Facile Solvothermal Method. *RSC Adv.* **2017**, *7*, 49273–49278. [CrossRef]
14. Xiao, B.; Zhao, Q.; Wang, D.; Ma, G.; Zhang, M. Facile Synthesis of Nanoparticle Packed In_2O_3 Nanospheres for Highly Sensitive NO_2 Sensing. *New J. Chem.* **2017**, *41*, 8530–8535. [CrossRef]
15. Prajapati, C.S.; Sahay, P.P. Influence of in Doping on the Structural, Optical and Acetone Sensing Properties of ZnO Nanoparticulate Thin Films. *Mater. Sci. Semicond. Process.* **2013**, *16*, 200–210. [CrossRef]
16. Aluri, G.S.; Motayed, A.; Davydov, A.V.; Oleshko, V.P.; Bertness, K.A.; Rao, M.V. Nitro-Aromatic Explosive Sensing Using GaN Nanowire-Titania Nanocluster Hybrids. *IEEE Sens. J.* **2013**, *13*, 1883–1888. [CrossRef]
17. Toda, K.; Furue, R.; Hayami, S. Recent Progress in Applications of Graphene Oxide for Gas Sensing: A Review. *Anal. Chim. Acta* **2015**, *878*, 43–53. [CrossRef] [PubMed]
18. Gawali, S.R.; Patil, V.L.; Deonikar, V.G.; Patil, S.S.; Patil, D.R.; Patil, P.S.; Pant, J. Ce Doped NiO Nanoparticles as Selective NO_2 Gas Sensor. *J. Phys. Chem. Solids* **2018**, *114*, 28–35. [CrossRef]
19. Walker, J.M.; Akbar, S.A.; Morris, P.A. Synergistic Effects in Gas Sensing Semiconducting Oxide Nano-Heterostructures: A Review. *Sens. Actuators B Chem.* **2019**, *286*, 624–640. [CrossRef]
20. Reddeppa, M.; Park, B.-G.; Murali, G.; Choi, S.H.; Chinh, N.D.; Kim, D.; Yang, W.; Kim, M.-D. NO_x Gas Sensors Based on Layer-Transferred N-MoS_2/P-GaN Heterojunction at Room Temperature: Study of UV Light Illuminations and Humidity. *Sens. Actuators B Chem.* **2020**, *308*, 127700. [CrossRef]
21. Khan, M.A.H.; Rao, M.V. Gallium Nitride (GaN) Nanostructures and Their Gas Sensing Properties: A Review. *Sensors* **2020**, *20*, 3889. [CrossRef] [PubMed]
22. Aluri, G.S.; Motayed, A.; Davydov, A.V.; Oleshko, V.P.; Bertness, K.A.; Sanford, N.A.; Rao, M.V. Highly Selective GaN-Nanowire/TiO_2-Nanocluster Hybrid Sensors for Detection of Benzene and Related Environment Pollutants. *Nanotechnology* **2011**, *22*, 295503. [CrossRef] [PubMed]
23. Shi, C.; Rani, A.; Thomson, B.; Debnath, R.; Motayed, A.; Ioannou, D.E.; Li, Q. High-Performance Room-Temperature TiO_2-Functionalized GaN Nanowire Gas Sensors. *Appl. Phys. Lett.* **2019**, *115*, 121602. [CrossRef]
24. Khan, M.A.H.; Thomson, B.; Yu, J.; Debnath, R.; Motayed, A.; Rao, M.V. Scalable Metal Oxide Functionalized GaN Nanowire for Precise SO_2 Detection. *Sens. Actuators B Chem.* **2020**, *318*, 128223. [CrossRef]
25. Liu, C.; Kuang, Q.; Xie, Z.; Zheng, L. The Effect of Noble Metal (Au, Pd and Pt) Nanoparticles on the Gas Sensing Performance of SnO_2-Based Sensors: A Case Study on the {221} High-Index Faceted SnO_2 Octahedra. *CrystEngComm* **2015**, *17*, 6308–6313. [CrossRef]
26. Addabbo, T.; Cappelli, I.; Fort, A.; Mugnaini, M.; Panzardi, E.; Vignoli, V.; Viti, C. The Effect of Au Nanoparticle Addition on Humidity Sensing with Ultra-Small TiO_2 Nanoparticles. *Chemosensors* **2021**, *9*, 170. [CrossRef]
27. Miyake, T.; Rolandi, M. Grotthuss Mechanisms: From Proton Transport in Proton Wires to Bioprotonic Devices. *J. Phys. Condens. Matter* **2015**, *28*, 023001. [CrossRef]
28. Munuera, G.; Rives-Arnau, V.; Saucedo, A. Photo-Adsorption and Photo-Desorption of Oxygen on Highly Hydroxylated TiO_2 Surfaces. Part 1.—Role of Hydroxyl Groups in Photo-Adsorption. *J. Chem. Soc. Faraday Trans. 1 Phys. Chem. Condens. Phases* **1979**, *75*, 736. [CrossRef]
29. Giberti, A.; Fabbri, B.; Gaiardo, A.; Guidi, V.; Malagù, C. Resonant Photoactivation of Cadmium Sulfide and Its Effect on the Surface Chemical Activity. *Appl. Phys. Lett.* **2014**, *104*, 222102. [CrossRef]
30. Zhao, D.; Huang, H.; Chen, S.; Li, Z.; Li, S.; Wang, M.; Zhu, H.; Chen, X. In Situ Growth of Leakage-Free Direct-Bridging GaN Nanowires: Application to Gas Sensors for Long-Term Stability, Low Power Consumption, and Sub-ppb Detection Limit. *Nano Lett.* **2019**, *19*, 3448–3456. [CrossRef]

31. Hwang, J.; Jung, H.; Shin, H.-S.; Kim, D.-S.; Kim, D.S.; Ju, B.-K.; Chun, M. The Effect of Noble Metals on CO Gas Sensing Properties of In_2O_3 Nanoparticles. *Appl. Sci.* **2021**, *11*, 4903. [CrossRef]
32. Gogurla, N.; Sinha, A.K.; Santra, S.; Manna, S.; Ray, S.K. Multifunctional Au-ZnO Plasmonic Nanostructures for Enhanced UV Photodetector and Room Temperature NO Sensing Devices. *Sci. Rep.* **2014**, *4*, 6483. [CrossRef]
33. Ueda, T.; Boehme, I.; Hyodo, T.; Shimizu, Y.; Weimar, U.; Barsan, N. Enhanced NO_2-Sensing Properties of Au-Loaded Porous In_2O_3 Gas Sensors at Low Operating Temperatures. *Chemosensors* **2020**, *8*, 72. [CrossRef]
34. Choi, S.-W.; Jung, S.-H.; Kim, S.S. Significant Enhancement of the NO_2 Sensing Capability in Networked SnO_2 Nanowires by Au Nanoparticles Synthesized via γ-Ray Radiolysis. *J. Hazard. Mater.* **2011**, *193*, 243–248. [CrossRef]
35. Mun, Y.; Park, S.; An, S.; Lee, C.; Kim, H.W. NO_2 Gas Sensing Properties of Au-Functionalized Porous ZnO Nanosheets Enhanced by UV Irradiation. *Ceram. Int.* **2013**, *39*, 8615–8622. [CrossRef]
36. Shim, Y.-S.; Moon, H.G.; Kim, D.H.; Zhang, L.; Yoon, S.-J.; Yoon, Y.S.; Kang, C.-Y.; Jang, H.W. Au-Decorated WO_3 Cross-Linked Nanodomes for Ultrahigh Sensitive and Selective Sensing of NO_2 and C_2H_5OH. *RSC Adv.* **2013**, *3*, 10452–10459. [CrossRef]
37. Chen, C.; Zhang, Q.; Xie, G.; Yao, M.; Pan, H.; Du, H.; Tai, H.; Du, X.; Su, Y. Enhancing Visible Light-Activated NO_2 Sensing Properties of Au NPs Decorated ZnO Nanorods by Localized Surface Plasmon Resonance and Oxygen Vacancies. *Mater. Res. Express* **2020**, *7*, 015924. [CrossRef]
38. Phani, A.R. X-ray Photoelectron Spectrscopy Studies on Pd-Doped SnO_2 Liquid Petroleum Gas Sensor. *Appl. Phys. Lett.* **1998**, *71*, 2358, Erratum in *Appl. Phys. Lett.* **1998**, *72*, 3383–3383. [CrossRef]
39. Reddeppa, M.; Mitta, S.B.; Park, B.-G.; Kim, S.-G.; Park, S.H.; Kim, M.-D. DNA-CTMA Functionalized GaN Surfaces for NO_2 Gas Sensor at Room Temperature under UV Illumination. *Org. Electron.* **2019**, *65*, 334–340. [CrossRef]
40. Reddeppa, M.; KimPhung, N.T.; Murali, G.; Pasupuleti, K.S.; Park, B.-G.; In, I.; Kim, M.-D. Interaction Activated Interfacial Charge Transfer in 2D G-C_3N_4/Gan Nanorods Heterostructure for Self-Powered UV Photodetector and Room Temperature NO_2 Gas Sensor at ppb Level. *Sens. Actuators B Chem.* **2021**, *329*, 129175. [CrossRef]
41. Sun, J.; Sokolovskij, R.; Iervolino, E.; Liu, Z.; Sarro, P.M.; Zhang, G. Suspended AlGaN/GaN HEMT NO_2 Gas Sensor Integrated with Micro-Heater. *J. Microelectromech. Syst.* **2019**, *28*, 997–1004. [CrossRef]
42. Shin, J.; Han, S.; Noh, S.; Yu, Y.-T.; Kim, J.S. Room-Temperature Operation of Light-Assisted NO_2 Gas Sensor Based on GaN Nanowires and Graphene. *Nanotechnology* **2021**, *32*, 505201. [CrossRef] [PubMed]
43. Chen, K.; Lu, H.; Li, G.; Zhang, J.; Tian, Y.; Gao, Y.; Guo, Q.; Lu, H.; Gao, J. Surface Functionalization of Porous In_2O_3 Nanofibers with Zn Nanoparticles for Enhanced Low-Temperature NO_2 Sensing Properties. *Sens. Actuators B Chem.* **2020**, *308*, 127716. [CrossRef]

chemosensors

MDPI

Article

Simultaneous Detection of CO_2 and CH_4 Using a DFB Diode Laser-Based Absorption Spectrometer

Xu Wu [1], Yulong Du [1], Shijian Shi [1], Cong Jiang [1], Xueliang Deng [2], Song Zhu [2], Xiaolong Jin [2] and Jingsong Li [1,*]

1 Laser Spectroscopy and Sensing Laboratory, Anhui University, Hefei 230601, China
2 Hefei Meteorological Bureau, Hefei 230041, China
* Correspondence: ljs0625@126.com

Abstract: In this paper, an attractive gas sensing technique based on fiber optical sensing ideal was developed for the detection of multi-gas species simultaneously. Calibration-free laser absorption spectroscopy (LAS) was used and combined with a multi-pass optical absorption cell, two fiber-coupled near-infrared (NIR) distributed feedback (DFB) diode lasers are used and coupled into a single optical path for measuring CO_2 and CH_4 absorption spectra simultaneously. The optimal sampling pressure, laser tuning characteristics, spectral sampling points, and potential optical interference are theoretically and experimentally investigated in detail. The results indicated that the proposed technique has good reliability and has been successfully demonstrated for ambient CO_2 and CH_4 detection by using a single sample cell and detector. An Allan–Werle deviation analysis shows that detection limits of 0.12 ppm for CH_4 and 35.97 ppm for CO_2 can be obtained with an integration time of 181 s and 166 s, respectively. The proposed technique can be expanded to measure more molecules simultaneously by combing laser array and may pave a new way for developing a low-cost and ultra-compact multi-gas laser spectroscopy sensing system.

Keywords: laser absorption spectroscopy; calibration-free; multi-gas detection; CO_2; CH_4

Citation: Wu, X.; Du, Y.; Shi, S.; Jiang, C.; Deng, X.; Zhu, S.; Jin, X.; Li, J. Simultaneous Detection of CO_2 and CH_4 Using a DFB Diode Laser-Based Absorption Spectrometer. *Chemosensors* **2022**, *10*, 390. https://doi.org/10.3390/chemosensors10100390

Academic Editor: Masanori Ando

Received: 6 September 2022
Accepted: 21 September 2022
Published: 24 September 2022

1. Introduction

Laser spectroscopy gas sensing technique plays a key role in various fields such as climate warming, Mars exploration, ocean exploration, bio-medical diagnosis, etc. [1–4]. With the continuous deterioration of global environmental ecology and energy problems, the ability to analyze multiple substances or even unknown components has been a highly technical challenge for developing optical gas sensors, considering the limitations of system volume, weight, power consumption, and cost. Mid-infrared (MIR) external cavity quantum cascade lasers (ECQCLs) are widely attractive for spectroscopic applications, which can be used for simultaneously measuring more than three gas components [5], but they are still not universal due to their high cost. With the development of optical communication technology, near-infrared lasers have been developed more maturely, especially for distributed feedback (DFB) diode lasers, which have been demonstrated for most atmospheric molecules by combing with various spectroscopy techniques, such as direct absorption spectroscopy (DAS), wavelength modulation spectroscopy (WMS), photoacoustic spectroscopy (PAS), off-axis integrated-cavity output spectroscopy (OA-ICOS), cavity ring-down spectroscopy (CRDS), or cavity-enhanced absorption spectroscopy (CEAS) [6–13].

In the case of multi-component analysis, a single broadband laser source or multi-laser array is usually adopted to cover the fingerprint spectral region of interest by combing with multiple photoelectric detectors and the time division multiplexing (TDM) or frequency-division multiplexing (FDM) technique [14,15]. For example, a single quartz crystal tuning fork (QCTF) detector and multiple diode lasers near 1391, 1574, and 1653 nm based compact multi-gas sensor system was developed for the simultaneous detection of H_2O, CO_2, and

CH_4, respectively, by Xu et al. [16]. Quartz-enhanced photoacoustic spectroscopy (QEPAS) sensor was reported by Wu et al. for detecting CH_4 and H_2O simultaneously with the FDM technique [17]. Jin et al. reported a robust FDM-based optical sensor for remote multi-species (CH_4 and C_2H_2) detection [18]. By combing the FDM technique with WMS and OA-ICOS techniques, a dual-gas NIR sensor was developed for sensing C_2H_2 and CH_4 molecules simultaneously [19]. As for the TDM detection technique, a TDM-based multi-gas sensor system was developed by combing a single broadband light source and multiple CO, CO_2, and CH_4 pyroelectric detectors by Dong et al. [20]. A multi-channel fiber surface plasmon resonance sensor was proposed by Liu et al. by using the multi-core fiber and TDM technique [21]. Moreover, a hybrid multi-component gas sensing system integrated with NIR diode laser and MIR quantum cascade laser (QCL) was most recently reported for atmospheric CO, N_2O, and CH_4 detection by Liu et al. [22]. Even though, atmospheric trace gases can also be detected by measuring the absorption spectroscopy in either NIR or MIR spectral bands. To meet all the requirements on cost, size, and power permissibility for field applications, the NIR spectral gas sensors are preferred.

However, the FDM technique is commonly combined with the WMS detection method because of the dependence of WMS on light intensity variation, which requires frequent corrections to ensure accuracy for real-time measurement applications [23]. Obviously, the TDM-based multi-laser array technique has one of the disadvantages of decreasing time resolution, while the simultaneous usage of multiple detectors or gas cells and other additional devices makes the gas sensor system more complicated. To resolve these issues mentioned above, a novel gas detection technique based on fiber optical sensing ideal and calibration-free DAS spectroscopy analysis technique is proposed for the detection of multi-component gas molecules simultaneously. To explore the characteristics of the proposed gas detection technique, two fiber-coupled NIR DFB diode lasers emitting near 1574 nm and 1653 nm, respectively, are used and coupled into a single optical path for simultaneously measuring CO_2 and CH_4 absorption spectra. The optimal sampling pressure, laser tuning characteristics, spectral sampling points, and potential optical interference are theoretically and experimentally investigated in detail. Finally, the developed sensor system is evaluated for ambient atmospheric measurement.

2. Theory Analysis

2.1. Absorption Spectroscopy Theory

The interaction between light and gas medium can be described by the well-known Lambert–Beer law, which is suitable for all electromagnetic radiation and absorbing medium, including gases, solids, liquids, molecules, atoms, and ions. Assuming a laser beam has a wavelength of λ and a light intensity of I_0, and after passing through a homogeneous medium, the output light intensity is I. The change of the light intensity after passing through the medium will satisfy the Beer–Lambert law

$$I(\lambda) = I_0(\lambda)e^{-\alpha(\lambda)L} \tag{1}$$

when it passes through a non-uniform gaseous medium, in addition to the absorption process of the interaction between light and matter, the change in light intensity will also be affected by other factors, such as scattering effects and instrument response characteristics, so the generalized Beer–Lambert law can be expressed as:

$$I(\lambda) = I_0(\lambda)e^{[-\alpha(\lambda)L+\beta(\lambda)L+\gamma L+\delta]A(\lambda)} \tag{2}$$

where $I(\lambda)$ and $I_0(\lambda)$ are the transmitted and initial light intensity, $\beta(\lambda)$ and γ are the Rayleigh scattering coefficient and Mie scattering coefficient, respectively, δ is the influencing factor of other effects, and $A(\lambda)$ is the response function of the instrument. Since the main medium studied in this paper is a gas molecule, the gas molecule can default to a uniform medium, so the interaction process between light and gas molecules is mainly absorption. The absorption coefficient $\alpha(\lambda)$ of gas molecules is related to the absorption lineshape, line strength, and molecular number of molecules, and the absorption of individual lines can be expressed as:

$$\alpha(\lambda) = \phi(\lambda - \lambda_0) \cdot S(T) \cdot N(T,P) \tag{3}$$

where $S(T)$ is the molecular absorption line strength, $N(T,P)$ is the molecular number density, $\phi(\lambda - \lambda_0)$ is an absorption line type centered on wavelength λ_0, and usually satisfies the normalization condition as follows:

$$\int_{-\infty}^{\infty} \phi(\lambda - \lambda_0) d\lambda = 1 \tag{4}$$

The number density of molecules is a function of temperature T and pressure p, which can be expressed as:

$$N(T,P) = \frac{P}{P_0} \cdot N_0 \cdot \frac{T_{ref}}{T} \tag{5}$$

where $N_0 = 2.6875 \times 10^{19}$ (mol/cm^3) is the molecular density of an ideal gas at a reference temperature $T_{ref} = 296$ K and a reference pressure $P_0 = 1$ atm. The strength of the molecular absorption process is usually measured by the absorption coefficient of the molecule and the product of the absorption path, that is, the absorption depth $\alpha(\lambda)L$. In summary, the integral area of a single absorption line of a molecule can be expressed as:

$$A = \int_{-\infty}^{\infty} \alpha(\lambda) L dv = \int_{-\infty}^{\infty} \phi(\lambda - \lambda_0) \cdot S(T) \cdot N(T,P) \cdot L dv \tag{6}$$

since the lineshape function satisfies the normalization condition, thus A can be simplified into:

$$A = S \cdot N \cdot L \tag{7}$$

Once the relevant experimental conditions (such as temperature, pressure, optical path, and spectral line parameters) are known, the number or concentration of the absorbing molecule is calculated by measuring the integral absorption area of the molecule, i.e., Equation (7). Conversely, the spectral line parameters of the interested molecule (e.g., line strength) can also be calculated.

2.2. Detection Principle

The illustration of three different detection schemes is presented and compared in Figure 1 in the time domain. The main principle of the multi-gas TDLAS sensor system based on the traditional time division multiplexing (TDM) technique is that different gas molecules are measured by driving each laser source successively, this approach presents a significant time lag in the time domain, as shown in Figure 1a. In the case of a frequency division multiplexing (FDM) detection scheme (Figure 1b), the key principle is combing modulation and demodulation techniques with a different frequency to realize signal separation, i.e., WMS detection, which deprives the most critical advantage of calibration-free nature. For these issues mentioned above, a new detection scheme was proposed for calibration-free detection of multi-gas species without any time lag in this study. The proposed detection scheme (Figure 1c) is based on calibration-free direct absorption spectroscopy and an effective combination of pressure broadening effect, laser tuning characteristics, and data acquisition mode. The fast current tuning characteristics and wide tuning range of diode lasers enable simultaneous measurement of multiple absorption peaks. Generally, the atmospheric small molecular absorption profile can be fully recovered using data sampling points less than 10^3. Obviously, as the sampling pressure decreases, the absorption profile will be narrower, the fewer sampling points are needed. Note that it is meaningless to record too much non-absorptive baseline background in the absorption spectrum. Moreover, the fewer sampling data points, the better for improving measurement time resolution. Due to the relatively high concentration of water vapor (H_2O) in ambient and strong NIR absorption characteristics (i.e., absorption saturation effect), H_2O detection is not experimentally demonstrated for the developed long path gas cell-based TDLAS sensor system. More details of theoretical simulation and experimental investigation will be described in the next section.

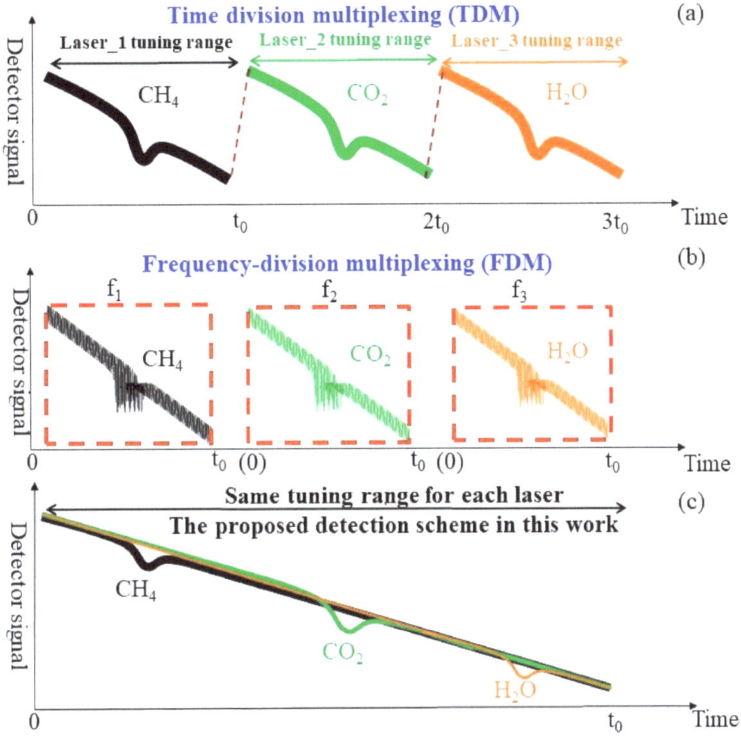

Figure 1. Illustration of different detection schemes: (**a**) Traditional TDM, (**b**) Traditional FDM, (**c**) The proposed detection scheme in this work.

3. Experimental Details

3.1. Absorption Line Selection

Atmospheric carbon dioxide (CO_2) and methane (CH_4) are two very important greenhouse gases, which play a key role in global warming and climate change. Therefore, the multi-gas sensing system was evaluated for simultaneous measuring CH_4 and CO_2 using a single photoelectric detector. Figure 2 shows the simulated absorption spectral line intensity (@296 K) of each molecule in the NIR spectral region (between 1 μm and 2.5 μm) based on the HITRAN database. For this mission, two NIR DFB diode lasers with a center wavelength near 1574 and 1653 nm are selected as excitation light sources, respectively. Table 1 summarizes the spectral parameters of the selected CH_4 and CO_2 molecular selection lines.

Table 1. Spectral Parameters for the Selected Two Species of Interest.

Species	Wavelength (nm)	Wavenumber (cm^{-1})	Line Intensity (cm/mol.)	γ_{air} (cm^{-1}·air $^{-1}$)	γ_{self} (cm^{-1}·air^{-1})
CO_2	1574.03395	6353.10312	1.122×10^{-23}	0.0822	0.112
CH_4	1653.72254 1653.72582 1653.7283	6046.96359 6046.9516 6046.9425	1.455×10^{-21} 9.277×10^{-22} 7.877×10^{-22}	0.0578 0.0774 0.0651	0.079

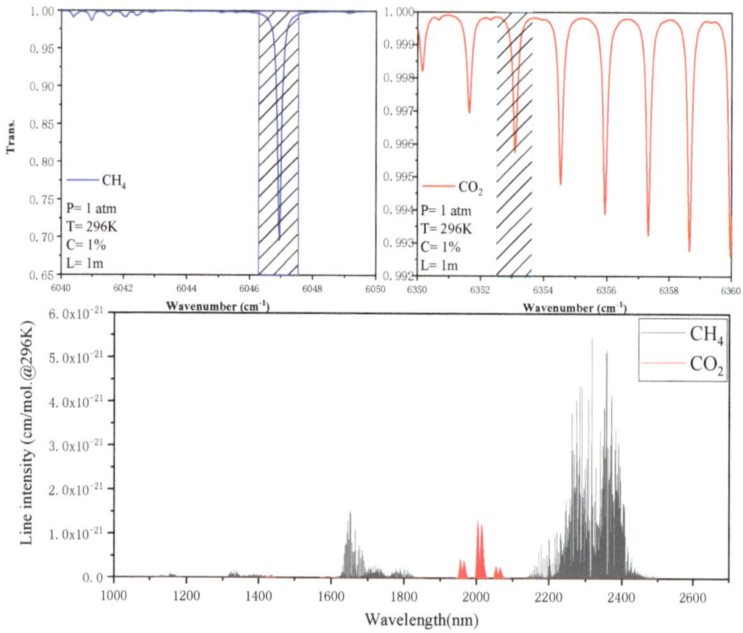

Figure 2. Spectral characteristic simulation of CH_4 and CO_2 based on HITRAN database.

3.2. DFB Diode Laser Characteristics

In general, the emitting wavelength of a diode laser can be tuned by scanning its operating temperature or driving current. For precisely controlling the laser wavelength and achieving coverage of the absorption spectra of the two analytes of interest, the characteristics of the two DFB lasers selected in this study were initially investigated. In this work, the relationship between the laser wavelength and driving voltage is recorded with a high-precision wavelength meter (HighFinesse GmbH, Tubingen, Germany, WS6-200) at a fixed operating temperature, as shown in Figure 3. According to the calibrated results, the operating temperature of the laser was determined at 30 °C, and the absorption line near 1574.03 nm for CO_2 and the absorption line near 1653.72 nm for CH_4 were finally selected, respectively.

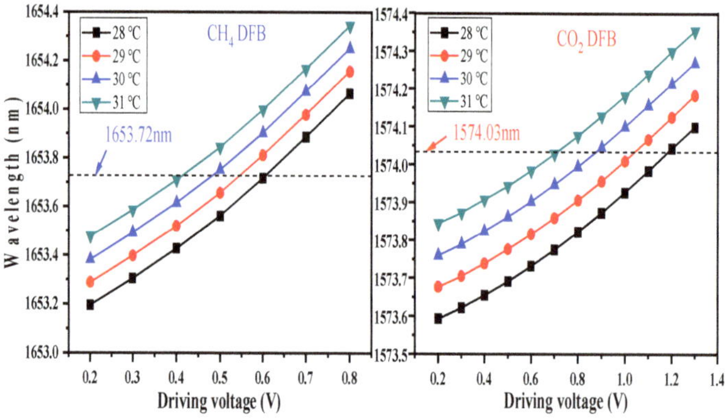

Figure 3. Wavelength tuning characteristic of the diode laser as a function of its operating temperature and driving voltage.

3.3. Sensor Platform Design and Data Processing

The configuration of the developed multi-gas sensor system is shown in Figure 4, which mainly consists of the following parts: optical module, system control, and data acquisition units. In the optical module, two fiber output DFB diode lasers with central wavelengths near 1574 nm and 1653 nm, respectively, are directly coupled into a single beam through a 2 × 1 fiber coupler and an optical fiber collimator, and the laser beam is carefully adjusted to enter a multi-pass gas sample Herriott cell (AMAC-76, Aerodyne Research Inc., Billerica, MA, USA). After exiting from the long-path gas cell, it was focused on a photodetector (InGaAs detector, New Focus 2053, Santa Clara, CA, USA) by a gold-plated parabolic mirror. All optical elements are mounted on an aluminum breadboard. The gas handling module mainly consists of an oil-type diaphragm pump, several two-way and three-way valves, two flow meters (MCR-2000 slpm ALICAT), and a pressure meter (PC3-Series ALICAT), several standard gas cylinders for system calibration. A personal computer (PC) and a data acquisition (DAQ) I/O card (National Instruments, Austin, TX, USA, NI USB-6259) were used for system control and data acquisition. Real-time signal processing and analysis were performed by a LabVIEW-based digital lock-in amplifier.

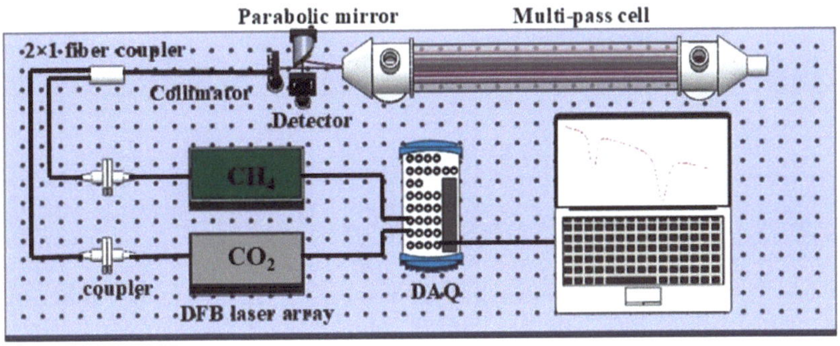

Figure 4. Configuration of the developed multi-component gas synchronous detection system.

3.4. Optimal Sampling Pressure

The key issue for simultaneous detection of multiple molecules is to ensure that there is no crosstalk effect (i.e., absorption overlapping interference between each molecule). Generally, diode lasers can be tuned with 3 cm^{-1} at a fixed operating temperature. The individual absorption profile for typical atmospheric molecules commonly covers 1 cm^{-1} even under the pressure of 1 atm. By decreasing gas sampling pressure, the molecular absorption profile can be significantly narrowed. However, the optical absorption depth will also gradually decrease with the decrease of gas pressure at a constant concentration, since the molecule number of the absorption medium is reduced, as demonstrated in Figure 5. Therefore, a trade-off should be carefully considered between sampling pressure, the number of analytes, and effective sampling points of the absorption spectrum for each analyte. To explore the capabilities of the proposed multi-gas detection technique, a theoretical simulation was first performed. For the CH_4 absorption line at 6046.9 cm^{-1} and CO_2 absorption line at 6353.1 cm^{-1}, 500 data sampling points are enough to cover individual spectral profiles (including absorption lineshape and non-absorption baseline) at a total pressure of 1 atm, as shown in Figure 5a, 1000 sampling points are selected for laser scanning ramp. By analogy, 300 data sampling points can be realized for simultaneous measuring three molecules (such as CH_4, CO_2, CO, or other atmospheric species) with a sampling pressure of less than 600 mbar. For more molecular analysis simultaneously, the related parameters should be carefully selected according to the available experimental conditions, especially for the selection of molecular absorption fingerprint spectral region and laser tuning characteristics, etc.

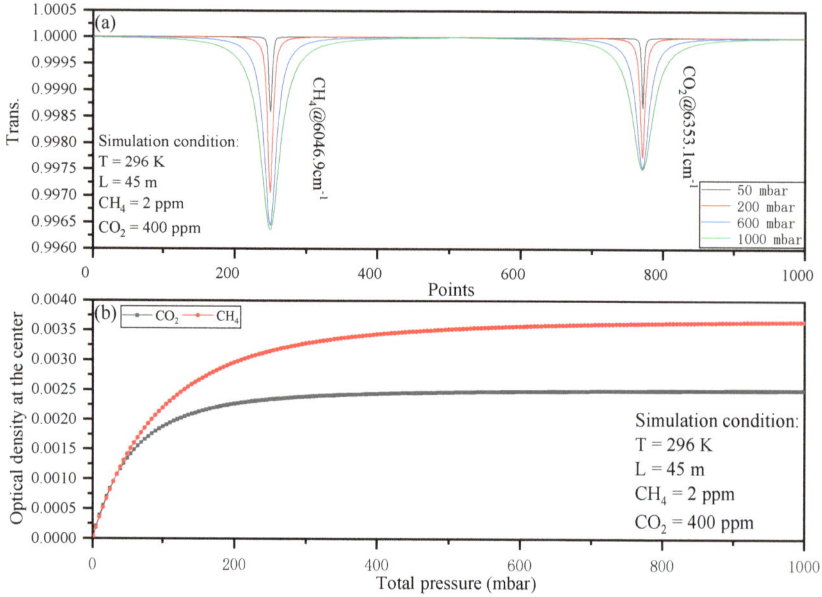

Figure 5. (**a**) Simulated CH_4 and CO_2 transmitted spectra under different pressures and (**b**) the optical density at absorption center as function of sample total pressure.

4. Results and Discussion

4.1. Stability Assessment

The proposed detection scheme was initially evaluated using a gas cylinder with known CH_4 (360 ppm) and CO_2 (29,270 ppm) concentrations before practical application. An alternate measurement between an individual laser and a dual laser was carried out to evaluate the potential crosstalk effect under the optimal sampling pressure of 200 mbar and a total of 1000 sampling points. First, the individual CH_4 laser is only coupled into a gas cell, and continuous measurements of CH_4 absorption spectral signal are made; for example, the normalized CH_4 absorption spectra are shown in Figure 6a. Then, the individual CO_2 laser is coupled into a gas cell without the CH_4 laser, and similar procedures are applied to CO_2 spectral signal measurements, as shown in Figure 6b. Finally, both lasers are coupled into a gas cell for recording CH_4 and CO_2 absorption signals simultaneously, as shown in Figure 6c. For comparison, the simultaneously measured CH_4 and CO_2 absorption signals are plotted in Figure 6c,d together with the manually superimposed spectrum from two individual CH_4 and CO_2 absorption signals. To evaluate the repeatability and stability of the sensor system, the experimental procedure described above was repeated several times. From this figure, a good agreement was obtained between the individual laser mode and dual laser measuring mode.

For quantitative comparison, the optical absorption depth for CH_4 and CO_2 are calculated from the continuous measurement of both spectral data, as demonstrated in Figure 7. The statistical standard deviations (1σ) with two successive cycles of alternating measurements are presented in the insert. The results indicated that the standard deviations for CH_4 are 0.002 and 0.0016, and the standard deviations for CO_2 are approximate 0.00052 and 0.00053, which shows good stability and repeatability for the proposed detection technique.

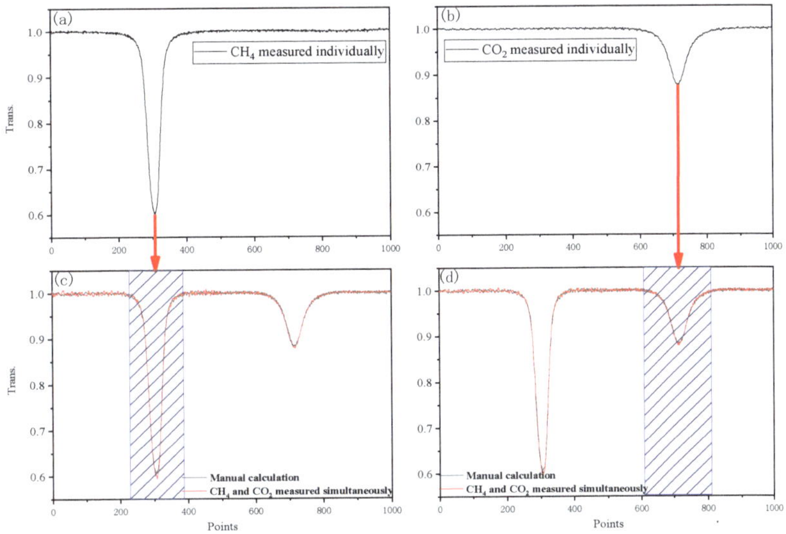

Figure 6. (**a**,**b**) CH_4 and CO_2 transmission spectra measured individually, (**c**,**d**) CH_4 and CO_2 transmission spectra measured simultaneously and the manually synthesized signals using individually measured spectra.

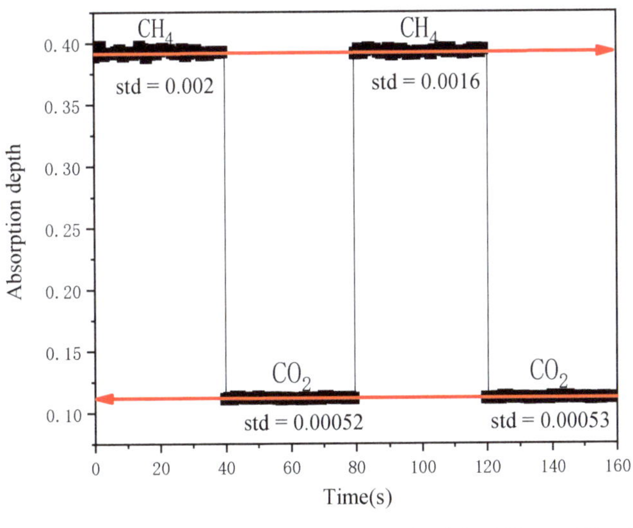

Figure 7. Experimentally measured absorption depth for CH_4 and CO_2 with two successive cycles.

4.2. Linear Calibration

Based on the optimal experimental conditions determined above, the concentration response characteristics of the gas sensor system were further investigated. Two gas cylinders with a certified concentration of 9000 ppm CH_4 and pure CO_2 were used and diluted with high pure nitrogen (N_2). The optical absorbances were extracted from measured CH_4 and CO_2 spectra to check their linear response. The relationship between absorbance and CH_4 and CO_2 concentrations was analyzed using a linear regression algorithm, shown in Figure 8. From this figure, we can see that good linearity with regression coefficients R^2 of 0.997 for CH_4 and 0.999 for CO_2, respectively, were obtained.

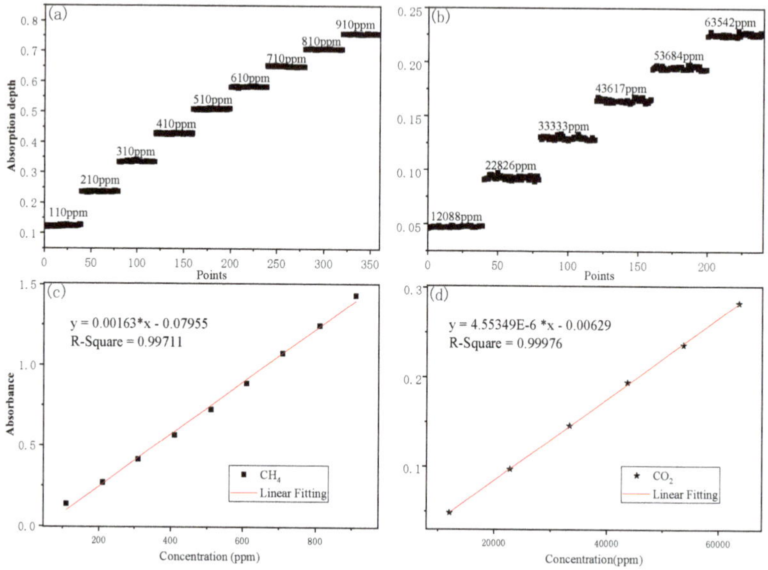

Figure 8. Measured absorption depth with different concentration gradients for (**a**) CH_4 and (**b**) CO_2 and linear calibrations (**c**,**d**).

4.3. Allan Deviation Analysis

Allan deviation analysis method was used to determine the sensitivity and long-time stability of the developed multi-gas sensing system. CH_4 and CO_2 standard cylinder gases with certified concentrations of 249 ppm and 120,600 ppm were measured continuously for 2 h. The time series concentration data are provided in Figure 9a,b. During the entire observation process, the CH_4 concentration varied in the range of 246 to 252 ppm, and the CO_2 concentration changed from 119,600 to 121,600 ppm. The Allan deviations were plotted on a log—log scale versus signal average time, as shown in Figure 9c,d, the results indicate that detection limits at a 1-second average time are 0.82 ppm for CH_4 and 447.68 ppm for CO_2, respectively. The measurement sensitivity can be improved to 0.12 ppm for CH_4 at an optimal averaging time of 181 s and 35.97 ppm for CO_2 at an optimal averaging time of 166 s. The results of Allan variance show that the gas sensing system has good stability.

4.4. Real-Time Measurement of Ambient CO_2 and CH_4

Finally, the developed dual-gas sensor system was applied for ambient CO_2 and CH_4 simultaneously. Lab indoor air was continuously sampled with a flow rate of 100 sccm and gas cell pressure of 200 mbar. The spectra for ambient CO_2 and CH_4 are repeatedly scanned one after the other at a total rate of 100 Hz and are subsequently averaged in 2 s to improve the signal-to-noise ratio. The finally measured CO_2 and CH_4 concentrations for several hours are shown in Figure 10. The result indicated that CO_2 concentration is stationary with an average of about 570 ppm within lunch break (13:30:00–14:30:00), after then the concentration showed an obviously increasing trend, with a maximum over 1000 ppm as shown in Figure 10a, which is mainly due to entry of laboratory personnel. In the case of CH_4, its concentration shows a stable during the entire measurement period as shown in Figure 10b, which indicated no correlation with human activity. The calculated average concentration of CH_4 is approximate 1.3 ppm, which is approximately consistent with its average atmospheric concentration level.

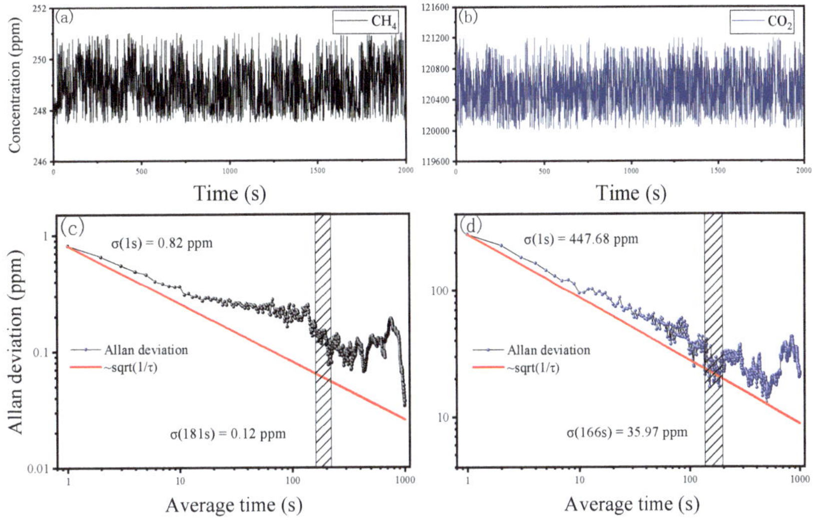

Figure 9. (**a**) and (**b**) Time series concentrations of CH_4 and CO_2 continuously measured standard gas samples, (**c**) and (**d**) Allan deviation of CH_4 and CO_2, respectively.

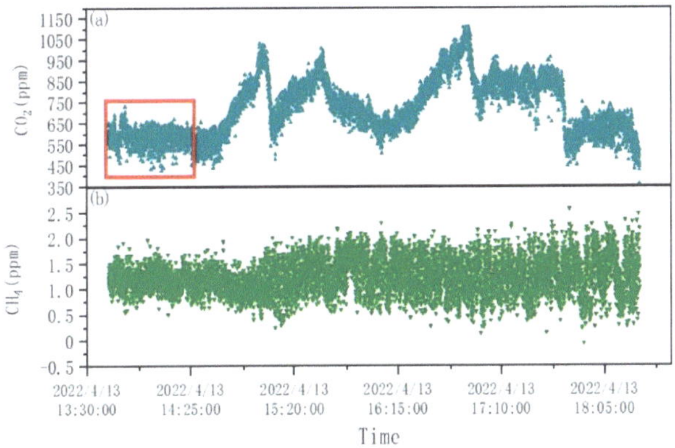

Figure 10. Time series of (**a**) CO_2 and (**b**) CH_4 were measured by using the gas sensing system from the indoor air for ~5 h on 13 April 2022.

For detailed analysis, the histogram analysis of the experimental data is also performed, as illustrated in Figure 11. The histogram plot of CO_2 is made from the relatively stable concentration intervals, as labeled with a rectangle in Figure 10a, while all data in Figure 10b were used for CH_4 histogram analysis. The results indicated that the full width half maximum (FWHM) of the Gaussian profile corresponds to the measurement precision of 76.73 ppm and 0.84 ppm for CO_2 and CH_4, respectively. The histogram plot analysis shows a good Gaussian distribution, which indicates that the developed dual-gas sensor system is mainly dominant with white noise.

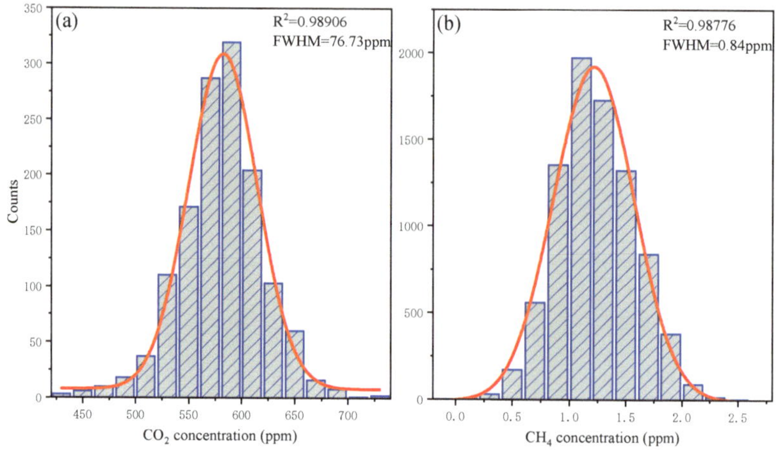

Figure 11. Histogram plot analysis of (**a**) CO_2 and (**b**) CH_4 concentration (details see text).

5. Conclusions

In this paper, an attractive gas sensing technique was proposed for the detection of multi-gas species simultaneously, which is based on the fiber optical sensing ideal. The details of parameters optimization of optimal sampling pressure, laser tuning characteristics, and spectral sampling points were theoretically and experimentally investigated. Two typical atmospheric greenhouse gases (CH_4 and CO_2) are selected as the analyte for simultaneous calibration-free gas detection without time delay, and two fiber-coupled NIR DFB diode lasers are used and coupled into a single optical path for measuring CO_2 and CH_4 absorption spectra simultaneously. The results indicated that the developed laser spectroscopy sensor has good reliability and has been successfully demonstrated for ambient CO_2 and CH_4 detection using a single detector without any time lag. The Allan–Werle deviation analysis indicated that detection limits at a 1-second averaging time are 0.82 ppm for CH_4 and 447.68 ppm for CO_2, respectively. The measurement sensitivity can be improved to 0.12 ppm for CH_4 at an optimal averaging time of 181 s and 35.97 ppm for CO_2 at an optimal averaging time of 166 s. detection limits of 0.12 ppm for CH_4 and 35.97 ppm for CO_2 can be obtained with an integration time of 181 s and 166 s, respectively. The proposed technique can be expanded to measure more molecules simultaneously by combing a laser array and may pave a new way for developing a low-cost and ultra-compact multi-gas laser spectroscopy sensing system.

Author Contributions: Conceptualization, X.W. and J.L.; methodology, X.W. and J.L.; software, Y.D.; validation, X.W., Y.D. and C.J.; formal analysis, C.J.; investigation, S.S.; resources, X.D.; data curation, S.Z. and X.J.; writing—original draft preparation, X.W.; writing—review and editing, X.W. and J.L. All authors have read and agreed to the published version of the manuscript.

Funding: The authors gratefully acknowledge the financial support from the National Natural Science Foundation of China (41875158 and 61675005), Research type for the operation project of Anhui Meteorological Bureau (YJG202102).

Conflicts of Interest: The authors declare no conflict of interest.

References

1. Le, L.D.; Tate, J.D.; Seasholtz, M.B.; Gupta, M.; Owano, T.; Baer, D.; Knittel, T.; Cowie, A.; Zhu, J. Development of a Rapid On-Line Acetylene Sensor for Industrial Hydrogenation Reactor Optimization Using Off-Axis Integrated Cavity Output Spectroscopy. *Appl. Spectrosc.* **2008**, *62*, 59–65. [CrossRef]
2. Durry, G.; Li, J.S.; Vinogradov, I.; Titov, A.; Joly, L.; Cousin, J.; Decarpenterie, T.; Amarouche, N.; Liu, X.; Parvitte, B.; et al. Near infrared diode laser spectroscopy of C_2H_2, H_2O, CO_2 and their isotopologues and the application to TDLAS, a tunable diode laser spectrometer for the martian PHOBOS-GRUNT space mission. *Appl. Phys. B* **2010**, *99*, 339–351. [CrossRef]

3. Zhang, X.; Kirkwood, W.J.; Walz, P.M.; Peltzer, E.T.; Brewer, P.G. A Review of Advances in Deep-Ocean Raman Spectroscopy. *Appl. Spectrosc.* **2012**, *66*, 237–249. [CrossRef] [PubMed]
4. Sigrist, M.W.; Bartlome, R.; Marinov, D.; Rey, J.M.; Vogler, D.E.; Wächter, H. Trace gas monitoring with infrared laser-based detection schemes. *Appl. Phys. B* **2008**, *90*, 289–300. [CrossRef]
5. Zhou, S.; Xu, L.; Zhang, L.; He, T.; Liu, N.; Liu, Y.; Yu, B.; Li, J. External cavity quantum cascade laser-based QEPAS for chlorodifluoromethane spectroscopy and sensing. *Appl. Phys. B* **2019**, *125*, 125. [CrossRef]
6. Ono, S.; Wang, D.T.; Gruen, D.S.; Lollar, B.S.; Zahniser, M.S.; McManus, B.J.; Nelson, D.D. Measurement of a Doubly Substituted Methane Isotopologue, 13CH3D, by Tunable Infrared Laser Direct Absorption Spectroscopy. *Anal. Chem.* **2014**, *86*, 6487–6494. [CrossRef]
7. Cai, W.; Kaminski, C.F. Multiplexed absorption tomography with calibration-free wavelength modulation spectroscopy. *Appl. Phys. Lett.* **2014**, *104*, 154106.
8. Ma, Y.; Yu, X.; Yu, G.; Li, X.; Zhang, J.; Chen, D.; Sun, R.; Tittel, F.K. Multi-quartz-enhanced photoacoustic spectroscopy. *Appl. Phys. Lett.* **2015**, *107*, 021106. [CrossRef]
9. Chen, H.; Winderlich, J.; Gerbig, C.; Hoefer, A.; Rella, C.W.; Crosson, E.R.; van Pelt, A.D.; Steinbach, J.; Kolle, O.; Beck, V.; et al. High-accuracy continuous airborne measurements of greenhouse gases (CO_2 and CH_4) using the cavity ring-down spectroscopy (CRDS) technique. *Atmos. Meas. Tech.* **2010**, *3*, 375–386. [CrossRef]
10. Zheng, K.; Zheng, C.; Zhang, Y.; Wang, Y.; Tittel, F.K. Review of Incoherent Broadband Cavity-Enhanced Absorption Spectroscopy (IBBCEAS) for Gas Sensing. *Sensors* **2018**, *8*, 3646. [CrossRef] [PubMed]
11. Lv, H.; Zheng, H.; Liu, Y.; Yang, Z.; Wu, Q.; Lin, H.; Montano, B.A.Z.; Zhu, W.; Yu, J.; Kan, R.; et al. Radial-cavity quartz-enhanced photoacoustic spectroscopy. *Opt. Lett.* **2021**, *46*, 3917–3920. [CrossRef] [PubMed]
12. Yin, X.; Gao, M.; Miao, R.; Zhang, L.; Zhang, X.; Liu, L.; Shao, X.; Tittel, F.K. Near-infrared laser photoacoustic gas sensor for simultaneous detection of CO and H_2S. *Opt. Express* **2021**, *29*, 34258–34268. [CrossRef]
13. Yin, X.; Dong, L.; Wu, H.; Gao, M.; Zhang, L.; Zhang, X.; Liu, L.; Shao, X.; Tittel, F.K. Compact QEPAS humidity sensor in SF_6 buffer gas for high-voltage gas power systems. *Photoacoustics* **2022**, *25*, 100319. [CrossRef]
14. Tucker, R.S.; Eisenstein, G.; Korotky, S.K. Optical time-division multiplexing for very high bit-rate transmission. *J. Lightwave Technol.* **1988**, *6*, 1737–1749. [CrossRef]
15. Oh, D.B.; Paige, M.E.; Bomse, D.S. Frequency modulation multiplexing for simultaneous detection of multiple gases by use of wavelength modulation spectroscopy with diode lasers. *Appl. Opt.* **1998**, *37*, 2499–2501. [CrossRef]
16. Xu, L.; Zhou, S.; Liu, N.; Zhang, M.; Liang, J.; Li, J. Multigas Sensing Technique Based on Quartz Crystal Tuning Fork-Enhanced Laser Spectroscopy. *Anal. Chem.* **2020**, *92*, 14153–14163. [CrossRef]
17. Wu, H.; Dong, L.; Yin, X.; Sampaolo, A.; Patimisco, P.; Ma, W.; Zhang, L.; Yin, W.; Xiao, L.; Spagnolo, V.; et al. Atmospheric CH_4 measurement near a landfill using an ICL-based QEPAS sensor with V-T relaxation self-calibration. *Sens. Actuators B Chem.* **2019**, *297*, 126753. [CrossRef]
18. Jin, W.; Zhang, H.; Hu, M.; Hu, M.; Wei, Y.; Liang, J.; Kan, R.; Wang, Q. A Robust Optical Sensor for Remote Multi-Species Detection Combining Frequency-Division Multiplexing and Normalized Wavelength Modulation Spectroscopy. *Sensors* **2021**, *21*, 1073. [CrossRef]
19. Zheng, K.; Zheng, C.; Yao, D.; Hu, L.; Liu, Z.; Li, J.; Zhang, Y.; Wang, Y.; Tittel, F.K. A near-infrared C_2H_2/CH_4 dual-gas sensor system combining off-axis integrated-cavity output spectroscopy and frequency-division-multiplexing-based wavelength modulation spectroscopy. *Analyst* **2019**, *144*, 2003–2010. [CrossRef]
20. Dong, M.; Zheng, C.; Miao, S.; Zhang, Y.; Du, Q.; Wang, Y.; Tittel, F.K. Development and Measurements of a Mid-Infrared Multi-Gas Sensor System for CO, CO2 and CH4 Detection. *Sensors* **2017**, *17*, 2221. [CrossRef]
21. Liu, Z.; Wei, Y.; Zhang, Y.; Wang, Y.; Zhao, E.; Zhang, Y.; Yang, J.; Liu, C.; Yuan, L. A multi-channel fiber SPR sensor based on TDM technology. *Sens. Actuators B Chem.* **2016**, *226*, 326–331. [CrossRef]
22. Liu, N.; Xu, L.; Zhou, S.; Zhang, L.; Li, J. Simultaneous Detection of Multiple Atmospheric Components Using an NIR and MIR Laser Hybrid Gas Sensing System. *ACS Sens.* **2020**, *5*, 3607–3616. [CrossRef] [PubMed]
23. Li, J.; Parchatka, U.; Königstedt, R.; Fischer, H. Real-time measurements of atmospheric CO using a continuous-wave room temperature quantum cascade laser based spectrometer. *Opt. Express* **2012**, *20*, 7590–7601. [CrossRef]

MDPI AG
Grosspeteranlage 5
4052 Basel
Switzerland
Tel.: +41 61 683 77 34

Chemosensors Editorial Office
E-mail: chemosensors@mdpi.com
www.mdpi.com/journal/chemosensors

www.ingramcontent.com/pod-product-compliance
Lightning Source LLC
Chambersburg PA
CBHW041348210926
51850CB00013B/3021
9783725824595